Proceedings in
Information and Communications Te

Shin-ya Nishizaki   Masayuki Numao   Jaime Caro
Merlin Teodosia Suarez (Eds.)

Theory and Practice of Computation

2nd Workshop on Computation:
Theory and Practice, Manila, The Philippines,
September 2012, Proceedings

 Springer

Volume Editors

Shin-ya Nishizaki
Tokyo Institute of Technology
Japan
E-mail: nisizaki@cs.titech.ac.jp

Masayuki Numao
Osaka University
Japan
E-mail: numao@sanken.osaka-u.ac.jp

Jaime Caro
University of the Philippines Diliman
The Philippines
E-mail: jdlcaro@dcs.upd.edu.ph

Merlin Teodosia Suarez
De La Salle University-Manila
The Philippines
E-mail: merlin.suarez@delasalle.ph

ISSN 1867-2914   e-ISSN 1867-2922
ISBN 978-4-431-54435-7   e-ISBN 978-4-431-54436-4
DOI 10.1007/978-4-431-54436-4
Springer Tokyo Heidelberg New York Dordrecht London

Library of Congress Control Number: 2013938943

CR Subject Classification (1998): D.3.1, F.3.2, J.3, I.5, H.5.2, H.5.1

Printed on acid-free paper

Springer is part of Springer Science+Business Media (www.springer.com)

Preface

Following the success of the Workshop on Computation: Theory and Practice-2011 (WCTP-2011), held at the University of the Philippines Diliman, Quezon City, September 2011, WCTP-2012 was held at De La Salle University–Manila, September 27–28, 2012. This volume is a collection of the papers that were presented at WCTP-2012 (The proceedings of WCTP 2011 was published as PICT 5 (2012)).

Computation should be a good blend of theory and practice. Researchers in the field should create algorithms to address real-world problems putting equal weight on analysis and implementation. Experimentation and simulation can be viewed as yielding to refined theories or improved applications. This workshop was organized by the Tokyo Institute of Technology, The Institute of Scientific and Industrial Research–Osaka University, the University of the Philippines Diliman and De La Salle University–Manila and was devoted to theoretical and practical approaches to computation. It aimed to present the latest developments by theoreticians and practitioners in academe and industry working to address computational problems that can directly impact the way we live in society.

The program of WCTP-2012 was a combination of an invited talk given by Dr. Ken-ichi Fukui (Osaka University) and selected research contributions. It included the most recent visions and research of the invited talk and the 26 contributions. We collected the original contributions after their presentation at the workshop and began a review procedure that resulted in the selection of the papers in this volume. They appear here in their final form.

WCTP-2012 required a lot of work that depended heavily on members of the program committee. Lastly, we owe a great debt of gratitude to the Tokyo Institute of Technology, specifically, its Philippines Office, which is managed by Ronaldo Gallardo, for sponsoring the workshop.

February 2013

Shin-ya Nishizaki
Masayuki Numao
Jaime Caro
Merlin Teodosia Suarez

Organization

WCTP-2012 was organized by the Philippines Office of the Tokyo Institute of Technology, the Institute of Scientific and Industrial Research of Osaka University, the University of the Philippines Diliman, and De La Salle University–Manila.

Executive Committee

Workshop Co-Chairs

Nobuaki Otsuki   Tokyo Institute of Technology
Kazuhiko Matsumoto   Osaka University

Program Committee

Program Co-Chairs

Jaime Caro   Univeristy of the Phlippines Dilima
Shin-ya Nishizaki   Tokyo Institute of Technology
Masayuki Numao   Osaka University
Merlin Suarez   De La Salle University–Manila

Program Committee

Ryutaro Ichise   National Institute of Informatics
Satoshi Kurihara   Osaka University
Koichi Moriyama   Osaka University
Ken-ichi Fukui   Osaka University
Mitsuharu Yamamoto   Chiba University
Hiroyuki Tominaga   Kagawa University
Naoki Yonezaki   Tokyo Institute of Technology
Takuo Watanabe   Tokyo Institute of Technology
Shigeki Hagihara   Tokyo Institute of Technology
Raymund Sison   De La Salle University–Manila
Jocelynn Cu   De La Salle University–Manila
Gregory Cu   De La Salle University–Manila
Rhia Trogo   De La Salle University–Manila

Table of Contents

On Generating Soft Real-Time Programs
for Non-Real-Time Environments

Ilankaikone Senthooran and Takuo Watanabe

Department of Computer Science, Tokyo Institute of Technology
2-12-1 Ookayama, Meguroku, Tokyo, 152-8552, Japan
senthooran@psg.cs.titech.ac.jp, takuo@acm.org

Abstract. Model based development of real-time embedded systems has actively been studied. Generating real-time code from a model is an important topic of these studies. In this paper, we briefly describe a code generation scheme for real-time programs for non-real-time environment. The generated code runs in a runtime environment without inherent real-time scheduling facilities. In our approach, models are formally written as systems of timed automata and are verified using UPPAAL model checker prior to be processed by our code generator. In order to make the generated code to meet the timing requirements in a non-real-time environment, the code generator weaves explicit timing checking code fragments in the code. We construct a simple two-wheeled robot as a case study of this approach.

Keywords: timed automata, model checking, UPPAAL, code generation, soft real-time systems.

1 Introduction

The primary motivation of this work is to provide an easy and systematic way of developing soft real-time systems. This paper presents a method for generating real-time programs running on top of non-real-time runtime environments. Overview of our method, which is an instance of model-based development, is described briefly in Fig. 1. The model is used in two ways: it is verified by the model verifier with respect to given properties and then is used as the source of the code generator. We adopt timed automata [2,4] for model description and thus use UPPAAL model checker [10,3] as the model verifier.

Code generation plays an important role in model-based development. We designed a code generation scheme that translates models written as a system of UPPAAL timed automata into plain Java programs. Based on this scheme, we implemented a code generator named TA2J. The generated code can be deployed to runtime environments that have no inherent real-time schedulers. The code explicitly checks the timing constraints by polling real-time clocks to meet the timing requirements specified in the model.

The rest of the paper is organized as follows. The next section briefly reviews the concepts of Timed Automata (TA) and real-time system modeling using

S. Nishizaki et al. (Eds.): WCTP 2012, PICT 7, pp. 1–12, 2013.
© Springer Japan 2013

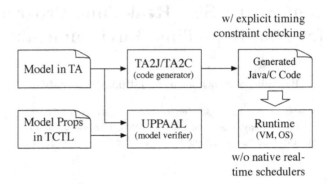

Fig. 1. Overview of Our Method

TAs. Section 3 presents our code generation scheme and Section 4 shows a case study — a small two-wheeled robot. Section 5 overviews related work. The final section concludes the paper.

2   Modeling Real-Time Systems Using Timed Automata

Modeling formalisms for real-time systems have to capture both quantitative and qualitative aspects of time. For modeling real-time embedded systems, UML StateCharts and Stateflow are gaining popularity these days. However, they are weak in their support for formal verification.

Several formalisms that support verification have been investigated. For example, timed extensions of Petri Nets and process algebras, real-time logics and timed automata [12]. In this work, we adopt timed automata [2] as the primary modeling formalism and use model checking for the verification purpose.

A timed automaton is an extension of a Büchi automaton enhanced with real-valued clock variables. Fig. 2 shows an example of timed automaton modeling a simple lamp with a button[3]. It has three locations[1] named off, dim and bright that represent the state of the lamp. In this example, off corresponds to the initial state. The label press? on each edge represents an action of pressing the button. Suppose that the lamp is off. When we press the button once, the lamp is turned on and becomes dim. But if we quickly press the button twice, the lamp is turned on and becomes bright.

To model such timing dependent behaviors using timed automata, we use *clock variables* and *clock constraints*. Clock variables x, y, ... are real-valued variables that increase continuously at the same rate with time. Clock constraints are expressions associated to edges or locations. They are defined using the abstract syntax $g ::= x < c \mid x \leq c \mid x = c \mid x \geq c \mid x > c \mid g \wedge g$ where x is a clock variable and c is a non-negative integer constant[2].

---

[1] In timed automata terminology, a state in an automaton is called a *location*.

[2] Constants in clock constraints are restricted to integers for model-checking purpose.

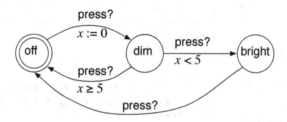

Fig. 2. A Timed Automaton Modeling a Lamp

The example in Fig. 2 has one clock variable x. When the first press occurs, the clock is reset to zero according to the assignment $x := 0$ associated to the edge from off to dim. Note that the values of clock variables increase with time. Thus after the reset of x, its value grows from 0. Now if the second press occurs before x reaches 5 (time units), the lamp becomes bright because the edge to the location has the clock constraint $x < 5$. As the other edge from dim has $x \geq 5$ as its clock constraint, the second press after 5 time units causes the lamp to be turned off.

Automatic verification of a real-time system modeled as a system of timed automata is possible by means of model checking. UPPAAL [10,3] is a popular model checker based on extended timed automata. It can verify temporal properties described as Timed Computation Tree Logic (TCTL) formulae.

In this work, we describe the model of a real-time system as a system of timed automata and verify it using UPPAAL. The verified model is used as the input of our code generator TA2J. The code generators actually generate skeletons of Java code that have some holes to be manually filled with platform specific code fragments.

3 Code Generation Rules

In this section, we describe our code generation rules that take UPPAAL timed automata as inputs.

3.1 A Single Automaton

In our code generation scheme, each timed automaton in a system is converted into a Java thread running a *state machine loop* (Fig. 3). A generated thread has a *location variable* of type `location` that represents the current location of the automaton. It is initialized to the initial location when the thread starts. The behavior of the automaton is represented as an infinite loop in the `run` method of the thread. The body of the loop is a switch sentence that selects a case for each location.

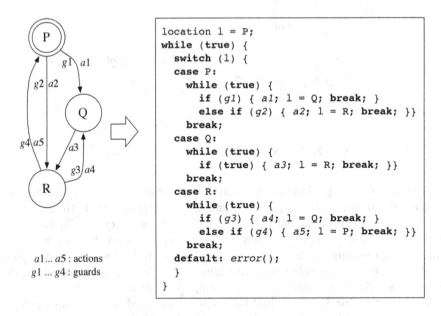

al... a5 : actions
g1 ... g4 : guards

Fig. 3. A State Machine Loop Representation for an Automaton

Local declarations in an UPPAAL template[3] are defined as the local variables in the thread. Template parameters are represented as the arguments of the thread constructor.

3.2 Clock Variables, Guards and Location Invariants

In a system of timed automata, the values of all clock variables increase at the same rate. Any subset of them can be set to constant values when a transition is executed. We use ordinary variables and a wall clock to simulate the behavior of clock variables.

In our translation scheme, a clock variable x is represented as a variable c_x of the type `time`[4] (Fig 4). An assignment $x := c$ within an action of the original (timed) automaton is translated to the assignment $c_x := \mathrm{ct}() + c$. Then, each occurrence of x in guards and location invariants is translated into $\mathrm{ct}() - c_x$. Note that the method `ct` is a synonym for a wall clock method such as `System.currentTimeMillis()` or `System.nanoTime()`.

The translation of location invariants needs a little twist. Let P be a location with an invariant l. To ensure that l holds at the moment the automaton comes to P, l is added to the guard of every incoming edge of P. This means that if g

---

[3] In UPPAAL, a *template* is a parameterized description of a Timed Automata.
[4] A synonym of a number type chosen dependent on the needed accuracy.

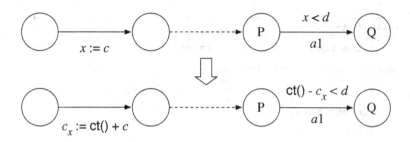

Fig. 4. Translation of a Clock Variable and a Guard

is such a guard, it is translated into $l \wedge g$. In addition, the body of the inner loop in the case clause corresponding to P should starts with the following sentence.

$$\texttt{if } (\neg l) \texttt{ \{ error \}}$$

3.3 Binary Synchronization

Channels are used to synchronize and communicate with processes. Binary synchronization can be categorized into two, internal and external. Internal synchronization is one that happens between processes running within one system/platform and external synchronization is between processes in different system/platform by means of a medium.

A class named `Channel` (Fig. 5) is introduced to model this scenario. The class has two thread variables `sender` and `receiver` to hold the information of its users. To ensure that only one sender or receiver becomes active at any given time, all the methods of the class are declared as synchronized. A boolean variable `message` to depict synchronous channels and six synchronized methods.

In UPPAAL, a channel involves at least two automata. To model this scenario in Java, the local and global channels are declared and passed to all the Java threads as constructor parameters. Normally, there might be more than one automaton (thread) uses a channel to synchronize, i.e. at a given time only one sender and only one receiver can communicate and synchronize via a channel. So, any sender or receiver first acquires permission to send or receive and when both acquired for a particular channel they use the channel.

Suppose that we have two communicating automata, partly shown in Fig. 6. Fragments of the generated Java code for these automata, corresponding to the communication via the channel a, are described in Fig. 7.

The automaton A, requires permission to get the lock for sending via channel a (line 4). When A gets permission it puts a message (line 5) and then waits until the message has been received (line 7) by the receiver (the automaton B). Finally, it releases the lock of both the sender and the receiver (line 8). Similarly, the automaton B acquires lock (line 4), checks whether there is a message to be received (line 5) and if so it proceeds with the transition (line 6) while resetting the message (line 7) to indicate the message read scenario.

```
 1 public class Channel {
 2   volatile Thread sender, receiver;
 3   volatile boolean message = false;
 4
 5   public synchronized boolean
 6     acquire_permission_to_send () {
 7     if (sender == null && receiver != null && !message) {
 8       sender = Thread.currentThread();
 9       return true;
10     }
11     return false;
12   }
13
14   public synchronized boolean
15     acquire_permission_to_receive() {
16     if (sender == null && receiver == null) {
17       receiver = Thread.currentThread();
18       return true;
19     }
20     return false;
21   }
22
23   public synchronized void release_send_and_receive () {
24     if (receiver != null &&
25         sender == Thread.currentThread())
26       receiver = sender = null;
27   }
28
29   public synchronized void release_receive () {
30     if (receiver == Thread.currentThread())
31       receiver = null;
32   }
33
34   public synchronized void set_message (boolean value) {
35     message = value;
36   }
37
38   public synchronized boolean
39     get_message () {
40     return message;
41   }
42 }
```

Fig. 5. Channel Class

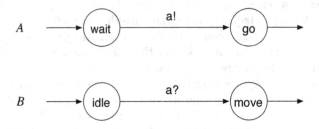

Fig. 6. Two Communicating Automata

```
 1  case LOC_WAIT:      // in automaton A
 2    while (true) {
 3      ...
 4      if (a.acquire_permission_to_send()) {
 5        a.set_message(true);
 6        location = LOC_GO;
 7        while (a.get_message());
 8        a.release_send_and_receive();
 9        break; }}
10    ...
```

```
 1  case LOC_IDLE:     // in automaton B
 2    while (true) {
 3      ...
 4      if (a.acquire_permission_to_receive()) {
 5        while (a.receiver != myThread || !a.get_message());
 6        location = LOC_MOVE;
 7        a.set_message(false);
 8        break; }}
 9    ...
```

Fig. 7. Code Fragments for the Communication Parts of A and B

3.4 Other Features of Timed Automata

External Binary Synchronization. Processes that are running in separate plat-forms communicate by means of a medium (ex. I/O pins, wireless, etc.). Transi-tion that involves a sending channel is in real scenario is an event sending signal. Since it is a binary synchronization, it also involves signal acknowledgment from receiving platform. Code conversion rule is not presented, as coding an external synchronous channel is straightforward.

Broadcast Channels. Broadcast channels allow one-to-many synchronizations. The code generation rules of broadcast channels are almost same as the internal binary synchronization except that the sender of the broadcast channel will not wait until the message is removed by the enabled receivers. In addition there are no receiver locks in the broadcast channel class. The senders of the broadcast channel might have to acquire a lock if more than one thread (automaton) shares the channel to broadcast.

Urgent Binary Synchronization and Urgent/Committed Locations (or States). Currently we have not dealt translating these features to Java code, as the time is not allowed to pass. Therefore, we have considered these features as normal binary synchronization or normal location and applied the normal conversion rules to generate Java code at the presence of these features in the model.

Global Variables. To translate global variables, excluding global channels, to Java, a class named Global is introduced.a class named Global is introduced. Global variables are declared as volatile static fields in the class.

3.5 Non-determinism

A location in an automaton may contain more than one outgoing edges with enabled guards. However, with an event that enables more than one edges, the translated program will always enter the first `if` statement even if other edges are enabled. This means that the translated code is deterministic. Following the perception that an implementation should be more deterministic, this implementation can be considered correct. We could not arrive at a general rule for ordering of such `if` statements, but we propose an ordering method for the guards on `clocks`. If there are more than one edges with enabled clock guards, the one with lowest upper bound or highest lower bound is given highest priority. To handle ordering of other scenarios, we directly use the ordering present in the model, i.e. we use the source file of the UPPAAL model.

4 Case Study: A Small Two-Wheeled Robot

We applied our method to a simple but non-trivial project: a small two-wheeled maze solving robot (Fig. 8). In the first run, the robot traverse a maze using right-hand (or left-hand) rule to find the way to the goal. In the second run, it will go directly to the goal with the map constructed on the first run.

Fig. 8. A Small Two-Wheeled Robot

4.1 The Robot Architecture

We use two commercially available products to construct the robot. It consists of a Java-based controller (SunSPOT[5]) and a two-wheeled small robot (Pololu 3pi[6]). The functions of the controller (SunSPOT) are learning the maze, finding a shortest route and directing the robot to the destination. 3pi robot has an 8-bit microcontroller (Atmel ATmega328P), two DC motors with gearboxes and five reflectance sensors for recognizing lines on the floor. These two components are connected via four GPIO pins to be able to communicate each other. The microcontroller in 3pi is bit small to run programs with full-fledged thread library such as Pthreads. This is the reason we use SunSPOT as the main controller. The C code running on 3pi is generated using a code generation scheme similar to the one for Java descrcibed above. Currently it is only applicable to a system with a single automaton. The resulting code generator targeted to C is called TA2C.

---

[5] http://sunspotworld.com
[6] http://www.pololu.com

4.2   The Robot Model and the Specifications of Desired Properties

The UPPAAL automata that model the controller and the robot body are shown in Fig. 9 and Fig. 10 respectively. The model of the entire system is the parallel composition of these two automata. Using UPPAAL, we verified the following properties of the composed automaton written in TCTL.

- The whole system is deadlock free: $\mathbf{A}\square\neg$deadlock
- Whenever the robot starts moving, it eventually reaches the goal unless it gets error[7]:

$$\text{controller.Ready} \longrightarrow \text{controller.ErrorState} \lor \text{controller.Goal},$$
$$\text{robot.Ready} \longrightarrow \text{robot.ErrorState} \lor \text{robot.Goal}.$$

- Whenever the robot starts a guided run (a run after the robot finished its maze learning run), it eventually reaches the goal:

$$\text{controller.GuildStart} \longrightarrow \text{controller.Final},$$
$$\text{robot.GuildeStart} \longrightarrow \text{robot.Final}.$$

4.3   Code Generation and Deployment

From the verified model of the robot, we generate Java and C code using TA2J and TA2C respectively. Actually, the generated code needs some manual code insertion for it to be able to deployable in the target environments: library calls, communications via channels and some miscellaneous code fragments. Of course these manually inserted code affects the execution timing of the resulting code. Thus we assign a worst-case execution time (WCET) to each code fragment. To determine the values for WCETs, we actually measured the execution time of the code fragments.

The model for the controller contains two timed automata with a total of 23 states and 38 transitions. The code generated from our code generator had two Java threads, around 500 lines of code and 8 channels. The final deployable code amounts to 800 lines.

5   Related Work

Model-based development of real-time and/or embedded systems has actively been studied. Generating real-time code from a verified model is an important topic of these studies. There have been several other approaches for generating real-time code from UPPAAL timed automata.

Hakimipour et al [6,7] presented an approach for generating real-time Java code from timed automata. Their approach uses native real-time features of RTSJ (Real-Time Specification for Java) [5], whereas our approach focuses on

---

[7] The formula $p \longrightarrow q$ means that whenever p holds, q eventually holds.

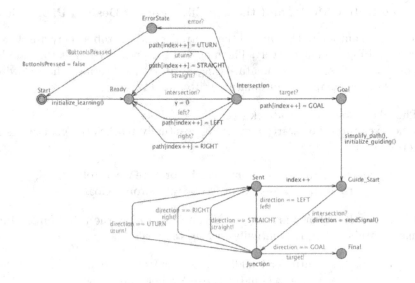

Fig. 9. The UPPAAL Model of the Robot Controller

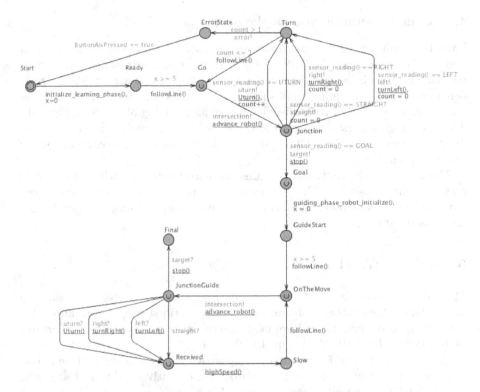

Fig. 10. The UPPAAL Model of the Robot Body

generating plain Java program running on environments without inherent real-time scheduling facilities. In their approach they map the binary channel synchronization feature of timed automata to two variables in a global class called environment. Their implementation will not make the sender to wait till there is a receiver, which we believe that it does not adhere to the concept of synchronization. In our approach we make the sender to wait till there is a receiver. Also, their approach resets the time during the transition starting from a committed location, due to the fact that a committed location cannot allow the time to pass. Even the time elapses due to the thread scheduling representing a false timing is not appropriate. We have not dealt this feature in our work as we generate non real-time Java and delays are inevitable to some extent.

Hendriks [8] presents translation from UPPAAL timed automata to NQC programs that run within LEGO RCX controllers. The generated code is single threaded and depends on the real-time features of RCX controllers.

Pajic et al. [11] present a model transformation method that adopts UPPAAL timed automata as models. In their approach, a model is translated into Stateflow model and then is used for simulation and code generation with Simulink. The generated code runs on top of real-time operating systems.

Altisen et al. [1] show a method to construct implementation models for timed automata. Their methodology allows to transform a timed automaton into a program and to check whether the execution of this program on a given platform satisfies a desired property.

6 Conclusion

In this paper, we present a code generation scheme that acquires UPPAAL timed automata and produces Java/C code with explicit timing checking. The primary contribution of this work is that using our scheme, we can generate correct code deployable to general runtime environments without inherent real-time schedulers; i.e., real-time operating systems are not required if we limit the scope of the applicable area to a class of soft real-time systems. We applied our approach to a simple two-wheeled robot, where we modeled its behavior as a system of timed automata and verified some important properties including deadlock freeness and some reachability.

The results seem promising based on our experience, but we have to take care of two things: (a) to show the correctness of our model-to-code translation rules and (b) to guarantee that the generated code can actually meets the timing requirements. For the latter, we are currently working on modeling the behavior of the generated code with timed automata and verifying the timing requirements using model checking. To achieve this, we have to perform WCET analysis for environment dependent code fragments within the generated code as [9].

Acknowledgments. This work is partly supported by JSPS KAKENHI Grant No. 24500033.

References

1. Altisen, K., Tripakis, S.: Implementation of timed automata: An issue of semantics or modeling? In: Pettersson, P., Yi, W. (eds.) FORMATS 2005. LNCS, vol. 3829, pp. 273–288. Springer, Heidelberg (2005)
2. Alur, R., Dill, D.L.: A theory of timed automata. Theoretical Computer Science 126(2), 183–235 (1994)
3. Behrmann, G., David, A., Larsen, K.G.: A tutorial on UPPAAL. In: Bernardo, M., Corradini, F. (eds.) SFM-RT 2004. LNCS, vol. 3185, pp. 200–236. Springer, Heidelberg (2004)
4. Bengtsson, J.E., Yi, W.: Timed automata: Semantics, algorithms and tools. In: Desel, J., Reisig, W., Rozenberg, G. (eds.) ACPN 2003. LNCS, vol. 3098, pp. 87–124. Springer, Heidelberg (2004)
5. Bollella, G., Gosling, J., Brosgol, B., Dibble, P., Furr, S., Hardin, D., Turnbull, M.: The Real-Time Specification for Java. Addison-Wesley (2000)
6. Hakimipour, N., Strooper, P., Wellings, A.: TART: Timed-automata to real-time Java tool. In: 8th IEEE International Conference on Software Engineering and Formal Methods (SEFM 2010), pp. 299–309. IEEE (2010)
7. Hakimipour, N., Strooper, P., Wellings, A.: A model-based development approach for the verification of real-time Java code. Concurrenty and Computation: Practice and Experience 23(13), 1583–1606 (2011)
8. Hendriks, M.: Translating Uppaal to Not Quite C. Tech. Rep. CSI-R0108, Computing Science Institute, Catholic University of Nijmegen (March 2001)
9. Kim, S., Patel, H.D., Edwards, S.A.: Using a model checker to determine worst-case execution time. Tech. Rep. CUCS-038-09, Department of Computer Science, Columbia University (2009)
10. Larsen, K.G., Pettersson, P., Yi, W.: UPPAAL in a nutshell. International Journal on Software Tools for Technology Transfer 1(1-2), 134–152 (1997)
11. Pajic, M., Jiang, Z., Lee, I., Sokolsky, O., Mangharam, R.: From verification to implementation: A model translation tool and a pacemaker case study. In: 18th IEEE Real-Time and Embedded Technology and Applications Symposium, RTAS 2012 (2012)
12. Wang, F.: Formal verification of timed systems: A survey and perspective. Proceedings of the IEEE 92(8), 1283–1305 (2004)

A Formal Ontology of Interactions with Intensional Quantitative Semantics

Takashi Tomita[1], Naoko Izumi[2], Shigeki Hagihara[1], and Naoki Yonezaki[1]

[1] Tokyo Institute of Technology, Tokyo, Japan
{tomita,hagihara,yonezaki}@fmx.cs.titech.ac.jp
[2] Jumomji University, Saitama, Japan
nizumi@jumonji-u.ac.jp

Abstract. The concepts *promotion* and *inhibition* are commonly used to explain the effects of interactions. These concepts are used with inference rules among them, such as "X *inhibits* Y, and Y *promotes* Z, therefore X *inhibits* Z." Even when considering highly complex systems such as biological processes, many experimental facts can be explained using a few critical chains of reactions and their *promotion/inhibition* effects. The overall interaction effect of paths can be determined by considering relative properties, for example when an interaction effect of certain paths is *stronger* than that of others. In this paper, we present a formal ontology of interactions by providing a set of rules for the quantitative relations of the interaction effects of paths. Quantitative relations can be used to infer the overall interaction effect of all paths in a reaction network. Additionally, we present denotational semantics based on mass action kinetics with linear approximation at a steady state and prove the soundness of the given rules.

Keywords: Deductive reasoning, formal ontology of interactions, intensional semantics, linear time-invariant systems.

1 Introduction

In an illustration of interactions on a system, we often use some concepts and reason about them. The concepts *promotion* and *inhibition* and the following inference rules among them are commonly used to explain the effects of interactions.

- X *promotes* Y, and Y *promotes* Z, therefore X *promotes* Z,
- X *promotes* Y, and Y *inhibits* Z, therefore X *inhibits* Z,
- X *inhibits* Y, and Y *promotes* Z, therefore X *inhibits* Z,
- X *inhibits* Y, and Y *inhibits* Z, therefore X *promotes* Z,

where X, Y, and Z are objects or reactions. This framework is also helpful for intuitively understanding facts of the interactions in highly complex systems such as biological processes.

S. Nishizaki et al. (Eds.): WCTP 2012, PICT 7, pp. 13–33, 2013.
© Springer Japan 2013

In [14], a formal ontology of the concepts *promotion* and *inhibition* was proposed with such inference rules. The basic ideas are as follows:

- a reaction is a consumption-production relation between sets of resource objects and product objects,
- objects (or reactions) interact with each other through a chain of reactions connected by the competition- or supply- relations of resource objects.

We refer to such a chain of reactions as a *path*. Additionally, [14] provided denotational semantics of these concepts on a qualitative level, i.e., they assumed that the presence/absence of resources change by the occurrence of reactions, which produce or consume the resources. [13] reported quantitative semantics based on *mass action kinetics* with linear approximation at a steady state. In this framework, a reaction is interpreted as a *linear time-invariant (LTI) system*. A path is interpreted as a composition of LTI systems of individual reactions on the path, and then the interaction effect (*promotion/inhibition*) of the path is given by a sign ($+/-$) of the integral of the impulse response of the system. Moreover, the rules are improved to be valid for this semantics. However, they capture the interaction effect of only a single path with simple feedback loops.

There are multiple paths between objects or reactions in general. However, even when considering highly complex systems, many experimental facts can be explained with the *promotion/inhibition* effects of a few critical paths. The overall interaction effect of critical paths can be determined by considering relative properties, for example when an interaction effect of certain paths is *stronger* than that of others. Consider two paths p_1 and p_2 in a reaction network, where p_1 and p_2 have a positive and a negative effect, respectively. In this case, it is generally impossible to infer whether the overall interaction effect of the two paths is positive (i.e., whether $effect(p_1 + p_2) > 0$), when considering *promotion* and *inhibition* (i.e., based on $effect(p_1) > 0$ and $effect(p_2) < 0$). However, it may be possible if we can recognize the quantitative relations among the local interaction effects of individual paths, e.g., $|effect(p_1)| > |effect(p_2)|$ and $effect(p_1 + p_2) = effect(p_1) + effect(p_2)$. Obviously, such linearity is not generally valid but it is also often assumed to approximate and intuitively understand the behavior of complex systems.

In this paper, we propose a formal ontology to capture the overall interaction effect of multiple paths. We provide axiomatic semantics for quantitative relations about the interaction effect of paths, by defining a set of rules among them. The rules consist of axioms of ordered commutative rings, and other axioms that consider reaction kinetics. A reaction network applied in our ontology must have no feedback loop paths constituted of only competitive chains. This is because of the trade-offs among derivability, rule simplicity, and adaptable kinetics. We provide denotational semantics based on mass action kinetics with linear approximation at a steady state, similar to [13]. Additionally, we prove the soundness of the given rules to justify the validity of derived properties for the denotational semantics. Finally, we show how to infer the overall interaction effect of multiple paths from their local interaction effects.

The remainder of this paper is organized as follows. In Section 2, we introduce some preliminary definitions of reactions and paths, as well as linear time-invariant (LTI) systems that capture the behavior of reactions obeying mass action kinetics with linear approximation at a steady state. In Section 3, we describe work related to our study before presenting our ontology. In Section 4, we present a formal ontology of interactions. In Section 5, we provide denotational semantics, and prove the soundness of the given rules. In Section 6, we present an example of the overall interaction effect of a reaction network, and demonstrate a method for finding a term that represents this effect. In Section 7, we conclude the paper and suggest some future research directions.

2   Preliminary Definitions

2.1   Reactions and Paths

We define a *reaction* as a consumption-production relation between resource objects and product objects and a *path* as a sequence of reactions connected by the competition- or supply- relations of resource objects. We assume that interactions among objects are caused through paths, and that all reactions involve two different resources and some products.

For each main component of the network (i.e., reactions and paths), we establish finite sets of objects (O), identifiers/names of reactions (ID), variables for coefficients (V).

First, we give definitions related to reactions.

Definition 1 (Reactions). *A* reaction *is a quadruple* $\langle \alpha, src, prd, c \rangle$, *where* $\alpha \in ID$ *is the name of a reaction,* $src \subseteq O$ *is a binary set of resource objects (i.e.,* $|src| = 2$), *prd* $\subseteq O$ *is a finite set of product objects such that* $src \cap prd = \emptyset$, *and* $c : prd \to \mathbb{R}_{>0} \cup V$ *is a mapping that results in a positive real coefficient in* $\mathbb{R}_{>0}$ *or a coefficient variable in* V *for each output object.*

For example, $\langle \text{Neutralization}_1, \{\text{HCl}, \text{NaOH}\}, \{\text{NaCl}, \text{H}_2\text{O}\} \rangle$ means that 1 molecule of sodium chloride (NaCl) and 1 molecule of water (H_2O) are produced by reacting 1 molecule of hydrogen chloride (HCl) with 1 molecule of sodium hydroxide (NaOH).

We often use the symbol R to represent a reaction, and abbreviate $\langle \alpha, src, prd, c \rangle$ as $\langle \alpha, src, \{c(z)z | z \in prd\} \rangle$ (if $c(z) = 1$, we omit 1 from $1z$), especially in examples.

In this paper, we use U to represent the *universe* of reactions, and use C as the set $\{c | \langle \alpha, src, prd, c \rangle \in U\}$ of coefficient functions for reactions in U. Additionally, we assume that U satisfies the following property.

Assumption 1. A name and a binary set of resource objects for reactions in U are unique, i.e., for all $\langle \alpha_1, src_1, prd_1, c_1 \rangle, \langle \alpha_2, src_2, prd_2, c_2 \rangle \in U$, $prd_1 = prd_2$ and $\alpha_1 = \alpha_2$ (resp. $src_1 = src_2$) if $src_1 = src_2$ (resp. $\alpha_1 = \alpha_2$).

We also assume that a reaction $\langle \alpha, \{x, y\}, \{z_1, \ldots, z_n\}, c \rangle$ obeys reaction kinetics based on the *law of mass action* with a rate-constant k_α as follows:

$$d[x]_t/dt = d[y]_t/dt = -k_\alpha \cdot [x]_t \cdot [y]_t, \tag{1}$$

$$d[z_i]_t/dt = k_\alpha \cdot c(z_i) \cdot [x]_t \cdot [y]_t \quad \text{for each } i, \tag{2}$$

where $[x]_t$ (resp. $[y]_t$ and $[z_i]_t$) is the concentration of the object x (resp. y and z_i) at time t. Note that c does not generally give stoichiometric coefficients.

Definition 2 (Paths). *A path p (in the universe U) is a finite sequence $\langle \alpha_1, x_1, y_1, prd_1, c_1 \rangle \ldots \langle \alpha_n, x_n, y_n, prd_n, c_n \rangle$ of quintuples, where $\langle \alpha_i, \{x_i, y_i\}, prd_i, c_i \rangle \in U$, $\alpha_i \neq \alpha_{i+1}$ and $x_{i+1} \in \{y_i\} \cup prd_i$ for $1 \leq i < n$.*

Intuitively, p represents a chain of reactions, where a reaction on the path is connected with the next reaction by competition or supply of resources, i.e., a reaction with the name α_i interacts with another reaction with the name $\alpha_{i+1} \neq \alpha_i$ by competition if $x_{i+1} = y_i$ or by supply if $x_{i+1} \in prd_i$. For such a path p, x_1 is called the *trigger* object, and α_n and $z \in prd_n$ are called the *target* reaction and object, respectively. The empty sequence is denoted by ϵ. We often use the symbol r to represent a quintuple $\langle \alpha, x, y, prd, c \rangle$, i.e., an atomic path. We use $|p|$ to denote the length n of $p = r_1 \ldots r_n$.

For an atomic path $r = \langle \alpha, x, y, prd, c \rangle$, we use $id(r)$ to denote the name α; $trigger(r)$ to denote the resource object x, which is a trigger or is directly influenced by the previous reaction; $products(r)$ as the set prd of product objects; $coeff(r)$ as the coefficient c; and r^{opp} to represent the opposite-directed path $\langle \alpha, y, x, prd, c \rangle$. A path $p = \langle \alpha_1, x_1, y_1, prd_1, c_1 \rangle \ldots \langle \alpha_n, x_n, y_n, prd_n, c_n \rangle$ is *competitive* if $y_i = x_{i+1}$ for $1 \leq i < n$ (i.e., each reaction on p competes with the next reaction for a resource object), and p is *cyclic* if $x_1 \in \{y_n\} \cup prd_n$ (i.e., x_1 regulates itself through p). For a competitive path $p = r_1 \ldots r_n$, we use p^{opp} to denote the opposite-directed path $r_n^{opp} \ldots r_1^{opp}$.

In a manner analogous to the notation of reactions, we abbreviate $\langle \alpha, x, y, prd, c \rangle$ as $\langle \alpha, x, y, \{c(z)z | z \in prd\} \rangle$ (if $c(z) = 1$, we omit 1 from $1z$), especially in examples.

We use P to denote the universe of paths in U, i.e., $p \in P$ is a path consisting of tuples $\langle \alpha, x, y, prd, c \rangle$ corresponding to reactions $\langle \alpha, \{x, y\}, prd, c \rangle \in U$.

We make the following assumption.

Assumption 2. The universe of paths in U, P, has no competitive cyclic path.

If U has such a path, it is difficult to define a set of rules among concepts that is sound for denotational semantics. If we assume that objects, which may exist massively in a given system, are not affected by supply or competition, then there may be no competitive feedback loop in the system. In this sense, Assumption 2 is a weak restriction.

Example 1. Consider a set of reactions $\{R_1, R_2, R_3, R_4\}$, where

$$R_1 = \langle \alpha_1, \{x_1, x_2\}, \{y_1\}\rangle,$$
$$R_2 = \langle \alpha_2, \{x_2, x_3\}, \{v_1 y_2, v_2 x_5\}\rangle,$$
$$R_3 = \langle \alpha_3, \{x_3, x_4\}, \{y_3\}\rangle,$$
$$R_4 = \langle \alpha_4, \{x_4, x_5\}, \{y_3\}\rangle.$$

Let $r_1, \ldots, r_4$ be the paths $\langle \alpha_1, x_1, x_2, \{y_1\}\rangle$, $\langle \alpha_2, x_2, x_3, \{v_1 y_2, v_2 x_5\}\rangle$, $\langle \alpha_3, x_3, x_4,$ $\{y_3\}\rangle$, and $\langle \alpha_4, x_4, x_5, \{y_3\}\rangle$, respectively. In this set of reactions, x_1 interacts with y_3 through paths $r_1 r_2 r_3$, $r_1 r_2 r_3 r_4$, $r_1 r_2 (r_4^{opp})$, and so on.

2.2 Linear Time-Invariant Systems

We assume that reaction kinetics is based on Equations (1) and (2), and that it is appropriate to apply linear approximation to ordinary differential equations of reactions at a steady state (i.e., a reaction system is interpreted as a *linear time-invariant (LTI) system*). Then, to capture the behavior of a reaction network, we define a model at a steady state as follows.

Definition 3 (Models at a Steady State). *A model Δ is a pair $\langle [\cdot]_0, K\rangle$, where*

- *$[\cdot]_0 : O \to \mathbb{R}_{>0}$ is a function that gives the concentration of an object at the steady state , and*
- *$K : ID \to \mathbb{R}_{>0}$ is a function that gives a reaction-rate constant of a reaction at the steady state .*

According to a model Δ and an assignment function $f : V \to \mathbb{R}_{>0}$ for coefficient variables, we capture the behavior of an individual reaction in U, which is linearized around a steady state given by Δ and f.

Definition 4 (Linearized Reaction Systems). *Let $\Delta = \langle [\cdot]_0, K\rangle$ be a model, f be an assignment function for coefficient variables, and $R = \langle \alpha, \{x, y\}, \{z_1, \ldots, z_n\}, c\}\rangle \in U$ be a reaction.*
A system $Sys_\Delta^f(R)$ of α on Δ and f is an LTI system with 2-inputs $\boldsymbol{u}_\alpha(t)$ and $(3 + n)$-outputs $\boldsymbol{y}_\alpha(t)$ such that:

$$\frac{d}{dt}\boldsymbol{x}_\alpha(t) = \boldsymbol{A}_\alpha \boldsymbol{x}_\alpha(t) + \boldsymbol{u}_\alpha(t), \tag{3}$$
$$\boldsymbol{y}_\alpha(t) = \boldsymbol{B}_\alpha \boldsymbol{x}_\alpha(t), \tag{4}$$

where $\boldsymbol{x}_\alpha(t)$ is a state vector that gives $([x]_t - [x]_0, [y]_t - [y]_0)^T$, $\boldsymbol{u}_\alpha(t) = (I_x^\alpha(t), I_y^\alpha(t))^T$ is an input vector, $\boldsymbol{y}_\alpha(t) = (O^\alpha(t), O_x^\alpha(t), O_y^\alpha(t), O_{z_1}^\alpha(t), \ldots, O_{z_n}^\alpha(t))^T$ is an output vector,

$$A_\alpha = K(\alpha) \begin{pmatrix} -[y]_0 & -[x]_0 \\ -[y]_0 & -[x]_0 \end{pmatrix} \text{ is a state matrix,}$$

$$B_\alpha = K(\alpha) \begin{pmatrix} [y]_0 & [x]_0 \\ -[y]_0 & -[x]_0 \\ -[y]_0 & -[x]_0 \\ c[V \mapsto f(V)](z_1) \cdot [y]_0 & c[V \mapsto f(V)](z_1) \cdot [x]_0 \\ \vdots & \vdots \\ c[V \mapsto f(V)](z_n) \cdot [y]_0 & c[V \mapsto f(V)](z_n) \cdot [x]_0 \end{pmatrix} \text{ is an output matrix,}$$

and $c[V \mapsto f(V)](z_i) = f(c(z_i))$ if $c(z_i) \in V$, otherwise $c[V \mapsto f(V)](z_i) = c(z_i)$. Additionally, a transfer function $\boldsymbol{G}_\alpha$ of $Sys_\Delta^f(R)$ is

$$G_\alpha(s) = B_\alpha(sI - A_\alpha)^{-1} = \frac{B_\alpha}{s + K(\alpha)([x]_0 + [y]_0)}. \tag{5}$$

Intuitively, $I_x^\alpha(t)$ (resp., $O^\alpha(t)$ and $O_{z_i}^\alpha(t)$) represents the external input for x (resp., the change of in the rate of the reaction α, and the output for z_i) at time t. The state-space model of R is given by the *state equation* (3) and the *output equation* (4). The characteristics of $Sys_\Delta^f(R)$ can be captured via the transfer function $\boldsymbol{G}_\alpha(s)$ given by (5). The transfer function $\boldsymbol{G}_\alpha(s)$ is the ratio $\boldsymbol{Y}_\alpha(s)/\boldsymbol{U}_\alpha(s)$, where $\boldsymbol{Y}_\alpha(s)$ and $\boldsymbol{U}_\alpha(s)$ are Laplace transforms of the output $\boldsymbol{y}_\alpha(t)$ and input $\boldsymbol{u}_\alpha(t)$, respectively. Then we can easily analyze systems on a reaction network based on the operations among their transfer functions. The system of a reaction network is a compositional LTI system in which each output O_x^α of a reaction α links to x-labeled inputs $I_x^{\alpha'}$ of other reactions α'. Therefore, each supply of a resource object can be viewed as a serial connection. Then we can easily capture supply connections based on the linearity of systems. Additionally, each competitor of a resource object acts as a feedback loop, because each system of α has an input I_x^α and an output O_x^α for a resource object x.

Example 2. Consider the compositional system of reactions R_1 and R_2 given in Example 1. This system is illustrated in Fig. 1. R_1 competes with R_2 for x_2, and hence the compositional system has a feedback loop. Let $U(s)$ be the Laplace transform of the input $\boldsymbol{u}(t) = (I_{x_1}(t), I_{x_2}(t), I_{x_3}(t))^T$, $\boldsymbol{Y}_{\alpha_1}(s)$ be the Laplace transform of the output $\boldsymbol{y}_{\alpha_1}(t) = (O^{\alpha_1}(t), O_{x_1}^{\alpha_1}(t), O_{x_2}^{\alpha_1}(t), O_{y_1}^{\alpha_1}(t))^T$, and $\boldsymbol{Y}_{\alpha_2}(s)$ be the Laplace transform of the output $\boldsymbol{y}_{\alpha_2}(t) = (O^{\alpha_2}(t), O_{x_2}^{\alpha_2}(t), O_{x_3}^{\alpha_2}(t),$

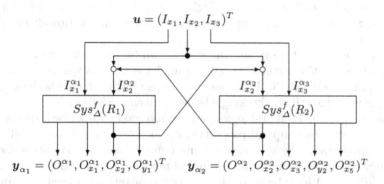

Fig. 1. The outline of the compositional system of reactions R_1 and R_2 in Example 1

$O_{y_2}^{\alpha_2}(t), O_{x_5}^{\alpha_2}(t))^T$. The behavior of the compositional system can be expressed by the following equations:

$$Y_{\alpha_1}(s) = G_{\alpha_1}(s)\left(\begin{pmatrix} 1 & 0 & 0 \\ 0 & 1 & 0 \end{pmatrix} U(s) + \begin{pmatrix} 0 & 0 & 0 & 0 & 0 \\ 0 & 1 & 0 & 0 & 0 \end{pmatrix} Y_{\alpha_2}(s)\right),$$

$$Y_{\alpha_2}(s) = G_{\alpha_2}(s)\left(\begin{pmatrix} 0 & 1 & 0 \\ 0 & 0 & 1 \end{pmatrix} U(s) + \begin{pmatrix} 0 & 0 & 1 & 0 \\ 0 & 0 & 0 & 0 \end{pmatrix} Y_{\alpha_1}(s)\right).$$

In our denotational semantics in Section 5, we interpret the integral of the unit impulse response of a system over time t from 0 to infinity as the interaction effect of the system. Regarding the integral of the unit impulse response of an LTI system, the following theorem holds.

Theorem 5 (Final Value Theorem). *Let Sys be a 1-input 1-output LTI system, $Sys(t)$ be the unit impulse response of Sys at time t, and G_{Sys} be a transfer function of Sys.*

Then the following equation can be derived:

$$\int_0^\infty Sys(t)dt = \lim_{s \to 0} G_{Sys}(s). \tag{6}$$

3   Related Work

Formal ontologies have recently been used in various fields. Usually, the semantics of a formal ontology is extensional, i.e., it is represented as a list of all valid relations among concepts. Each ontology organizes the knowledge of a considering domain. One of the most successful ontologies is the Gene Ontology [2] (GO) used in bioinformatics. GO provides a controlled vocabulary to describe characteristics of gene products and annotation data.

In contrast, the semantics of the ontologies in this paper and [14,13] is intensional, i.e., it is presented based on the definitions of properties of relations. Such ontologies formalize the general properties of relations. The ontologies of [14,13] focused on the interaction effect of a single path at the *promotion/inhibition* level, whereas ours considers the quantitative relations of the interaction effects of multiple paths.

To analyze a complex system, the system can be abstracted and verified on several levels. At the qualitative level, a model can be viewed as a transition system, process algebra [3], Petri net [10], and so forth. Model checking [5] is a technique for exhaustively checking whether a given model satisfies a given specification (property). Temporal logics [6] are often used to describe a specification that reflects various temporal properties, e.g., safety, liveness, reachability and progress properties. The verification using temporal logics is traditionally used to analyze of various systems, and standard model-checking tools (e.g., NuSMV [9] and SPIN [12]) have been developed. At the quantitative level, a model cab be viewed not only as a hybrid or probabilistic extension of a transition system, process algebras, Petri nets, and so forth, but also as a system of ordinary differential equations (ODEs). Methods of quantitative analysis (especially for a system of ODEs) are often based on simulation.

In systems biology, several tools (e.g., BIOCHAM [8,7]) are used for multi-level analysis of systems. BIOCHAM can analyze models at the Boolean, differential, and stochastic levels. A model on BIOCHAM consists of a set of reaction rules among objects with kinetic expressions. It can be checked whether the model satisfies various types of properties represented in some temporal logics. At the Boolean level, the presence/absence of objects is switched by reaction rules. BIOCHAM can be interfaced with NuSMV, and then a model can be verified by classical model-checking. At the differential level, the concentrations of objects are changed continuously according to kinetics described flexibly in the modeling language of BIOCHAM. In general, it is difficult to verify a model with a continuous state-space. In BIOCHAM, a model must be defined completely, i.e., the behavior of the model is deterministic. Then, the model is verified with only a finite-time trace, i.e., via simulation-based quantitative verification. A BIOCHAM model is compatible with the systems biology markup language [1] (SBML), which is used as a standard modeling format in systems biology. Hence, analysis tools for SBML can also be used for verifications of BIOCHAM models. At the stochastic level, a model is verified with only a certain number of sampled finite traces, i.e., using Monte-Carlo methods. A probabilistic model-checker (e.g., PRISM[11]) is often used for general quantitative verification for probabilistic systems. In [4], the authors conducted a quantitative analysis of discrete concentration levels using PRISM, although it is intractable to verify a model with an infinite state-space.

In contrast, our ontology and that set out in [13] (resp., [14]) enable analysis using deductive reasoning for a model at the differential (resp., Boolean) level. Our ontology (resp., [13]) can only handle properties of quantitative (resp., qualitative) relations among the interaction effects of paths in a reaction network. We also assume that each reaction requires two types of resources and that each reaction-rate is based on simple mass action kinetics. Hence, from properties in the known range, we can derive conclusions that are valid for any model with those properties. Additionally, our semantics captures the approximate behavior of a system with a continuous state-space.

4   A Formal Ontology of Interactions

In this section, we introduce a formal ontology of interactions.

First, we provide sets of symbols that is used in the language of the ontology, as follows:

- $\mathcal{O} \stackrel{\text{def}}{=} \{\hat{x}|x \in O\}$ is a set of symbols representing objects,
- $\mathcal{ID} \stackrel{\text{def}}{=} \{\hat{\alpha}|\alpha \in ID\}$ is a set of symbols representing names of reactions,
- $\mathcal{V} \stackrel{\text{def}}{=} \{\hat{v}|v \in V\}$ is a set of symbols representing variables for coefficients,
- $\mathcal{C} \stackrel{\text{def}}{=} \{\hat{c}|c \in C\}$ is a set of coefficient functions for symbols of objects, where $\hat{c}(\hat{x}) = \hat{v}$ if $c(x) = v \in V$, otherwise $\hat{c}(\hat{x}) = c(x)$ for $x \in \text{Dom}(c)$,
- $\mathcal{P} \stackrel{\text{def}}{=} \{\hat{p}|p \in P\}$, where $\hat{p} = p[x/\hat{x}$ for all $x \in O, \alpha/\hat{\alpha}$ for all $\alpha \in ID, c/\hat{c}$ for all $c \in C]$.

Then, we define a language for representing properties of the structure and behavior of a system as follows.

Definition 6 (Syntax). *The language $\mathcal{L}$ of the formal ontology of interactions is a kind of first-order logic that has: standard logical connectives $\neg$ and $\wedge$, a quantifier $\forall$, a relation $\precsim$, function symbols $\oplus$, $\otimes$, $\ominus$ and $^{\circledast}$, constant symbols in $\mathbb{R} \cup \mathcal{P}$, and finite variable symbols in $\mathcal{V}$, where a variable $v \in V$ is assigned only a real constant in $\mathbb{R}$.*

For formulae φ and ψ, and terms τ, τ' and τ'', we use the following abbreviations:

$$\varphi \vee \psi \equiv \neg(\varphi \wedge \psi),$$
$$\varphi \leftrightarrow \psi \equiv (\varphi \rightarrow \psi) \wedge (\psi \rightarrow \varphi),$$
$$\tau \approx \tau' \equiv \tau \precsim \tau' \wedge \tau' \precsim \tau,$$
$$\tau \succsim \tau' \equiv \tau' \precsim \tau,$$
$$\tau \, \Box \, \tau \, \Diamond \, \tau'', \equiv \tau \, \Box \, \tau' \wedge \tau' \, \Diamond \, \tau'',$$
$$\tau^i \equiv \overbrace{\tau \otimes \ldots \otimes \tau}^{i},$$

$$\varphi \rightarrow \psi \equiv \neg\varphi \vee \psi,$$
$$\exists v.\varphi(v) \equiv \neg\forall v.\neg\varphi(v),$$
$$\tau \prec \tau' \equiv \tau \precsim \tau' \wedge \neg(\tau' \precsim \tau),$$
$$\tau \succ \tau' \equiv \tau' \prec \tau,$$
$$\tau \ominus \tau' \equiv \tau \oplus (\ominus\tau'),$$
$$abs(\tau) = \begin{cases} \tau & \text{if } 0 \precsim \tau, \\ \ominus\tau & \text{otherwise,} \end{cases}$$

where $i \in \mathbb{N}_{>0}$, and $\Box, \Diamond \in \{\precsim, \prec, \approx, \succ, \succsim\}$.

Intuitively, a term τ represents:

- a constant-multiplier that responds a times of the input if $\tau = \hat{a} \in \mathbb{R} \cup V$,

$$In \rightarrow \boxed{a} \rightarrow Out \ (= a \cdot In)$$

- a system corresponding to a path $p = r_1 \ldots r_n$ if $\tau = \hat{p} \in \mathcal{P}$, where the system responds to the change in the rate of a target reaction $id(r_n)$ through a path p, for the input to a trigger object $trigger(r_1)$.

$$In \text{ for } trigger(r_1) \rightarrow \boxed{\text{The system for } p} \rightarrow Out \text{ for } id(r_n)$$

- a serial connection system of τ_1 to τ_2 if $\tau = \tau_1 \otimes \tau_2$,

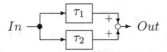

- a parallel connection system of τ_1 and τ_2 if $\tau = \tau_1 \oplus \tau_2$,

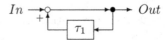

- an inverter of τ_1 (i.e., it gives -1 times of the output of τ_1, for the input) if $\tau = \ominus \tau_1$,
- a feedback-loop system of τ_1 if $\tau = \tau_1^{\circledast}$.

$$In \rightarrow \begin{array}{c} + \\ \end{array} \rightarrow Out$$
$$\boxed{\tau_1}$$

A predicate $\tau_1 \precsim \tau_2$ means that the interaction effect of τ_1 is less (weaker) than or equal to that of τ_2. Intuitively, $\hat{0} \prec \tau$ (resp., $\tau \prec \hat{0}$) means that the interaction effect of τ is positive (resp., negative).

We intend to use a formula $\tau_1 \precsim \tau_2$ to represent a quantitative relation between terms τ_1 and τ_2 with "well-connected" structures. However, in the process of deriving $\tau_1 \precsim \tau_2$ from a hypothesis set Γ of formulae (or deriving another formula φ with $\tau_1 \precsim \tau_2$), it is not necessary to restrict terms to being well connected. Therefore, note the following points.

- A term $p_1 \otimes p_2$ is well connected if $p_1 p_2 \in \mathcal{P}$, but it is not generally required. In addition, $\hat{p}_1 \hat{p}_2$ may be not equivalent to $\hat{p}_1 \otimes \hat{p}_2$. If the last reaction of p_1 competes with the first reaction of p_2 for a resource object, it is not appropriate to interpret $p_1 p_2$ as $p_1 \otimes p_2$ in our formalism. Even if the last reaction of p_1 supplies a resource object of the first reaction of p_2, we must consider a coefficient for the resource object.
- A term $\hat{p}_1 \oplus \hat{p}_2$ is well connected if trigger objects (and target reaction/object) of p_1 and p_2 are the same, but it is also not required.
- For a competitive path p, a term $\hat{p}^{opp}$ is not generally equivalent to $\ominus \hat{p}$. Consider an atomic path $r = \langle \alpha, x, y, prd, c \rangle$. A term $\hat{r}$ (resp., $\hat{r}^{opp} = \langle \hat{\alpha}, \hat{y}, \hat{x}, \widehat{prd}, \hat{c} \rangle$) represents a system that transfers the interaction effect from the trigger x (resp., y) to the target α through r. We can intuitively find that both r and r^{opp} have a positive effect.
- A term $\hat{p}^{\circledast}$ is well connected if a trigger object of p is one of target objects, but it is not required. In general, $\hat{p}^{\circledast}$ is not equivalent to $\hat{p}$. Furthermore, $\tau^{\circledast}$ may be incomparable to any other term, i.e., the value of the interaction effect of $\tau^{\circledast}$ may be undefined.

In the following, we use p as $\hat{p}$ (i.e., x as $\hat{x} \in \mathcal{O}$, α as $\hat{\alpha} \in \mathcal{ID}$, v as $\hat{v} \in \mathcal{V}$ and c as $\hat{c} \in \mathcal{C}$) if we can clearly recognize that these symbols are in syntax.

Next we give a set of rules among formulae in $\mathcal{L}$. $\mathcal{L}$ is a kind of first-order logic, and hence the set of rules for $\mathcal{L}$ contains the ones of the first-order logic. For formula φ and a set Γ of formulae, we denote $\Gamma \vdash \varphi$ if φ is derivable from Γ with the derivation rules in first-order logic, e.g., $\{\varphi_1, \varphi_2\} \vdash \varphi_1 \wedge \varphi_2$ and $\{\varphi, \varphi \to \psi\} \vdash \psi$. Additionally, the ontology includes the following rules; they reflect the denotational semantics given in Section 5, therefore some of them may not seem intuitive.

Definition 7 (Additional Rules). *The additional rules in the ontology consist of*

- *(A0) a variant of axioms of an ordered commutative ring, and*
- *(A1-A10) rules that reflect the kinetics and structure of a reaction network.*

A0 (Axioms of an ordered commutative ring on comparable terms). These rules consist of axioms of an ordered commutative ring with some conditions. The equivalent relations (and also $\tau_1 \square \tau_2$) on the left side of the implications are conditions to guarantee the comparability of terms.

(a) : $\vdash a \precsim a$, where $a \in \mathbb{R} \cup \mathcal{V}$

(b) : $\vdash \tau_1 \square \tau_2 \to \tau_1 \precsim \tau_1$, where $\square \in \{\precsim, \prec, \approx, \succ, \succsim\}$,

(c) : $\vdash (\tau_1 \precsim \tau_2 \precsim \tau_3) \to \tau_1 \precsim \tau_3$,

(d) : $\vdash (\bigwedge_{i=1,2} \tau_i \approx \tau_i) \to \tau_1 \precsim \tau_2 \vee \tau_2 \precsim \tau_1$,

(e) : $\vdash \tau_1 \approx \tau_1 \to 0 \oplus \tau_1 \approx \tau_1$,

(f) : $\vdash \tau_1 \approx \tau_1 \to \tau_1 \oplus (\ominus \tau_1) \approx 0$,

(g) : $\vdash (\bigwedge_{i=1,2} \tau_i \approx \tau_i) \to \tau_1 \oplus \tau_2 \approx \tau_2 \oplus \tau_1$,

(h) : $\vdash (\bigwedge_{i=1,2,3} \tau_i \approx \tau_i) \to (\tau_1 \oplus \tau_2) \oplus \tau_3 \approx \tau_1 \oplus (\tau_2 \oplus \tau_3)$,

(i) : $\vdash (\tau_1 \approx \tau_1 \wedge \tau_2 \precsim \tau_3) \to \tau_1 \oplus \tau_2 \precsim \tau_1 \oplus \tau_3$,

(j) : $\vdash \tau_1 \approx \tau_1 \to 1 \otimes \tau_1 \approx \tau_1$,

(k) : $\vdash (\bigwedge_{i=1,2} \tau_i \approx \tau_i) \to \tau_1 \otimes \tau_2 \approx \tau_2 \otimes \tau_1$,

(l) : $\vdash (\bigwedge_{i=1,2,3} \tau_i \approx \tau_i) \to (\tau_1 \otimes \tau_2) \otimes \tau_3 \approx \tau_1 \otimes (\tau_2 \otimes \tau_3)$,

(m) : $\vdash (\bigwedge_{i=1,2,3} \tau_i \approx \tau_i) \to (\tau_1 \oplus \tau_2) \otimes \tau_3 \approx (\tau_1 \otimes \tau_3) \oplus (\tau_2 \otimes \tau_3)$,

(n) : $\vdash (0 \prec \tau_1 \wedge 0 \prec \tau_2) \to 0 \prec \tau_1 \otimes \tau_2$.

A1 (Empty path axiom). The interaction effect of the empty path ϵ is equal to the given effect (i.e., 1).

$$\vdash \epsilon \approx 1.$$

A2 (Positive coefficient axiom). Every coefficient variable $v \in \mathcal{V}$ is positive.

$$\vdash 0 \prec v.$$

A3 (Non-competitive path axiom). The interaction effect of a path $r_1 \ldots r_n \in \mathcal{P}$ with a supply connection between r_i and r_{i+1} $(1 \leq i < n)$ is equivalent to the multiplication of that of paths $r_1 \ldots r_i$ and $r_{i+1} \ldots r_n$, considering the coefficient $coeff(r_i)(trigger(r_{i+1}))$ of the supplied object $trigger(r_{i+1})$.

$$\vdash r_1 \ldots r_n \approx r_1 \ldots r_i \otimes coeff(r_i)(trigger(r_{i+1})) \otimes r_{i+1} \ldots r_n,$$

where $1 \leq i < n$ and $trigger(r_{i+1}) \in products(r_i)$.

A4 (Competitive path sign axiom). For a competitive and nonempty path $p \in \mathcal{P}$, the interaction effect of p is *promotion* if p has an odd length, and is *inhibition* if p has an even length.

$$\vdash 0 \prec (\ominus 1)^{|p|-1} \otimes p,$$

where p is competitive and nonempty.

A5 (Competitive suffix path axiom). The absolute value of the interaction effect of a competitive path $p_1 p_2 \in \mathcal{P}$ is equal to that of the nonempty suffix p_2.

$$\vdash abs(p_1 p_2) \approx abs(p_2),$$

where $p_1 p_2$ is competitive and p_2 is nonempty.

A6 (Competitive prefix path axiom). The absolute value of the interaction effect of a competitive path $p_1 p_2 \in \mathcal{P}$ is less than that of the proper prefix p_1.

$$\vdash abs(p_1 p_2) \prec abs(p_1),$$

where $p_1 p_2$ is competitive and p_2 is nonempty.

A7 (Shared trigger axiom). If atomic paths $r_1, \ldots, r_n \in \mathcal{P}$ have the same trigger object, then the summation of the absolute values of their interaction effects is less than the given effect 1 on their trigger object.

$$\vdash \bigoplus_{i=1}^{n} r_i \prec 1,$$

where $trigger(r_i) = trigger(r_j)$ and $id(r_i) \neq id(r_j)$, for all $i \neq j$.

A8 (Dual trigger axiom). For an atomic path $r = \langle \alpha, x, y, prd, c \rangle$, if we give the same input effect to both resources x and y of the reaction α, then the interaction effect $r \oplus r^{opp}$ on α is exactly equal to that of the given effects.

$$\vdash r \oplus r^{opp} \approx 1.$$

A9 (Feedback loop convergence axiom). The interaction effect of a feedback loop of τ is considered as an infinite geometric series $\epsilon \oplus \tau_1 \oplus \tau_1^2 \oplus \ldots$ of τ. In illegal notation, $\tau^{\circledast}$ is equivalent to $(1 \ominus \tau)^{-1}$ if the absolute value of τ is less than 1 and not equal to 0.

$$\vdash (\ominus 1 \prec \tau \prec 0 \lor 0 \prec \tau \prec 1) \rightarrow (1 \ominus \tau) \otimes \tau^{\circledast} \approx 1.$$

A9' (Feedback loop comparability axiom). A term $\tau^{\circledast}$ is comparable if the absolute value of τ is less than 1 and not equal to 0.

$$\vdash (\ominus 1 \prec \tau \prec 0 \vee 0 \prec \tau \prec 1) \rightarrow \tau^{\circledast} \approx \tau^{\circledast}.$$

A10 (Feedback loop divergence axiom). The interaction effect of a feedback loop of τ_1 is undefined if the absolute value of τ_1 is greater than or equal to 1.

$$\vdash 1 \precsim abs(\tau_1) \rightarrow \neg(\tau_1^{\circledast} \,\square\, \tau_2),$$

where $\square \in \{\precsim, \prec, \approx, \succ, \succsim\}$.

As a result, we can define axiomatic semantics of $\mathcal{L}$ as follows.

Definition 8 (Axiomatic Semantics). *Axiomatic semantics of $\mathcal{L}$ is defined by the derivation rules of first-order logic and rules A0–A10.*

Example 3. With the additional rules in Definition 7, we can derive some properties on the reaction network in Example 1. Let $p_1 = r_1 r_2$, $p_2 = r_1 r_2 r_3$, $p_3 = r_1 r_2 r_3 r_4$, $p_4 = r_4^{opp}$, $p_5 = (r_3 r_4)^{opp}$ and $p_6 = (r_2 r_3 r_4)^{opp}$. The derivation $\Gamma \vdash \varphi$ with a rule An is denoted by $\dfrac{\Gamma}{\varphi} An$.

1. The object x_1 has a negative interaction effect on α_4 through $p_1 p_4$ ($= r_1 r_2 (r_4^{opp})$), i.e., $p_1 p_4 \prec 0$.

$$\dfrac{\dfrac{}{p_1 p_4 \approx p_1 \otimes v_2 \otimes p_4} A3 \qquad \dfrac{\dfrac{}{0 \prec v_2} A2 \qquad \dfrac{\dfrac{}{0 \prec p_4} A4 \qquad \dfrac{0 \prec \ominus p_1}{} A4}{0 \prec (\ominus p_1) \otimes v_2 \otimes p_4} A0}{p_1 p_4 \prec 0} A0$$

2. The term $p_2 \oplus p_3$ indicates the overall interaction effect of p_2 and p_3, and we can derive $0 \prec p_2 \oplus p_3$ by A6, because p_2 is a proper prefix of p_3. Therefore, x_1 promotes y_3 through p_2 and p_3.

$$\dfrac{\dfrac{}{\ominus p_3 \prec p_2} A6}{0 \prec p_2 \oplus p_3 \quad (= \varphi_1)} A0$$

3. Using the following derivation, we obtain the quantitative relation "the total interaction effect of p_2 and p_3 is equal to that of p_4 and p_5."

$$\dfrac{\dfrac{r_3 \approx p_2, r_3^{opp} \approx \ominus p_5}{p_2 \oplus (\ominus p_5) \approx 1} A5 \quad \dfrac{r_3 \oplus r_3^{opp} \approx 1}{} A8 \quad A0 \qquad \dfrac{\dfrac{r_4 \approx \ominus p_3}{(\ominus p_3) \oplus p_4 \approx 1} A5 \quad \dfrac{r_4 \oplus p_4 \approx 1}{} A8}{} A0}{p_2 \oplus p_3 \approx p_4 \oplus p_5 \quad (= \varphi_2)} A0$$

4. The reaction α_2 produces x_5, and then this reaction network has the feedback-loop of the cyclic path p_6. The interaction effect from x_5 to x_5 through the feedback loop is $(p_6 \otimes v_2)^\circledast$, because $coeff(last(p_6)) = v_2$. Regarding this feedback loop, the following derivation can be obtained with hypothesis $v_1 \oplus v_2 \approx 1$.

$$\cfrac{\cfrac{}{0 \prec p_6 \prec 1}\,A4 \qquad \cfrac{v_1 \oplus v_2 \approx 1 \quad \cfrac{\overline{0 \prec v_1}}{0 \prec v_2 \prec 1} \,A2 \quad \cfrac{\overline{0 \prec v_2}}{(=\varphi_3'')}\,\begin{matrix}A2\\A0\end{matrix}}{0 \prec v_2 \prec 1 \quad (=\varphi_3'')}\,A0}{\cfrac{0 \prec p_6 \otimes v_2 \prec 1 \quad (=\varphi_3')}{(1 \ominus p_6 \otimes v_2) \otimes (p_6 \otimes v_2)^\circledast \approx 1 \quad (=\varphi_3)}\,A9}$$

5   Denotational Semantics

In this section, we consider how to denotationally interpret a formula in the language, and justify the validity of properties derived from the given rules.

The behavior of a composition of LTI systems of individual reactions is defined in Subsection 2.2, but it is troublesome to explicitly consider all feedback loops among the LTI systems for competitive connections. Hence, we define a system of a competitive path as follows, to effectively derive quantitative relations on a whole system.

Definition 9 (Competitive Path Systems). *Let Δ be a model, f be a variable assignment for V, and $p = r_1 \dots r_n \in P$ be a nonempty and competitive path such that $r_n = \langle \alpha, x, y, \{z_1, \dots, z_m\}, c \rangle$.*

A competitive path system $CompSys_\Delta^f(p)$ of p on Δ and f is a compositional LTI system that considers all competitive connections of reaction systems $Sys_\Delta^f(R)$ for $R \in U$ but does not consider supply connections and has 1-input $I_{trigger(r_1)}$ and $(m+1)$-output $(O^\alpha, O_{z_1}^\alpha, \dots, O_{z_m}^\alpha)^T$, where $I_{trigger(r_1)}$ is connected with all x-labeled inputs.

Example 4. Consider a competitive path system $CompSys_\Delta^f(p_2)$ of path p_2 in Examples 1 and 3. If the universe of reactions, U, consists only of reactions $R_1, \dots, R_4$, the outline of $CompSys_\Delta^f(p_2)$ is illustrated in Fig. 2. The reaction R_1 competes with the reaction R_2 for the resource x_2. Therefore, the output $O_{x_2}^{\alpha_1}$ of $Sys_\Delta^f(R_1)$ (resp., $O_{x_2}^{\alpha_2}$ of $Sys_\Delta^f(R_2)$) is connected with the inputs $I_{x_2}^{\alpha_2}$ of $Sys_\Delta^f(R_2)$ (resp., $I_{x_2}^{\alpha_1}$ of $Sys_\Delta^f(R_1)$). The reaction R_2 supplies the reaction R_4 with the resource x_5, but this supply relation (i.e., $O_{x_5}^{\alpha_2}$ and $I_{x_5}^{\alpha_4}$) is not connected on $CompSys_\Delta^f(p_2)$.

A competitive path system for a competitive path (e.g., p_2 in Example 4) reflects the competition in surrounding reactions (e.g., α_4) which do not appear in the competitive path. It is difficult to capture behavioral properties in any network

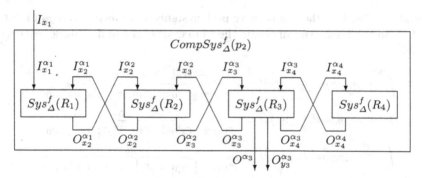

Fig. 2. The outline of the competitive path system $CompSys^f_\Delta(p_2)$ of Example 1

structure, but we can easily capture the interaction effect of competitive paths on network structures satisfying Assumption 2 (i.e., without competitive cyclic paths). For a competitive path $p = r_1 \ldots r_n$, we use $CompSys^f_\Delta(p)(t)$ to denote the unit impulse response of the output $O^{id(r_n)}$ of $CompSys^f_\Delta(p)$ at time t. Then, the following lemma holds.

Lemma 10 (The Integral of the Unit Impulse Response of $CompSys^f_\Delta(p)$).
Let $\Delta = \langle [\cdot]_0, K \rangle$ be a model, f be a variable assignment for V, and $p = r_1 \ldots r_n \in P$ be a nonempty and competitive path such that $trigger(r_1) = x$ and $id(r_n) = \alpha$.
Then we can derive the following equation:

$$\int_0^\infty CompSys^f_\Delta(p)(t)dt = (-1)^{|p|+1} \cdot \left(\frac{\displaystyle\sum_{x' \in CompTriggers(\{r_n^{opp}\})} [x']_0}{\displaystyle\sum_{x' \in CompTriggers(\{r_n, r_n^{opp}\})} [x']_0} \right), \quad (7)$$

where, for a set S of atomic paths, $CompTriggers(S) \overset{def}{=} \{ trigger(r'_1) | r'_1 \ldots r'_m \in P$ such that it is competitive and $r'_m \in S \}$.

Proof. (Sketch) From Assumption 2 in Section 2, there is no competitive cyclic path in U. Due to the simplicity in the system structure, we can easily confirm Equation (7), using Theorem 5 (Equation (6)) and the limit solution of the equations of transfer functions.

Based on Equation (7), if the sum of the concentrations of resources containing x in $CompTriggers(\{r_n\})$ is relatively large (resp., small) compared to that of resources in $CompTriggers(\{r_n^{opp}\})$, the unit impulse input for x has a relatively weak (resp., strong) effect on r_n.

Example 5. Consider the competitive path systems of paths $p_1, \ldots, p_6$ in Examples 1, 3, and 4. Then, we can obtain their integrals of the unit impulse responses as follows.

$$\int_0^\infty CompSys_\Delta^f(p_1)(t)dt = \frac{-([x_3]_0 + [x_4]_0 + [x_5]_0)}{[x_1]_0 + [x_2]_0 + [x_3]_0 + [x_4]_0 + [x_5]_0},$$

$$\int_0^\infty CompSys_\Delta^f(p_2)(t)dt = \frac{[x_4]_0 + [x_5]_0}{[x_1]_0 + [x_2]_0 + [x_3]_0 + [x_4]_0 + [x_5]_0},$$

$$\int_0^\infty CompSys_\Delta^f(p_3)(t)dt = \frac{-[x_5]_0}{[x_1]_0 + [x_2]_0 + [x_3]_0 + [x_4]_0 + [x_5]_0},$$

$$\int_0^\infty CompSys_\Delta^f(p_4)(t)dt = \frac{[x_1]_0 + [x_2]_0 + [x_3]_0 + [x_4]_0}{[x_1]_0 + [x_2]_0 + [x_3]_0 + [x_4]_0 + [x_5]_0},$$

$$\int_0^\infty CompSys_\Delta^f(p_5)(t)dt = \frac{-([x_1]_0 + [x_2]_0 + [x_3]_0)}{[x_1]_0 + [x_2]_0 + [x_3]_0 + [x_4]_0 + [x_5]_0},$$

$$\int_0^\infty CompSys_\Delta^f(p_6)(t)dt = \frac{[x_1]_0 + [x_2]_0}{[x_1]_0 + [x_2]_0 + [x_3]_0 + [x_4]_0 + [x_5]_0}.$$

Next, we give the denotational semantics of $\mathcal{L}$. In our semantics, a competitive path is interpreted as a competitive path system of the path, and a non-competitive path is interpreted as a serial connection system of individual competitive path systems of competitive subpaths.

Definition 11 (Interpretation of Paths). *Let Δ be a model, and f be a variable assignment for $\mathcal{V}$.*

An interpretation $\langle\!\langle \cdot \rangle\!\rangle_\Delta^f$ of paths is defined as follows:

- $\langle\!\langle \hat{\epsilon} \rangle\!\rangle_\Delta^f$ *is the 1-input 1-output LTI system responding the input-self,*
- $\langle\!\langle \hat{p} \rangle\!\rangle_\Delta^f$ *is a competitive path system $CompSys_\Delta^f(p)$, for a competitive and nonempty path p,*
- $\langle\!\langle \hat{p}_1\hat{p}_2 \rangle\!\rangle_\Delta^f$ *is a serial connection system of $\langle\!\langle \hat{p}_1 \rangle\!\rangle_\Delta^f$ and $\langle\!\langle \hat{p}_2 \rangle\!\rangle_\Delta^f$ (Fig. 3), where the input of $\langle\!\langle \hat{p}_1\hat{p}_2 \rangle\!\rangle_\Delta^f$ is the input of $\langle\!\langle \hat{p}_1 \rangle\!\rangle_\Delta^f$, the output $O_{z_i}^\alpha$ of $\langle\!\langle \hat{p}_1 \rangle\!\rangle_\Delta^f$ is connected with the input I_x of $\langle\!\langle \hat{p}_2 \rangle\!\rangle_\Delta^f$, and the output of $\langle\!\langle \hat{p}_1\hat{p}_2 \rangle\!\rangle_\Delta^f$ is the output of $\langle\!\langle \hat{p}_2 \rangle\!\rangle_\Delta^f$, for $\langle\!\langle \hat{p}_1 \rangle\!\rangle_\Delta^f$ such that the output is $(O^\alpha, O_{z_1}^\alpha, \ldots, O_{z_n}^\alpha)^T$, and $\langle\!\langle \hat{p}_2 \rangle\!\rangle_\Delta^f$ such that the input is (I_x) and $x = z_i$ for some i.*

A term is interpreted as a 1-input 1-output system obtained by composition operations over systems for constants, variables, and paths.

Definition 12 (Interpretation of Terms). *Let Δ be a model, and f be a variable assignment for $\mathcal{V}$.*

An interpretation $[\![\cdot]\!]_\Delta^f$ of terms is defined as follows:

- $[\![a]\!]_\Delta^f$ *(where $a \in \mathbb{R}$) is an LTI system responding a times of the input,*
- $[\![v]\!]_\Delta^f$ *(where $v \in \mathcal{V}$) is $[\![f(v)]\!]_\Delta^f$,*
- $[\![\hat{\epsilon}]\!]_\Delta^f$ *is $\langle\!\langle \hat{\epsilon} \rangle\!\rangle_\Delta^f$,*

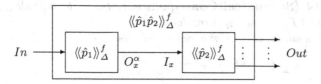

Fig. 3. The system represented by $\langle\langle\hat{p}_1\hat{p}_2\rangle\rangle_\Delta^f$ for a non-competitive path $\hat{p}_1\hat{p}_2$, where the last reaction of $\hat{p}_1$ supplies the resource x of the first reaction of $\hat{p}_2$

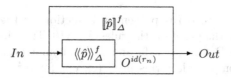

Fig. 4. The system represented by $[\![\hat{p}]\!]_\Delta^f$ for a nonempty path $\hat{p}$

- $[\![\hat{p}]\!]_\Delta^f$ *(where $p = r_1 \ldots r_n$ is a nonempty path) is a 1-input 1-output LTI system (Fig. 4), which is obtained by eliminating outputs other than $O^{id(r_n)}$ from $\langle\langle\hat{p}\rangle\rangle_\Delta^f$,*
- $[\![\tau_1 \otimes \tau_2]\!]_\Delta^f$ *is a serial connection system of $[\![\tau_1]\!]_\Delta^f$ and $[\![\tau_2]\!]_\Delta^f$,*
- $[\![\tau_1 \oplus \tau_2]\!]_\Delta^f$ *is a parallel connection system of $[\![\tau_1]\!]_\Delta^f$ and $[\![\tau_2]\!]_\Delta^f$,*
- $[\![\ominus\tau]\!]_\Delta^f$ *is an inverter of $[\![\tau]\!]_\Delta^f$,*
- $[\![\tau^{\circledast}]\!]_\Delta^f$ *is a feedback-loop system of $[\![\tau]\!]_\Delta^f$.*

Then, a quantitative relation $\tau_1 \precsim \tau_2$ is defined as a relation between the integrals of the unit impulse responses of the systems for τ_1 and τ_2.

Definition 13 (Denotational Semantics of $\mathcal{L}$). *Let f be a variable assignment for V.*

For a model Δ and a formula φ in $\mathcal{L}$, we define a satisfaction relation $\models_f$ among them as follows:

$$\Delta \models_f \tau_1 \precsim \tau_2 \overset{def}{=} \int_0^\infty [\![\tau_1]\!]_\Delta^f(t)dt \in \mathbb{R}, \int_0^\infty [\![\tau_2]\!]_\Delta^f(t)dt \in \mathbb{R} \text{ and}$$

$$\int_0^\infty [\![\tau_1]\!]_\Delta^f(t)dt \le \int_0^\infty [\![\tau_2]\!]_\Delta^f(t)dt,$$

$$\Delta \models_f \neg\varphi \overset{def}{=} \Delta \not\models_f \varphi,$$

$$\Delta \models_f \varphi \wedge \psi \overset{def}{=} \Delta \models_f \varphi \text{ and } \Delta \models_f \psi,$$

$$\Delta \models_f \forall v.\varphi(v) \overset{def}{=} \forall v' \in \mathbb{R}_{>0}.(\Delta \models_{f[v \mapsto v']} \varphi(v)),$$

where $[\![\tau_i]\!]_\Delta^f(t)$ is the unit impulse response of $[\![\tau_i]\!]_\Delta^f$ at time t, and $f[v \mapsto v'](v'') = v'$ if $v'' = v$, otherwise $f[v \mapsto v'](v'') = f(v'')$.

Finally, we define semantical consequences and show the soundness of the given rules in Section 4.

Definition 14 (Semantical Consequences). *Let Δ be a model, φ be a formula on $\mathcal{L}$, and Γ be a set of formulae on $\mathcal{L}$.*

We define semantical consequences as follows:

$$\Delta \models_f \Gamma \stackrel{def}{=} \forall \varphi \in \Gamma.(\Delta \models_f \varphi),$$
$$\Gamma \models \varphi \stackrel{def}{=} \forall \Delta, f.(\text{if } \Delta \models_f \Gamma, \text{then } \Delta \models_f \varphi).$$

Theorem 15 (Soundness). *Let φ be a formula, and Γ be a set of formulae on $\mathcal{L}$. If $\Gamma \vdash \varphi$, then $\Gamma \models \varphi$.*

Proof. (Sketch) This theorem is proven by induction on the derivations. This paper only represents the proof for the rules A0-A10. Trivially, rules A1 and A2 are sound. Based on Definition 4, 9, and 11-13, homomorphism appears from the closure set of comparable terms in $\mathcal{L}$ (except for terms whose values are undefined) to a real number field $\mathbb{R}$, for each model Δ and variable assignment f. Therefore, rules A0 and A9 (and also A9') are sound. However, for a term τ such that $abs(\tau) \gtrsim 1$, the value $\int_0^\infty [\![\tau^\circledast]\!]_\Delta^f(t)dt$ is undefined, and then $\tau^\circledast$ is incomparable to any other term. Therefore, the rule A10 is sound. The soundness of the rule A4-A8 is derived from Lemma 10. Finally, the rule A3 is sound because of Lemma 10 and Definition 11.

Due to the linearity of systems, this definition is sufficient to capture the overall interaction effects of some/all paths in a reaction network. From Definition 9 and Lemma 10, the term p denotes the interaction effect of a competitive path p, considering reactions that do not appear on p. Therefore, to some degree we can avoid the cumbersome inferences and expressions on feedback loops caused by competition among resources. Then, we can represent the overall interaction effect between objects and reactions by the summation of paths between them.

In this semantics, any term in $\mathcal{L}$ is interpreted as a certain kind of LTI system, and then the rules defining the ontology are specialized for them. The definition of this semantics can easily be extended not only to linearized reaction systems but also to other types of reaction systems at a steady state. However, it is not obvious that the representation with serial connections $\otimes$, parallel connections $\oplus$, and feedback loops $\circledast$ is appropriate for describing a complex structure that consists of other types of systems (e.g., nonlinear systems).

6   Overall Interaction Effect of Reaction Networks

To determine the overall interaction effect of a reaction network (or subnetwork), we must find a certain quantitative relation that represents a property of the overall interaction effect of all paths on the network. In this section, we demonstrate how to do this, with the following property, for example.

- The overall interaction effect from the trigger object x_1 to the target object y_3 in the reaction network in Example 1 (and 3) is positive, i.e., x_1 *promote* y_3 in this reaction network.

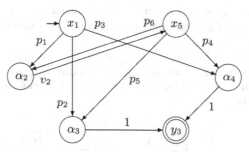

Fig. 5. A DFA recognizing $\mathcal{P}(x_1, y_3)$

Let U be a set $\{R_1, \ldots, R_4\}$ of reactions in Example 1 (and 3), and $\mathcal{P}(x_1, y_3)$ be a set of paths $r_1 \ldots r_n$ such that $x_1 = trigger(r_1)$ and $y_3 \in products(r_n)$. Intuitively, a term τ that represents the overall interaction effect from x_1 to y_3 is equivalent to the sum of all terms $\hat{r}_1 \ldots \hat{r}_n \otimes coeff(\hat{r}_n)(y_3)$ for paths $\hat{r}_1 \ldots \hat{r}_n \in \mathcal{P}(x_1, y_3)$. The reaction network has a feedback loop p_6, and hence we cannot finitely enumerate all paths belonging to $\mathcal{P}(x_1, y_3)$. However, a set of paths with feedback loops can be represented by the Kleene star "*" in regular expression. Therefore, we represent $\mathcal{P}(x_1, y_3)$ in a regular expression, and then translate it into τ, by substituting the concatenation".", the choice "+" and the Kleene star "*" in the regular expression with "$\otimes$", "$\oplus$" and "$\circledast$" in $\mathcal{L}$, respectively. Note rule A3, i.e., we must consider the coefficients of connecting objects for every supply connection on a path. Therefore, to describe the intended term τ, we use a regular expression over the alphabet $\mathbb{R} \cup \mathcal{V} \cup \{p \in \mathcal{P}(x_1, y_3) | p \text{ is competitive}\}$.

We can obtain a regular expression representing the intended term by using a *deterministic finite automaton* (DFA) $(Q, \Sigma, \delta, I, F)$ (Fig. 5) as follows:

- the set Q of states is $\mathcal{O} \cup \mathcal{ID}$,
- the alphabet Σ is $\mathbb{R} \cup \mathcal{V} \cup \{p \in \mathcal{P}(x_1, y_3) | p \text{ is competitive}\}$
- the transition relation $\delta \subseteq Q \times \Sigma \times Q$ is $\{\langle trigger(r_1), p, id(r_n) \rangle | r_1 \cdots r_n \in \mathcal{P}(x_1, y_3)$ such that it is competitive$\} \cup \{\langle id(r), coeff(r)(z), z \rangle | r \in \mathcal{P}(x_1, y_3)$ and $z \in products(r)\}$,
- the set I of initial states is $\{x_1\}$,
- the set F of final states is $\{y_3\}$.

This automaton accepts paths belonging to $\mathcal{P}(x_1, y_3)$ if real constants and co-efficient variables are substituted by ϵ. Then, we translate DFA into a regular expression that represents an acceptance language of DFA. This expression has a finite size, and is not redundant, i.e., it does not have the form $p + p$. This is because the transitions for the alphabet Σ are unique. Finally, we translate this expression into a term in $\mathcal{L}$, and obtain the intended term $\tau = p_2 \oplus p_3 \oplus p_1 \otimes v_2 \otimes (p_6 \otimes v_2)^{\circledast} \otimes (p_4 \oplus p_5)$.

In a manner analogous to this translation, we can obtain a term representing several kinds of the overall interaction effect of multiple paths. This type of translation can be summarized into the following three steps: (i) constructing a

DFA, considering a trigger and a target; (ii) translating from the DFA into a regular expression; and (iii) translating the regular expression into a term in $\mathcal{L}$.

Example 6. We show that x_1 promotes y_3 (i.e., $0 \prec p_2 \oplus p_3 \oplus p_1 \otimes v_2 \otimes (p_6 \otimes v_2)^{\circledast} \otimes (p_4 \oplus p_5)$) in the reaction network in Example 1 (and 3) if $v_1 \oplus v_2 \approx 1$. Then we can derive the conclusion, with lemmas $\vdash \varphi_1$, $\vdash \varphi_2$ and $v_1 \oplus v_2 \approx 1 \vdash \varphi_3, \varphi_3', \varphi_3''$ in Example 3, as follows:

$$
\cfrac{
  \overline{\varphi_2} \quad
  \cfrac{
    \overline{\varphi_1} \quad
    \cfrac{
      \cfrac{v_1 \oplus v_2 \approx 1}{\varphi_3'} \quad
      \cfrac{
        \cfrac{v_1 \oplus v_2 \approx 1}{\varphi_3} \quad
        \cfrac{
          \cfrac{
            \cfrac{v_1 \oplus v_2 \approx 1}{\varphi_3''} \quad \cfrac{p_6 \oplus (\ominus p_1) \approx 1}{\quad} \, A5, A8
          }{0 \prec 1 \ominus p_6 \otimes v_2 \oplus p_1 \otimes v_2} \, A0
        }{0 \prec 1 \ominus p_6 \otimes v_2 \oplus p_1 \otimes v_2 \otimes (p_6 \otimes v_2)^{\circledast} \otimes (1 \ominus p_6 \otimes v_2)} \, A0
      }{0 \prec (1 \oplus p_1 \otimes v_2 \otimes (p_6 \otimes v_2)^{\circledast}) \otimes (1 \ominus p_6 \otimes v_2)} \, A0
    }{0 \prec 1 \oplus p_1 \otimes v_2 \otimes (p_6 \otimes v_2)^{\circledast}} \, A0
  }{0 \prec (1 \oplus p_1 \otimes v_2 \otimes (p_6 \otimes v_2)^{\circledast}) \otimes (p_2 \oplus p_3)} \, A0
}{0 \prec p_2 \oplus p_3 \oplus p_1 \otimes v_2 \otimes (p_6 \otimes v_2)^{\circledast} \otimes (p_4 \oplus p_5)} \, A0
$$

Because of the denotational semantics in Section 5, this conclusion is also valid for a set of reactions larger than $\{R_1, \ldots, R_4\}$ if there is no additional path with a trigger object x_1 and the target object y_3.

7   Conclusions and Future Directions

We proposed a formal ontology of interactions. We considered the quantitative relations among the interaction effects of paths as basic properties that is used to infer the overall interaction effect in the reaction network. We gave their axiomatic semantics by a set of rules among them, which reflect the kinetics and structure of reaction networks. The applicability of our ontology to reaction networks is limited, but it allow us to simply represent the overall interaction effects of multiple paths, and to infer their properties. Additionally, we presented the denotational semantics based on mass action kinetics with linear approximation at a steady state. Then we proved the soundness of the given rules for the denotational semantics to justify the validity of the derived proprieties.

Our ontology enables formal analysis of quantitative processes using deductive reasoning applied to properties associated with denotational semantics. Such analyses may be useful for elucidating the robust properties of the processes, because all derived properties have concrete meanings that are guaranteed at a certain abstraction level. There are, however, gaps between theory and practice. Our ontology captures a certain aspect of behavioral properties, which is related to the integral of the transient response change at a steady state. The properties in the ontology are not appropriate to capture other aspects of behavior, e.g., changes in steady states. Such an aspect is also important to understand interactions with temporal contexts. To provide sophisticated reasoning in such cases, more research will be required to capture properties of considering phenomena at an appropriate abstraction level and to give rules and denotational semantics associated with them.

References

1. The Systems Biology Markup Language, http://sbml.org/
2. Ashburner, M., Ball, C.A., Blake, J.A., Botstein, D., Butler, H., Cherry, J.M., Davis, A.P., Dolinski, K., Dwight, S.S., Eppig, J.T., Harris, M.A., Hill, D.P., Issel-Tarver, L., Kasarskis, A., Lewis, S., Matese, J.C., Richardson, J.E., Ringwald, M., Rubin, G.M., Sherlock, G.: Gene ontology: tool for the unification of biology. Nature Genetics 25(1), 25–29 (2000)
3. Bergstra, J.A.: Handbook of Process Algebra. Elsevier Science Inc. (2001)
4. Calder, M., Vyshemirsky, V., Gilbert, D., Orton, R.: Analysis of signalling pathways using the prims model checker. In: Proceedings of Computational Methods in Systems Biology (CMSB 2005), pp. 179–190 (2005)
5. Clarke, E.M., Grumberg, O., Peled, D.A.: Model Checking. MIT Press (1999)
6. Emerson, E.A.: Temporal and modal logic. In: Handbook of Theoretical Computer Science, pp. 995–1072. Elsevier (1995)
7. Fages, F., Soliman, S.: Formal cell biology in biocham. In: Bernardo, M., Degano, P., Zavattaro, G. (eds.) SFM 2008. LNCS, vol. 5016, pp. 54–80. Springer, Heidelberg (2008)
8. Fages, F., Solimman, S., Chabrier-River, N.: Modelling and querying interaction networks in the biochemical abstract machine biocham. Journal of Biological Physics and Chemistry 4(2), 64–73 (2004)
9. NuSMV, http://nusmv.irst.itc.it/
10. Peterson, J.L.: Petri Net Theory and the Modeling of Systems. Prentice Hall PTR (1981)
11. PRISM, http://www.prismmodelchecker.org/
12. SPIN, http://spinroot.com/
13. Tashima, K., Izumi, N., Yonezaki, N.: A quantitative semantics of formal ontology of drug interaction. In: BIOCOMP 2008, pp. 760–766 (2008)
14. Yonezaki, N., Izumi, N., Akiyama, T.: Formal ontology of object interaction. In: Proceedings of International Symposium on Large-scale Knowledge Resources, pp. 15–20 (2006)

An Object Calculus with Remote Method Invocation

Shohei Matsumoto and Shin-ya Nishizaki

Department of Computer Science, Tokyo Institute of Technology,
2-12-1-W8-69, O-okayama, Meguro-ku, Tokyo, 152-8552, Japan

Abstract. Recently, object-oriented programming languages have been the fundamental infrastucture in software development. Several theoretical frameworks for object-oriented programming languages have been developed. The object calculus, proposed by Abadi and Cardelli, models various styles of object-oriented programming languages, and the sigma calculus is the simplest among the several systems of the object calculus.

Distributed computation is also a fundamental technology in information technology. The remote procedure call (RPC) is a popular technology for distributed computation. The object-oriented version of RPC is called remote method invocation (RMI). Cooper and Wadler proposed the RPC calculus, which is a formal computational system for modeling remote procedure calls.

In this paper, we propose the RMI calculus, which is an extension of the sigma calculus by adding a remote method invocation in the Cooper–Wadler style. We investigate the translation of the RPC calculus into the RMI calculus and prove its soundness with respect to the operational semantics.

1 Introduction

1.1 Remote Procedure Calls and Remote Procedure Invocations

In a distributed system, there are multiple autonomous computers that communicate through a network. The computers interact with each other in order to achieve a common computational goal. Several kinds of programming constructs have been proposed for distributed computation: the *remote procedure call* (RPC) is one of such constructs. It is an inter-process communication that allows a program to cause a procedure to be executed on another computer, in exactly the same manner as the usual procedure call. RPC lightens the programmer's burden by making the transport layer of the network transparent [3].

The RPC calculus [4] (in short, λrpc), proposed by Cooper et al., is an extension of the lambda calculus that incorporates the concepts of location and remote procedure call. The terms of the calculus are defined by the grammar.

S. Nishizaki et al. (Eds.): WCTP 2012, PICT 7, pp. 34–49, 2013.

$$
\begin{array}{lll}
a, b ::= & & \text{locations} \\
& \mathbf{c} & \text{client} \\
& \mid \mathbf{s} & \text{server} \\
M ::= & & \text{terms} \\
& c & \text{constants} \\
& \mid x & \text{variables} \\
& \mid \lambda^a x.\, M & \text{lambda abstraction} \\
& \mid (MN) & \text{function application}
\end{array}
$$

The operational semantics $\Downarrow_a$ is defined by the following rules:

$$
\frac{}{V \Downarrow_a V}\ \text{Value}
$$

$$
\frac{L \Downarrow_a \lambda^b x.\, N \quad M \Downarrow_a W \quad N\{W/x\} \Downarrow_b V}{(LM) \Downarrow_a V}\ \text{Beta,}
$$

where L, M are terms and V, W values, in other words, terms which are not function application. The expression $M \Downarrow_a V$ is a big-step reduction relation, and is read "the term M, evaluated at location a, results in the value V." Each evaluation is connected with a location where it is processed. The term $\lambda^b x.\, N$ is called b-*annotated abstraction* and its body N is evaluated at location b. This is obtained by formalizing the remote procedure call. Cooper and Wadler proposed the client–server calculus λ_{CS} [4], which defines a state-transition machine for the operational semantics of the RPC calculus. A state in the state-transition machine denotes a client and server configuration. The calculus only formalizes *sequential* computations with remote procedure calls.

1.2   The Object Calculus

The *object calculus* [1,2] was proposed by Abadi and Cardelli in order to formalize various kinds of object-oriented computations. Several systems have been proposed in the object calculus. The ς-calculus is the simplest system of the object calculus, which formulates imperative object-based object-oriented computation. Although it has simple syntax and operational semantics, the ς-calculus also involves the principal features of object-oriented computation, such as method invocation, method override, and late binding of self parameters.

The syntax of the ς-calculus is defined by the following grammar.

$$
\begin{array}{lll}
a, b ::= & & \text{term} \\
& x & \text{variable} \\
& \mid [l_i = \varsigma(x_i)b_i{}^{i \in 1..n}] & \text{object} \\
& \mid a.l & \text{method invocation, field selection} \\
& \mid a.l \Leftarrow \varsigma(x)b & \text{method override, field update}
\end{array}
$$

The notation $1..n$ represents the set $\{1, \ldots, n\}$. The notation $[l_i = \varsigma(x_i)b_i{}^{i \in 1..n}]$ is an abbreviation for $[l_1 = \varsigma(x_1)b_1, \cdots, l_n = \varsigma(x_n)b_n]$. l_i and l are referred to as methods, $\varsigma(x_i)b_i$ is said to be the body of a method l_i and $\varsigma(x_i)$ denotes a binder to the self-parameter x_i.

The operational semantics of the ς-calculus is provided by the reduction defined inductively by the following rules.

$$o.l_j \to b_j\{x_j \leftarrow o\} \quad (\text{select}),$$

where o is $[l_i = \varsigma(x_i)b_i^{i \in 1..n}]$ and the notation $b_j\{x_j \leftarrow o\}$ is a substitution of the term o for the variable x_j in the term b_j. The syntax $\varsigma(x_j)$ is a binder designating a self parameter and the whole object is bound to the self parameter x_j.

$$[l_i = \varsigma(x_i)b_i^{i \in 1..n}].l_j \Leftarrow \varsigma(y)b \to \begin{array}{l} [l_j = \varsigma(y)b, \\ l_i = \varsigma(x_i)b_i^{\;i \in (1..n)\backslash\{j\}}] \end{array} \quad (\text{update})$$

This rule provides semantics for method override. The body $\varsigma(x_j)b_j$ of a method l_j is replaced by a new body $\varsigma(y)b$.

In the ς-calculus, a method whose body does not refer to its self parameter is considered to be a field. Hence, field selection and field update are special cases of method invocation and method override, respectively. When we perform a field update, we use := instead of $\Leftarrow$.

1.3   Research Motivation

In this paper, we will formalize the notion of remote method invocation in the framework of the sigma calculus, which is a theoretical model of object-based and non-imperative object-oriented programming languages. The formalization proposed in this paper could be the starting point for the theoretical study of distributed object-oriented programming languages.

2   The RMI-calculus ςrmi

The syntax of the RMI-calculus ςrmi consist of locations, terms, and values. We define them in this section.

Definition 1 (Locations of ςrmi). *Locations* m are either c (a *client*) or s (a *server*):

$$m ::= \text{c} \mid \text{s}.$$

The location means the names for identifying the computer nodes where the terms are evaluated. In this paper, we restrict the locations to the two names c and s.

Definition 2 (Terms of ςrmi). *Terms* of ςrmi are defined inductively as follows:

$$
\begin{array}{llr}
a, b, c ::= & & \text{term} \\
& c & \text{constant} \\
\mid & x & \text{variable} \\
\mid & [l_1 = \varsigma(m_1, x_1)b_1, \ldots, l_n = \varsigma(m_n, x_n)b_n] & \text{object} \\
\mid & a.l & \text{method invocation,} \\
& & \text{field selection} \\
\mid & a.l \Leftarrow \varsigma(m, x)b & \text{method override,} \\
& & \text{field update}
\end{array}
$$

The difference between the terms of the ςrmi-calculus and of those of the ς-calculus is the number of parameters in the ς-binder. In the ςrmi-calculus, ς takes two parameters: the first parameter is the location where its body is to be evaluated and the second is the self-parameter, which is similar to that of the ς-calculus. An object $[l_1 = \varsigma(m_1, x_1)b_1, \ldots, l_n = \varsigma(m_n, x_n)b_n]$ is sometimes abbreviated as

$$[l_i = \varsigma(m_i, x_i)b_i^{i \in 1..n}]$$

similar to the ς-calculus [1], or more simply, as

$$\overline{[l_n = \varsigma(m_n, x_n)b_n]}.$$

In both the ς-calculus and the ςrmi-calculus, a method whose body does not have any free occurrence of its self-parameter, that is, a method

$$[\ldots, l = \varsigma(m, x)b, \ldots]$$

such that the self-parameter x does not occur freely in b, is called a *field* and is written as

$$[\ldots, l =_m b, \ldots].$$

The notion of free variable occurrence in the ςrmi-calculus will be defined in the next section. The method override

$$a.l \Leftarrow \varsigma(m, x)b$$

where x does not occur freely in b, is called a *field update*, and written as

$$a.l :=_m b.$$

Definition 3 (Values). *Values* of the ςrmi-calculus, a subset of the terms, are defined inductively by the grammar

$$v, w ::= c \mid x \mid \overline{[l_i = \varsigma(m_i, x_i)b_i]}.$$

A *value* of the ςrmi-calculus is a syntactic characterization of a result of evaluation, similarly to the evaluation strategies in the lambda calculus.

Example 1. A term

$$[l_1 = \varsigma(\mathsf{c}, s)[\ ]].l_1 \tag{1}$$

represents an object which has a method l_1. If the method l_1 returns, an empty object is created at client node c.

A term

$$[l_1 = \varsigma(\mathsf{c}, s)s.l_1].l_1 \Leftarrow \varsigma(\mathsf{s}, s)[\ ] \tag{2}$$

is an example of overriding the method l_1.

The following term represents an interaction between a client and a server node via the message "hello."

$$
\begin{aligned}
[\ &message =_\mathsf{c} \text{ "hello"}, \\
&server =_\mathsf{c} [message =_\mathsf{s} \text{ " "}, echo = \varsigma(\mathsf{s}, s)s.message], \\
&echo = \varsigma(\mathsf{c}, s)(s.server.message :=_\mathsf{s} s..message).echo\ ].echo
\end{aligned}
\tag{3}
$$

3   Operational Semantics of the RMI Calculus

We give an operational semantics for the RMI calculus, in the style of Kahn's *natural semantics* [7], or the *bigstep* reduction.

Before giving a definition of the operational semantics for the RMI calculus, we would like to prepare by introducing several notions.

Definition 4 (Free Variable). We define the *free variables* in a term a of the ςrmi-calculus inductively by the following equations.

$$
\begin{aligned}
\mathsf{FV}(c) &= \emptyset, \\
\mathsf{FV}(x) &= \{x\}, \\
\mathsf{FV}([l_i = \varsigma(m_i, x_i)b_i^{i\in 1..n}]) &= \bigcup_{i=1,\ldots,n} \mathsf{FV}(b_i) - \{x_i\}, \\
\mathsf{FV}(a.l) &= \mathsf{FV}(a), \\
\mathsf{FV}(a.l \Leftarrow \varsigma(m, x)b) &= \mathsf{FV}(a) \cup (\mathsf{FV}(b) - \{x\}).
\end{aligned}
$$

Definition 5 (Substitution). We define a *substitution* $a\{x \leftarrow b\}$ of term b for variable x in term a inductively by the following equations.

$$
\begin{aligned}
c\{x \leftarrow b\} &= c, \\
x\{x \leftarrow b\} &= b, \\
y\{x \leftarrow b\} &= y, \quad \text{where } y \neq x \\
([l_i = \varsigma(m_i, x_i)b_i^{i\in 1..n}])\{x \leftarrow b\} &= [l_i = \varsigma(m_i, x_i')b_i\{x \leftarrow b\}^{i\in 1..n}] \\
&\qquad \text{where } x_i \notin \mathsf{FV}(b) \cup \{x\}, \\
([l_i = \varsigma(m_i, x_i)b_i^{i\in 1..n}])\{x \leftarrow b\} &= [l_i = \varsigma(m_i, x_i')b_i\{x_i \leftarrow x_i'\}\{x \leftarrow b\}^{i\in 1..n}] \\
&\qquad \text{where } x_i' \notin \mathsf{FV}(b_i) \cup \mathsf{FV}(b) \cup \{x\}, \\
(a.l)\{x \leftarrow b\} &= (a\{x \leftarrow b\}).l, \\
(a.l \Leftarrow \varsigma(m, x)b)\{x \leftarrow b\} &= (a\{y \leftarrow b\}).l \Leftarrow \varsigma(m, y)(a\{x \leftarrow b\}) \\
&\qquad \text{where } y \notin \mathsf{FV}(b) \cup \{x\}, \\
(a.l \Leftarrow \varsigma(m, x)b)\{x \leftarrow b\} &= (a\{y \leftarrow b\}).l \Leftarrow \varsigma(m, y')(a\{y \leftarrow y'\}\{x \leftarrow b\}) \\
&\qquad \text{where } y' \notin \mathsf{FV}(a) \cup \mathsf{FV}(b) \cup \{x\},
\end{aligned}
$$

Definition 6 (Operational Semantics of the RMI Calculus). A ternary relation $a \Downarrow_m v$ between the term a, location m, and value v, is called an *evaluation relation*, and is defined inductively by the following rules.

$$\frac{}{v \Downarrow_m v} \text{ Value,}$$

$$\frac{a \Downarrow_m w \quad b_j\{x_j \leftarrow w\} \Downarrow_{m_j} v}{a.l_j \Downarrow_m v} \text{ Select}$$

$$\text{where } w \text{ is } [l_i = \varsigma(m_i, x_i')b_i\{x \leftarrow x_i'\}^{i \in 1..n}],$$

$$\frac{a \Downarrow_m [l_i = \varsigma(m_i, x_i)b_i^{i \in 1..n}]}{a.l_j \Leftarrow \varsigma(m', x)b \Downarrow_m [l_j = \varsigma(m', x)b, \; l_i = \varsigma(m_i, x_i)b_i^{i \in \{1,...,n\}-\{j\}}]} \text{ Override.}$$

In the rule Select, a method l_j is invoked in an object a. The self parameter x_j in the method's body b_j is bound to the whole object $w = [l_i = \varsigma(m_i, x_i')b_i\{x \leftarrow b\}^{i \in 1..n}]$ itself and then the body $b_j\{x_j \leftarrow w\}$ is evaluated.

Example 2. We will give the evaluation of the examples in Example 1. The evaluation of the term (1) is given by the following derivation.

$$\frac{\dfrac{}{[l_1 = \varsigma(c, s)[\;]] \Downarrow_c [l_1 = \varsigma(c, s)[\;]]} \text{ Value} \quad \dfrac{}{[\;] \Downarrow_c [\;]} \text{ Value}}{[l_1 = \varsigma(c, s)[\;]].l_1 \Downarrow_c [\;]} \text{ Select}$$

Similar to the other big-step operational semantics, or natural semantics [7], if a term a and a location m are given as inputs, then the resulting value v is determined satisfying $a \Downarrow_c v$.

The evaluation derivation of the term (2) in the previous section is as follows.

$$\frac{\dfrac{}{[l_1 = \varsigma(c, s)s.l_1] \Downarrow_c [l_1 = \varsigma(c, s)s.l_1]} \text{ Value}}{[l_1 = \varsigma(c, s)s.l_1].l_1 \Leftarrow \varsigma(s, s)[\;]] \Downarrow_c [l_1 = \varsigma(c, s)[\;]]} \text{ Override}$$

Next we will give the evaluation of the term (3) representing the echo server. For convenience, we refer to the subterms as follows.

$$\text{Server}_1 = [message = \text{``''}, echo = \varsigma(s, s)s.message],$$
$$\text{Server}_2 = [message = \text{Client}.message, echo = \varsigma(s, s)s.message],$$
$$\text{Client} = [message = \text{``hello''}, server = \text{Server}_1,$$
$$echo = \varsigma(c, s)(s.server.message :=_s s.message).echo].$$

The following is a derivation of the term's evaluation.

$$\frac{\dfrac{}{\text{Client} \Downarrow_c \text{Client}} \text{ Value} \quad \begin{array}{c} \vdots \; \Sigma_1 \\ (\text{Client}.server.message \\ :=_c \text{Client}.message).echo \Downarrow_c \text{``hello''} \end{array}}{\text{Client}.echo \Downarrow_c \text{``hello''}} \text{ Select}$$

The subtree Σ_1 is

$$\frac{\begin{array}{cc} \vdots\ \Sigma_2 & \vdots\ \Sigma_3 \\ \text{Client.}server.message :=_c \text{Client.}message \Downarrow_c \text{Server}_2 & \text{Server}_2.message \Downarrow_s \text{``hello''} \end{array}}{(\text{Client.}server.message :=_c \text{Client.}message).echo \Downarrow_c \text{``hello''}}$$

which is derived by the rule Select. Here the upper right evaluation relation $\text{Server}_2.message$ is derived from

$$s.message\{s \leftarrow \text{Server}_2.message\}.$$

The subtree Σ_2 is

$$\cfrac{\cfrac{\cfrac{}{\text{Client} \Downarrow_c \text{Client}}\text{Value} \quad \cfrac{}{\text{Server}_1 \Downarrow_c \text{Server}_1}\text{Value}}{\text{Client.}server \Downarrow_c \text{Server}_1}\text{Select}}{\text{Client.}server.message :=_c \text{Client.}message \Downarrow_c \text{Server}_2}\text{Override}$$

The subtree Σ_3 is

$$\cfrac{\cfrac{}{\text{Server}_2 \Downarrow_s \text{Server}_2}\text{Value} \quad \cfrac{\cfrac{\cfrac{}{\text{Client} \Downarrow_s \text{Client}}\text{Value} \quad \cfrac{}{\text{``hello''} \Downarrow_c \text{``hello''}}\text{Value}}{\text{Client.}message \Downarrow_s \text{``hello''}}\text{Select}}{}}{\text{Server}_2.message \Downarrow_s \text{``hello''}}\text{Select}$$

4   Theoretical Study of the RMI Calculus

Abadi and Cardelli [1] give a translation of the lambda-calculus into the ς-calculus. If we extend the translation to the RMI calculus, we should study a translation of the RPC calculus [4,9] into the RMI calculus.

4.1   Attempt at Interpreting the RPC Calculus in the RMI Calculus

In this section, we try to develop a translation of Cooper–Wadler's RPC calculus (the λrpc-calculus) into the RMI calculs (the ςrmi-calculus). It is based on the translation given by Abadi and Cardelli [1] of the λ-calculus into the ς-calculus.

Definition 7 (Translation of the λrpc-calculus into the ςrmi-calculus).
For a term M of the λrpc-calculus and a location m, a term $\langle M \rangle_m$ is defined inductively by the following equations.

$$\begin{aligned} \langle x \rangle_m &= x, \\ \langle c \rangle_m &= c, \\ \langle (MN) \rangle_m &= (\langle M \rangle_m.arg :=_m \langle N \rangle_m).val, \\ \langle \lambda^n x.M \rangle_m &= [arg = \varsigma(n,x)x.arg, val = \varsigma(n,x)\langle M \rangle_n\{x \leftarrow x.arg\}] \end{aligned}$$

The translated term $\langle M\rangle_m$ intuitively means that the subterms of the term M should be evaluated at the location m if the location is not specified by remote procedure calls. The term $\langle \lambda^n x.M\rangle_m$ means that if the remote procedure call is issued, then the body M is evaluated at location n and it returns the result to the location m. This translation is a naive extension of the translation of the lambda-calculus terms into terms of the ς-calculus [1].

Example 3. We will give some examples of this translation.

$$\langle \lambda^c x.x\rangle_c = [arg = \varsigma(c,x)x.arg, val = \varsigma(c,x)x.arg],$$
$$\langle \lambda^c x.xx\rangle_c = [arg = \varsigma(c,x)x.arg, val = \varsigma(c,x)(x.arg.arg :=_c x.arg).val],$$
$$\langle (\lambda^c x.x)(\lambda^s y.y)\rangle_s = ([arg = \varsigma(c,x)x.arg, val = \varsigma(c,x)x.arg]$$
$$.arg :=_s [arg = \varsigma(s,y)y.arg, val = \varsigma(s,y)y.arg]).var$$

It is known that the translation of the λ-calculus into the ς-calculus given by Abadi and Cardelli does not satisfy soundness with respect to the operational semantics of the ς-calculus [5]. Similar to this result, our translation does not satisfy soundness, that is, $M \Downarrow_m V$ in the λrpc-calculus does not necessarily imply $\langle M\rangle_m \Downarrow_m \langle V\rangle_m$ in the ςrmi-calculus. For example, consider a λrpc-term

$$M = (\lambda^c x.(\lambda^c y.x))((\lambda^c x.x)(\lambda^c z.z))$$

is evaluated to

$$V = \lambda^c y.(\lambda^c z.z)$$

at location c. These terms M and V are translated into

$$\langle M\rangle_c = \langle (\lambda^c x.(\lambda^c y.x))((\lambda^c x.x)(\lambda^c z.z))\rangle_c$$
$$= ([arg = \varsigma(c,x)x.arg, val = \varsigma(c,x)\langle \lambda^c y.x\rangle_c \{x \leftarrow x.arg\}]$$
$$.arg :=_c \langle ((\lambda^c x.x)(\lambda^c z.z))\rangle_c).val$$

$$\langle V\rangle_c = [arg = \varsigma(c,y).y.arg,$$
$$val = \varsigma(c,y)].$$

The term $\langle M\rangle_c$ is evaluated at the location c as follows.

$$\langle (\lambda^c x.(\lambda^c y.x))((\lambda^c x.x)(\lambda^c z.z))\rangle_c$$
$$\Downarrow_c ([\ arg = \varsigma(c,x)x.arg,$$
$$val = \varsigma(c,x)[\ arg =_c \langle (\lambda^c x.x)(\lambda^c z.z)\rangle_c$$
$$val = \varsigma(c,x)\langle \lambda^c y.x\rangle_c \{x \leftarrow x.arg\}] \tag{4}$$

However, the result of the evaluation is not equal to $\langle \lambda^c y.(\lambda^c z.z)\rangle_c$, which shows that the translation is not sound.

The reason why soundness does not hold is as follows.

The gap between the evaluation strategies of the two calculi. The evaluation of the RPC calculus λrpc [4, 9] is based on the call-by-*value* strategy [10]. The actual parameter is evaluated before binding it to the formal parameter. On the other hand, in the RMI calculus ςrmi, the actual parameter is not evaluated before binding it to the formal parameter. We may therefore say that the evaluation of the RMI calculus ςrmi is based on the call-by-*name* strategy [10], unlike the RPC calculus.

The gap between the methods of access to variables. In the RPC calculus λrpc, variable reference is similar to the usual λ-calculus. In the *lambda*-calculus and the λrpc-calculus, the reference to an actual parameter is made through the corresponding formal parameter. This is made possible by the *substitution* mechanism: the formal parameter is replaced by the actual parameter by β-reduction.

$$(\lambda x.M)N \rightarrow_\beta M\{x \leftarrow N\}$$

On the other hand, in the ςrmi-calculus, the varable reference of the λrpc-calculus is translated as the method invocation. For example, a λrpc-term $\lambda^c x.N$ is translated to

$$\langle \lambda^c x.N \rangle_c = [arg = \varsigma(c, x)x.arg, val = \varsigma(c, x)\langle N \rangle_c\{x \leftarrow x.arg\}].$$

The subterm

$$\langle N \rangle_c\{x \leftarrow x.arg\}$$

corresponding to the body N of the lambda-abstraction can be accessed via invocation of the method *arg*. Substitution of the actual parameter $x.arg$ for the formal parameter x is delayed when the method *val* is invoked. The difference can be seen in that although the actual parameter is substituted for the formal parameter just after the function call in the λrpc-calculus, this substitution is postponed in the ςrmi-calculus.

The term M in the previous example is evaluated to

$$\lambda^c y.((\lambda^c x.x)(\lambda^c x.x))$$

in the λrpc-calculus, whose translated image is, however, evaluated to

$$\langle \lambda^c y.((\lambda^c x.x)(\lambda^c z.z)) \rangle_c = [\ arg = \varsigma(c, y)y.arg, \qquad (5)$$
$$val = \varsigma(c, y)\langle ((\lambda^c x.x)(\lambda^c z.z)) \rangle_c]$$

Such disagreement is caused by the fact that the subterm in the term, (4)

$$[arg =_c \langle (\lambda^c x.x)(\lambda^c z.z) \rangle_c, val = \varsigma(c, x)\langle \lambda^c y.x \rangle_c\{x \leftarrow x.arg\}].arg,$$

is not evaluated. If it were evaluated and the method *arg* were invoked, the results of (4) and (5) would be equivalent.

4.2   The RMI Calculus for Translation of the RPC Calculus

In the previous section, the naive translation of the RPC calculus λrpc into the RMI calculus ςrmi was not sound. In this section, we propose a variation of the RMI calculus, ςrmi-let, by adding let-bindings, and we give a sound translation of the RPC calculus into the extended RMI calculus.

First, we define the terms of the ςrmi-let.

Definition 8 (Terms of ςrmi-let). We define a *term* of the ςrmi-let-calculus inductively by the following grammar.

$a, b, c ::=$		term
	c	constant
	x	variable
	$[l_1 = \varsigma(m_1, x_1)b_1, \ldots, l_n = \varsigma(m_n, x_n)b_n]$	object
	$a.l$	method invocation, field selection
	$a.l := \varsigma(m, x)b$	method override, field update
	$\mathsf{let}^m\ x = a\ \mathsf{in}\ b$	let-expression

The operational semantics of the ςrmi-let-calculus is given as an extension of that of the ςrmi-calculus by adding the following rule to the those of ςrmi.

Definition 9 (Operational Semantics of ςrmi-let). The evaluation relation $a \Downarrow_m v$ of ςrmi is extended to the ςrmi-let-calculus by adding the following rule to the rules of the ςrmi-calculus: Value, Select, and Override. The extended relation is written as $a \Downarrow_m^{\mathsf{let}} v$.

$$\frac{}{v \Downarrow_m^{\mathsf{let}} v}\ \text{Value,}$$

$$\frac{a \Downarrow_m^{\mathsf{let}} w \quad b_j\{x_j \leftarrow w\} \Downarrow_{m_j}^{\mathsf{let}} v}{a.l_j \Downarrow_m^{\mathsf{let}} v}\ \text{Select}$$

$$\text{where } w \text{ is } [l_i = \varsigma(m_i, x_i')b_i\{x \leftarrow b\}^{i \in 1..n}],$$

$$\frac{a \Downarrow_m^{\mathsf{let}} [l_i = \varsigma(m_i, x_i)b_i^{i \in 1..n}]}{a.l_j \Leftarrow \varsigma(m', x)b \Downarrow_m^{\mathsf{let}} \left[l_j = \varsigma(m', x)b,\ l_i = \varsigma(m_i, x_i)b_i^{i \in (1..n) - \{j\}}\right]}\ \text{Override}$$

$$\frac{a \Downarrow_m^{\mathsf{let}} w \quad b\{x \leftarrow w\} \Downarrow_m^{\mathsf{let}} v}{\mathsf{let}^m\ x = a\ \mathsf{in}\ b \Downarrow_m^{\mathsf{let}} v}\ \text{Let.}$$

Next we define a translation of the λrpc-calculus to the ςrmi-let-calculus. We obtain the translation by improving the translation in the previous section, using let-expressions.

Definition 10 (Translation of the λrpc-calculus to the ςrmi-let-calculus).
For a λrpc-term M and a location m, the translation $\langle M \rangle_m^{let}$ which maps them to a ςrmi-term is defined inductively by the following equations.

$$\langle x \rangle_m^{let} = x,$$
$$\langle c \rangle_m^{let} = c,$$
$$\langle (MN) \rangle_m^{let} = \mathsf{let}^m \ y = \langle N \rangle_m^{let} \ \mathsf{in} \ (\langle M \rangle_m^{let}.arg :=_m y).val,$$
$$\langle \lambda^n x.N \rangle_m^{let} = [arg = \varsigma(n,x)x.arg, \ val = \varsigma(n,x) \ \mathsf{let}^n \ x = x.arg \ \mathsf{in} \ \langle N \rangle_n^{let}]$$

Function application (MN) was translated in the previous translation (Definition 7) by

$$(\langle M \rangle_m.arg :=_m \langle N \rangle_m).val,$$

On the other hand, it is now translated in this translation (Definition 10) by

$$\mathsf{let}^m \ y = \langle N \rangle_m^{let} \ \mathsf{in} \ (\langle M \rangle_m^{let}.arg := \varsigma(m,x)y).val.$$

In this translation, the subterm $\langle N \rangle_m^{let}$ is folded as a let-expression, which makes it possible for it to be evaluated before parameter passing.

In the translation of lambda-abstraction $\lambda^n N.$, we find the subterm

$$\mathsf{let}^n \ x = x.arg \ \mathsf{in} \ \langle N \rangle_n^{let}$$

in place of the subterm

$$\langle N \rangle_n \{x \leftarrow x.arg\}$$

in the naive translation of Definition 7. The defect of the naive translation was that evaluation of the actual parameter is made later than the formal parameter's reference. The improvement is to make the evaluation of the actual parameter earlier than the access to the formal parameter's reference.

Example 4. We present several examples of the improved translation.

$$\langle \lambda^c x.x \rangle_c^{let} = [arg = \varsigma(c,x)x.arg, \quad val = \varsigma(c,x) \ \mathsf{let}^c \ x = x.arg \ \mathsf{in} \ x],$$
$$\langle \lambda^c x.xx \rangle_c^{let} = [\ arg = \varsigma(c,x)x.arg,$$
$$val = \varsigma(c,x) \ \mathsf{let}^c \ x = x.arg$$
$$\mathsf{in} \ (\mathsf{let}^c \ y = x \ \mathsf{in} \ (x.arg :=_c y).val)]$$
$$\langle (\lambda^c x.x)(\lambda^s y.y) \rangle_s^{let} = \mathsf{let}^s \ y_1 = [\ arg = \varsigma(s,y).arg,$$
$$val = \varsigma(c,y) \ \mathsf{let}^s \ y = y.arg \ \mathsf{in} \ y]$$
$$\mathsf{in} \ ([\ arg = \varsigma(c,x)x.arg,$$
$$val = \varsigma(c,x) \ \mathsf{let}^c \ x = x.arg \ \mathsf{in} \ x]$$
$$.arg :=_s y_1).val$$

The improved translation satisfies soundness with respect to the operational semantics of the ςrmi-let-calculus.

To prepare for the proof of Theorem 1, we first prove the following lemma.

Lemma 1 (Substitution Lemma). For a term M, a value V, and locations m, n of the λrpc-calculus,

$$\langle M \rangle_m^{let} \{ x \leftarrow \langle V \rangle_n^{let} \} = \langle M\{ x \leftarrow V \} \rangle_m^{let}$$

Proof. We prove this lemma by induction on the structure of the term M.

Case 1: M is a constant c. Then we have

$$\langle c \rangle_m^{let} \{ x \leftarrow \langle V \rangle_n^{let} \} = c \{ x \leftarrow \langle V \rangle_n^{let} \} = c$$

and

$$\langle c \{ x \leftarrow \langle V \rangle_n^{let} \} \rangle_m^{let} = \langle c \rangle_m^{let} = c.$$

Case 2: M is a variable x. Then we have

$$\langle x \rangle_m^{let} \{ x \leftarrow \langle V \rangle_n^{let} \} = x \{ x \leftarrow \langle V \rangle_n^{let} \} = \langle V \rangle_n^{let}$$

and

$$\langle x \{ x \leftarrow \langle V \rangle_n^{let} \} \rangle_m^{let} = \langle V \rangle_n^{let}.$$

Case 3: M is a variable y where $x \neq y$. Then we have

$$\langle y \rangle_m^{let} \{ x \leftarrow \langle V \rangle_n^{let} \} = y \{ x \leftarrow \langle V \rangle_n^{let} \} = y$$

and

$$\langle y \{ x \leftarrow \langle V \rangle_n^{let} \} \rangle_m^{let} = \langle y \rangle_m^{let} = y.$$

Case 4: $M = \lambda^l y.N$. By the induction hypothesis,

$$\langle N \rangle_l^{let} \{ x \leftarrow \langle V \rangle_n^{let} \} = \langle N\{ x \leftarrow V \} \rangle_l^{let}.$$

We may assume that the bound variable y is sufficiently fresh, in other words, y does not occur freely in V and $y \neq x$, without loss of generality. Then we have

$$\langle \lambda^l y.N \rangle_m^{let} \{ x \leftarrow \langle V \rangle_n^{let} \}$$
$$= [\ arg = \varsigma(l, y)y.arg,$$
$$\quad val = \varsigma(l, y) \ \mathsf{let}^l \ y = y.arg \ \mathsf{in} \ \langle N \rangle_l^{let}] \{ x \leftarrow \langle V \rangle_n^{let} \},$$
$$= [\ arg = \varsigma(l, y)y.arg,$$
$$\quad val = \varsigma(l, y) \ \mathsf{let}^l \ y = y.arg \ \mathsf{in} \ \left(\langle N \rangle_l^{let} \{ x \leftarrow \langle V \rangle_n^{let} \} \right)]$$
$$= [\ arg = \varsigma(l, y)y.arg,$$
$$\quad val = \varsigma(l, y) \ \mathsf{let}^l \ y = y.arg \ \mathsf{in} \ \underline{\langle N\{ x \leftarrow \langle V \rangle_n^{let} \} \rangle_l^{let}}]$$

(by the induction hypothesis).

On the other hand,

$$\langle \lambda^l y.(N\{ x \leftarrow V \}) \rangle_m^{let} = [\ arg = \varsigma(l, y)y.arg,$$
$$\quad val = \varsigma(l, y) \ \mathsf{let}^l \ y = y.arg \ \mathsf{in} \ \langle N\{ x \leftarrow \langle V \rangle_n^{let} \} \rangle_l^{let}].$$

Hence we know that

$$\langle \lambda^l y.N \rangle_m^{\text{let}} \{x \leftarrow \langle V \rangle_n^{\text{let}}\} = \langle \lambda^l y.(N\{x \leftarrow V\}) \rangle_m^{\text{let}}.$$

Case 5: $M = (LN)$. By the induction hypothesis, we have

$$\langle L \rangle_m^{\text{let}} \{x \leftarrow \langle V \rangle_n^{\text{let}}\} = \langle L\{x \leftarrow V\} \rangle_m^{\text{let}},$$

$$\langle N \rangle_m^{\text{let}} \{x \leftarrow \langle V \rangle_n^{\text{let}}\} = \langle N\{x \leftarrow V\} \rangle_m^{\text{let}}.$$

Then

$$\langle (LN) \rangle_m^{\text{let}} \{x \leftarrow \langle V \rangle_n^{\text{let}}\}$$
$$= \left(\mathsf{let}^m \; y = \langle N \rangle_m^{\text{let}} \; \mathsf{in} \; (\langle L \rangle_m^{\text{let}}.arg :=_m y).val \right) \{x \leftarrow \langle V \rangle_n^{\text{let}}\}$$
$$= \mathsf{let}^m \; y = \langle N \rangle_m^{\text{let}} \{x \leftarrow \langle V \rangle_n^{\text{let}}\} \; \mathsf{in} \; \left(\left((\langle L \rangle_m^{\text{let}}.arg :=_m y).val \right) \{x \leftarrow \langle V \rangle_n^{\text{let}}\} \right)$$
$$= \mathsf{let}^m \; y = \langle N \rangle_m^{\text{let}} \{x \leftarrow \langle V \rangle_n^{\text{let}}\} \; \mathsf{in} \; \left((\langle L \rangle_m^{\text{let}} \{x \leftarrow \langle V \rangle_n^{\text{let}}\}).arg :=_m y \right).val$$
$$= \mathsf{let}^m \; y = \langle N\{x \leftarrow V\} \rangle_m^{\text{let}} \; \mathsf{in} \; \left(\langle L\{x \leftarrow V\} \rangle_m^{\text{let}}.arg :=_m y \right).val$$

(by the induction hypothesis).

On the other hand, we have

$$\langle (LN)\{x \leftarrow V\} \rangle_m^{\text{let}}$$
$$= \langle L\{x \leftarrow V\} \; N\{x \leftarrow V\} \rangle_m^{\text{let}}$$
$$= \mathsf{let}^m \; y = \langle N\{x \leftarrow V\} \rangle_m^{\text{let}} \; \mathsf{in} \; \left(\langle L\{x \leftarrow V\} \rangle_m^{\text{let}}.arg :=_m y \right).val.$$

Hence, we obtain

$$\langle (LN) \rangle_m^{\text{let}} \{x \leftarrow \langle V \rangle_n^{\text{let}}\} = \langle (LN)\{x \leftarrow V\} \rangle_m^{\text{let}}.$$

Q.E.D.

We will now give the proof of the soundness theorem.

Theorem 1 (Soundness of $\langle M \rangle_m^{\text{let}}$). For a term M, a value V, and a location m of the λrpc-calculus, if $M \Downarrow_m V$, then $\langle M \rangle_m^{\text{let}} \Downarrow_m^{\text{let}} \langle V \rangle_m^{\text{let}}$.

Proof. We prove the soundness theorem proceeds by induction on the structure of the derivation of $M \Downarrow_m V$. We carry out an anlysis by cases for the last rule deriving $M \Downarrow_m V$.

Case 1: rule Value of λrpc. We assume that $M = V$ (and $V \Downarrow_m V$). By the definition of the translation, $\langle V \rangle_m^{\text{let}}$ is a value of the çrmi-let-calculus and therefore

$$\langle V \rangle_m^{\text{let}} \Downarrow_m^{\text{let}} \langle V \rangle_m^{\text{let}}$$

by rule Value of the çrmi-let-calculus.

Case 2: rule Beta. We assume that M is (KL) and the last rule of the derivation is

$$\frac{K \Downarrow_m \lambda^n x.N \quad L \Downarrow_m W \quad N\{x \leftarrow W\} \Downarrow_n V.}{(KL) \Downarrow_m V} \text{ Beta}$$

By the induction hypothesis, we have

$$\langle K \rangle_m^{let} \Downarrow_m^{let} \langle \lambda^n x.N \rangle_m^{let}, \; \langle L \rangle_m^{let} \Downarrow_m^{let} \langle W \rangle_m^{let}, \text{ and } \langle N\{x \leftarrow W\} \rangle_n^{let} \Downarrow_n^{let} \langle V \rangle_n^{let}.$$

The first evaluation relation means that

$$\langle K \rangle_m^{let} \Downarrow_m^{let} [arg = \varsigma(n,x)x.arg, val = \varsigma(n,x) \text{ let}^n x = x.arg \text{ in } \langle N \rangle_n^{let}].$$

The third evaluation relation implies that

$$\langle N \rangle_n^{let} \{x \leftarrow W\} \Downarrow_n^{let} \langle V \rangle_n^{let}$$

by the substitution lemma. Moreover, we have

$$\langle KL \rangle_m^{let} = \text{let}^m \; y = \langle L \rangle_m^{let} \text{ in } (\langle K \rangle_m^{let}.arg :=_m y).val$$

where y is a fresh variable. Since

$$\langle \lambda^n x.N \rangle_m^{let} = [arg = \varsigma(n,x)x.arg, val = \varsigma(n,x) \text{ let}^n x = x.arg \text{ in } \langle N \rangle_n^{let},]$$

we have

$$\frac{\langle K \rangle_m^{let} \Downarrow_m^{let} [arg = \varsigma(n,x)x.arg, val = \varsigma(n,x) \text{ let}^n x = x.arg \text{ in } \langle N \rangle_n^{let}]}{\langle K \rangle_m^{let}.arg :=_m \langle W \rangle_m^{let} \Downarrow_m^{let} o} \text{ Override}$$

where $o = [arg =_m \langle W \rangle_m^{let}, val = \varsigma(n,x) \text{ let}^n x = x.arg \text{ in } \langle N \rangle_n^{let}].$

Moreover, we have $\langle W \rangle_m^{let} \{z \leftarrow o\} = \langle W \rangle_m^{let}$, and by the substitution lemma

$$\langle N \rangle_n^{let} \{x \leftarrow \langle W \rangle_m^{let}\} = \langle N\{x \leftarrow W\} \rangle_n^{let}.$$

Then we have

$$\frac{\dfrac{o \Downarrow_n^{let} o \quad \langle W \rangle_m^{let} \{z \leftarrow o\} \Downarrow_n^{let} \langle W \rangle_m^{let}}{o.arg \Downarrow_n^{let} \langle W \rangle_m^{let}} \quad \langle N \rangle_n^{let} \{x \leftarrow \langle W \rangle_m^{let}\} \Downarrow_n^{let} \langle V \rangle_n^{let}}{\text{let}^n x = o.arg \text{ in } \langle N \rangle_n^{let} \Downarrow_n^{let} \langle V \rangle_n^{let}}$$

Therefore,

$$\frac{\langle K \rangle_m^{let}.arg :=_m \langle W \rangle_m^{let} \Downarrow_m^{let} o \quad \text{let}^n x = o.arg \text{ in } \langle N \rangle_n^{let} \Downarrow_n^{let} \langle V \rangle_n^{let}}{(\langle K \rangle_m^{let}.arg :=_m \langle W \rangle_m^{let}).val \Downarrow_m^{let} \langle V \rangle_n^{let}}.$$

Accordingly,

$$\frac{\langle L\rangle_m^{let}\ \Downarrow_m^{let}\ \langle W\rangle_m^{let}\quad \langle K\rangle_m^{let}.arg :=_m \langle W\rangle_m^{let}.val \Downarrow_m^{let} \langle V\rangle_n^{let}}{let^m\ y = \langle L\rangle_m^{let}\ in\ (\langle K\rangle_m^{let} :=_m y).val \Downarrow_m^{let} \langle V\rangle_n^{let}}\quad,$$

which means that

$$\langle KL\rangle_m^{let}\ \Downarrow_m^{let}\ \langle V\rangle_n^{let}.$$

Q.E.D.

5   Conclusion

In this paper, we proposed the ςrmi-calculus as a framework for the theoretical study of remote method invocation in object-oriented programming. This calculus is obtained by adding the notion of location to the ς-calculus and labeling the evaluation relation of its operational semantics with labels of the location. As well as the translation of the λ-calculus into the ς-calculus, we studied the translation of the λrpc-calculus into the ςrmi-calculus and proved the soundness of the translation.

6   Related Works

Our paper studied the distributed object-oriented calculus from the approach of the object calculus and remote method invocation. Besides our approach, there could be several alternatives to the distributed oriented calculus. The most distinguished study on this area is the object-oriented concurrent system ABCL [11] based on the actor model [6] proposed by Hewitt et al. The process calculi [8] proposed by Milner et al. are also related to our work. Such theoretical frameworks are provided for formalizing concurrent systems, but the difference between our framework and theirs is that our framework explicitly handles the locations where the processes are executed, whereas their frameworks handle these locations implicitly and transparently. For the abstract investigation of the concurrent systems, their approaches seem to be appropriate. On the other hand, our approach is suitable for describing a distributed computation which is aware of the locations.

References

1. Abadi, M., Cardelli, L.: A Theory of Objects. Springer, Berlin (1996)
2. Abadi, M., Cardelli, L.: A Theory of Primitive Objects — Untyped and First-Order Systems. In: Hagiya, M., Mitchell, J.C. (eds.) TACS 1994. LNCS, vol. 789, pp. 296–320. Springer, Heidelberg (1994)
3. Birrell, A.D., Nelson, B.J.: Implementing remote procedure calls. ACM Transactions on Computer Systems 2(1), 39–59 (1984)

4. Cooper, E., Wadler, P.: The RPC calculus. In: Proceedings of the 11th ACM SIG-PLAN Conference on Principles and Practice of Declarative Programming, PPDP 2009. ACM Press (2009)
5. Gordon, A., Rees, G.: Bisimilarity for a first-order calculus of objects with sub-typing. In: Proceedings of the 23rd Annual ACM Symposium on Principles of Programming Languages. ACM Press (1996)
6. Hewitt, C., Bishop, P., Steiger, R.: A universal modular actor formalism for ar-tificial intelligence. In: IJCAI 1973: Proceedings of the 3rd International Joint Conference on Artificial Intelligence, pp. 235–245. Morgan Kaufmann Publishers Inc. (1973)
7. Kahn, G.: Natural semantics. In: Brandenburg, F.J., Wirsing, M., Vidal-Naquet, G. (eds.) STACS 1987. LNCS, vol. 247, pp. 22–39. Springer, Heidelberg (1987)
8. Milner, R., Parrow, J., Walker, D.: A calculus of mobile processes, Part I and Part II. Information and Computation 100(1), 1–77 (1992)
9. Narita, K., Nishizaki, S.: A Parallel Abstract Machine for the RPC Calculus. In: Abd Manaf, A., Sahibuddin, S., Ahmad, R., Mohd Daud, S., El-Qawasmeh, E. (eds.) ICIEIS 2011, Part III. CCIS, vol. 253, pp. 320–332. Springer, Heidelberg (2011)
10. Plotkin, G.: Call-by-name, call-by-value, and the λ-calculus. Theor. Comput. Sci. 1, 125–159 (1975)
11. Yonezawa, A. (ed.): ABCL: An Object-Oriented Concurrent System. The MIT Press, Cambridge (1990)

A Metric for User Requirements Traceability in Sequence, Class Diagrams, and Lines-Of-Code via Robustness Diagrams

Jasmine A. Malinao[1,2], Kristhian B. Tiu[2], Louise Michelle O. Lozano[2],
Sonia M. Pascua[2], Richard Bryann Chua[2],
Ma. Sheila A. Magboo[2], and Jaime D.L. Caro[1,2]

[1] Department of Computer Science (Algorithms and Complexity Lab),
College of Engineering, University of the Philippines, Diliman,
Quezon City 1101, Metro Manila, Philippines
[2] UP Information Technology Development Center (UPITDC), Vidal Tan Hall,
Quirino Avenue corner Velasquez Ave., University of the Philippines, Diliman,
Quezon City 1101, Metro Manila, Philippines
{jamalinao,jdlcaro}@dcs.upd.edu.ph

Abstract. In this work, we propose a metric based on the ICONIX paradigm to calibrate the consistency, completeness, and correctness of commonly used dynamic and static models of software design with a pre-specified set of user requirements expressed as Use Case Texts. A depth-first search-based algorithm is presented to extract scenarios and describe the temporal aspect of software development-related tasks embedded in the Robustness Diagram of ICONIX to derive results needed to perform further verification of these models. A procedure to perform a similar verification of a software's set of Lines-of-Codes is also proposed. Finally, we perform empirical tests on real-world data and report the results.

Keywords: ICONIX, Metric, Robustness Diagram, UML.

1 Introduction

In the domain of software quality assurance, the value of a software product is mainly measured by how it caters to the needs of its users. These users may be of varied types and skill sets, each with varying satisfaction levels in terms of how they perceive a software functionalities match with what they need to perform their tasks. In the process of making sure that software products deliver what they are supposed to provide to these users, software designers and developers are burdened with the following tasks, 1) effectively capture user requirements from requirements elicitation stage, 2) effectively and, if possible, completely design all of the user requirements in the form of formal models by using standard modeling tools, 3) continually assess and quantitatively measure satisfiability of the formal models with the user requirements, 4) develop fully implemented designs as executable and usable software, and 5) test for usability for different kinds of users.

S. Nishizaki et al. (Eds.): WCTP 2012, PICT 7, pp. 50–63, 2013.

In this work, we shall focus on two of the four viewpoints of software quality as discussed in [1,2]. These are manufacturing and product view of software quality. The manufacturing view focuses on the production aspect of the software itself. It emphasizes on implementing optimal processes and decisions at the initial periods of software development rather than addressing defects and/or bugs at the latter stage and waste time and incur unnecessary testing costs[3] in doing so. Meanwhile, the product view looks at ensuring soundness of the features and functionalities a software product offers by scrutinizing its more detailed internal components. The principle that is followed is that by controlling internal product quality indicators, software products exhibit positive external behavior that gives user satisfaction. Even when metrics are formulated for black box testing[4] for requirements traceability checking, positive external behavior may not always imply sound internal product quality when some tests are bypassed. It is also notable that both of the indicated viewpoints are coherent with principles in the paradigm of Design Driven Testing (DDT)[5]. DDT also recognizes that to develop effective, efficient, bug-free and usable codes, these codes should match up with the designs, and designs fulfill user requirements.

In the software design and analysis phase, many software products are built from constructions of Sequence Diagrams (SD) and Class Diagrams (CD), all of which emanates from semiformal expositions of captured user requirements, i.e. from Use Case Texts (UCT). One important concern is realizing whether these constructions match up to their corresponding UCTs, i.e. User Requirements Traceability, to effectively build quality software codes. A study conducted in [7] performs requirements traceability by using two concepts, namely, 1) Requirements Coverage Metrics, and 2) Design Compliance Metrics. It checks existence of a use case for each requirement, a sequence diagram for each use case, a class diagram containing the classes and methods modeled in the sequence diagram. However, using the methods in this study, one can never ascertain how complete a use case model is with respect to the minute details in all the requirements of its corresponding UCT; how complete a sequence diagram in capturing the intricate details of all components in one use case model; how consistent is the ordering of activities in the sequence diagram with the ordering expressed in the UCT; how assignments of methods to classes are to be accomplished; which are input and output parameters are manipulated among these methods in the class diagrams, etc. In order to correctly measure these aspects of traceability from UCTs to their designs, we shall formulate and propose a metric that is powered by the use of Robustness Diagrams in the ICONIX process[5,6,11]. This metric is aligned with the underlying principles of DDT paradigm and the two viewpoints of software quality.

Among the three types of software metrics described in [9], the proposed metric is within the realm of both product and project metrics. In coherence with the basic principles of these two sets of metrics, our proposed metric is able to check whether key attributes and process flows mentioned in the UCTs are manifested in the intended software products. However, we improve our metric's capability by fully calibrating the amount of completeness, consistency, and correctness in both static and temporal representations in the SDs and CDs as compared with the user requirements in the UCT. This can be done proposing a set of guidelines in Section 2.2 to transform UCT to entities in the Robustness Diagram and verifying the veracity of the attributes' representation in the SDs and CDs. As for the verification of the correctness of representation of process flows, two algorithms shown in Sections 2.3 and 2.4 to extract scenarios from RDs and SDs, respectively. Full calibration for User Requirements Traceability is discussed in Section 2.5. A discussion of how to check for LOC compliance with the design is discussed in Section 2.7. In Section 3, we provide one real-world test data that represents one functionality in a software of Fujitsu Ten Limited, and show user requirement satisfaction with the functionality's corresponding UML model. Finally, we present our conclusions in Section 4.

2   Methodology for Metric on Requirements Traceability

2.1   The Robustness Diagram

Formally, a Robustness Diagram (RD) is a "pictorial representation of the behavior described by a use case, showing both participating classes and software behavior, although it intentionally avoids showing which class is responsible for which bits of behavior"[8]. The components of a robustness diagram are objects, i.e. either boundary or entity objects, controllers, and arcs that shows interaction between these objects and controllers. Boundary objects act as an interface between the system and an external user, e.g. human, other system, or subsystem within a software. Entity objects are classes found in the domain model while controllers are the "glue" between the boundary and entity objects[8]. Roughly speaking, controllers represent computations and I/O operations implemented by the boundary and entity objects in the diagram.

2.2   From Use Case Texts to Robustness Diagrams

An RD can be drawn directly from the UCT. This requires the latter to be as brief and as clear as possible to effectively and easily layout the former. The process of converting UCT to RD involves translating UCT words into their corresponding RD components. The initial step is to identify all nouns and verbs in the UCT, denoted by UCT.Noun and UCT.Verb, respectively. Each UCT.Noun is then represented as either boundary or entity classes or attributes of an entity class while each UCT.Verb is represented as either controllers (e.g. get, set, show and other methods of an entity class) or user actions (e.g. clicking a button). In [8], it was emphasized that there is direct one-to-one correlation between each step in the UCT and flow of action in the RD. Thus, each step in the UCT must be visibly represented in the RD. An arc (A, B) is drawn in the RD to represent the flow of actions from controllers A to B, or ownership relationship between a class A and a controller B. These arc may contain labels, denoted as $(A, B).label$, with the "send" word and an I/O value/s, that is usually from an attribute/s of A. Note that there may exist a UCT.Noun reflected in this label, thus, we access them using the notation $(A, B).label.noun$.

Guidelines for UCT to RD Conversion. The guidelines in converting UCTs to RDs are summarized below.

- Each UCT.Noun is represented as either candidate I/O variable or class or attribute in the RD
- Each UCT.Verb is represented as either Set or Get or Show Controller containing UCT.Noun where UCT.Noun is either a candidate I/O Variable or a class attribute

  Remark 1. *All other verbs can either be classified as methods of a class or user actions.*

- Controllers (i.e. methods) with outgoing arcs pointing to entity objects (i.e. classes) MUST be instantiators of them.
- Controller interaction, i.e. an arc from controllers A to B, is possible even if A and B are owned by different entity classes provided that the B's entity class has an object that has already been instantiated.
- A controller that requires n inputs MUST have n incoming arcs from other controllers, if the arc represents one value passed.
- Ownership arcs (i.e. arcs without labels from a class C to another controller) are present for non-boundary classes because it is not certain that for a specific time, C had an object instantiation already.

  Remark 2. *A controller D that has no ownership arc is a method of a boundary class B. D is unreachable without instantiating B through a user action.*

2.3   Extracting Scenarios and Process Flow in Robustness Diagrams

A UCT may contain many scenarios which are not readily traceable from input to output. Alternate flows from one task specification in the UCT may exist, entirely disjoint from one another, but may have one junction point later on. Flows in the list of task specifications in the UCT may also be influenced by satisfaction of some conditions in some tasks, system and/or user tasks, before processing of other tasks are pushed through.

In order to establish an effective means of checking whether all of these circumstances in the UCT are addressed and, later on, modeled in other UML diagrams and in LOCs, it is important that we can extract each of the complete scenarios, i.e. from a given the input specification of the UCT, the appropriate and desired output is obtained. We introduce Algorithm 1 that extracts one scenario of the UCT. This algorithm may be implemented again until all scenarios are extracted from the UCT.

Roughly speaking, Algorithm 1 employs a Depth-First-Search strategy by treating the RD as a directed graph. However, we set rules in backtracking from the input to the output stages. A path is established using this strategy, and in doing so, update the information contained in the graph. This information pertains to the time-of-execution of a task. When a final task is reached, the

algorithm stops and outputs the set of tasks and the time of execution of each
of them to form one scenario of a UCT.

Algorithm 1
Input: An RD of a UCT that adheres to the UCT to RD conversion guidelines
Output: A scenario described in the UCT from an input to a desired output

1. Label all the boundary, entity classes and controllers in the RD with $X_i = 0$,
   $1, \ldots, n-1$, where n is the total number of these components in RD.
2. For an arc q emanating from a class or a controller that needs to send an I/O
   value/s to a class or a controller of RD, assign a value for $q.label$ containing
   the "send" and the I/O variable/s corresponding to a UCT.Noun/s.
3. Whenever there are arcs E having a common source and destination, merge
   them as one arc e in RD. Then, set the value of $e.label$ as the concatenation
   of the labels of each element of E.
4. As an initial step to all scenarios modeled by RD, label its boundary class
   with X_0. This boundary class can be thought of the interface between the
   user and the system or a subsystem interacting as an API to another subsystem. By using RD's corresponding UCT, determine the final arc f that
   indicates the entry to the last process of a scenario in RD.
5. To model the actual series of processes from a starting point (i.e. X_0) to the
   exit point X_i (which is incident to f), $i \in 0, 1, \ldots, n-1$, select one arbitrary
   arc q extending out of X_0. To indicate that this arc q is already explored,
   mark it with its Time-of-Execution(TE), where $q.TE = 0$ to indicate the
   first time step in one scenario within RD.
   TE shall be incremented each time for newly explored arcs in RD. Note that
   for each unexplored arc q, $q.TE = nil$.
6. Implement a Depth First Search (DFS) in traversing RD from X_0, thereby,
   discovering new unexplored arcs until f is explored and marked with its TE.
   Note that each arc must be labeled with a number representing the appropriate time step as it is being traversed.
       In this paper, we modify DFS's backtracking criterion. The algorithm
   backtracks from a node X_i to its predecessor node X_j (which may be a
   boundary/entity objects or a controller) using an arc q if there exists an
   indegree arc r of X_i where $r.TE = nil$, $q.label \neq r.label$, and $q.label$ is not a
   substring of $r.label$.
   Whenever the newly explored arc q is the last explored (indegree) arc of X_i,
   the algorithm performs the following:
   (a) collects the set of all indegree arcs W of X_i (inclusive of q),
   (b) determines $v = \max_{\forall e, e \in W} e.TE$,
   (c) and updates $e.TE = v, \forall e, e \in W$.
   In backtracking, if DFS reaches X_0 again is some point in time and f is still
   not explored, the algorithm will determine the maximum TE assigned to all
   explored arcs in its implementation. Then, from X_0, DFS will traverse an
   unexplored arc and will update its TE with the obtained maximum plus one.
   Otherwise, if during the backtrack DFS reaches X_k, $k \in \{1, 2, \ldots, n-1\}$,

and finds an unexplored outgoing arc w of X_k, it sets $w.TE = p$, where p is the TE of an indegree arc of X_k.

7. When f is explored, the DFS traverses it and goes back to X_0. This accounts for one complete scenario embedded in RD that accomplishes the desired output requirement specified in the UCT from the initiation of the user action or API invocation.

Remark 3. *There may exist many scenarios embedded in RD that are derived from the UCT. Implicitly, one RD can represent multiple SDs.*

2.4   Extracting Process Flows in Sequence Diagrams

For an SD that has been previously developed out of the user requirements modeled in the UCT, we need to check if the SD has completely modeled one scenario in the UCT. Apart from checking if all the needed components, i.e. classes, methods, appropriateness of interaction by message passing, are present and consistent in the SD as specified in a scenario of a UCT, the temporal aspect of the SD should be checked. This aspect must be consistent to the ordering of task execution of the UCT. With this, we build Algorithm 2 that initializes values in the SD pertaining to this aspect. Then, we shall formally specify how validation of all of these aspects in these UML models, based on completeness, correctness of component interaction, and consistency of task execution modeling amongst these models via RD.

Algorithm 2
Input: Sequence Diagram of a UCT
Output: Updated Sequence Diagram with methods assigned with TE values

1. Label all classes and methods in the SD by following the labeling of the boundary/entity objects and controllers in RD.
2. Label arcs (representing method invocation in a class X_i by X_j) with increasing TE values using a left-right and top-bottom traversal of the diagram.
3. Instantiation time of an object of a class is equal to the time of the first use of a method of that class.

2.5   Algorithm for UCT to Sequence and Class Diagrams Requirements Traceability via Robustness Diagrams

Using the RD constructed from the UCT representing one software requirement specification, we validate the completeness of the representation of its corresponding SD and CD. First, we check the static aspect of the UML studies under study, i.e. classes, and methods including the I/O values associated to the attributes of the classes. This aspect is quantitatively measured by the completeness-of-class-representation V_c and completeness-of-method-representation V_m, respectively. Secondly, we check for the temporal aspect of these models, i.e. correctness of the ordering of method execution as specified in the UCT, as represented by V_t. Finally, we determine and measure if these UML models are consistent and complete with respect to the overall user requirements specified by the UCT via RD, as measured by the Requirements Traceability Value(RTV).

1. **Components Capture and Transformation**

   - Computation for V_c.
     For each $X_i \in C$, where C is the set of classes in RD represented as entity/boundary objects, determine the following:
     (a) $\alpha = \{X_j | \exists (X_j, X_i) \text{ in RD}\}$.
     (b) $\alpha' = \{X_j | X_j \text{ is an outdegree arc of } X_i \text{ in SD}\}$.
     (c) Compute for the completeness-of-class-representation $I(X_i)$ of X_i,
        as follows, $I(X_i) = \frac{|\alpha \cap \alpha'|}{|\alpha|}$.
     Therefore,
     $$V_c = \frac{\sum\limits_{\forall X_i, X_i \in C} I(X_i)}{|C|}.$$

     In these computations, it can be easily checked that a class is not considered present in SD, therefore not satisfying the user requirements indicated in the UCT, when all of its listed method in the RD is not included.

   - Computation for V_m.
     For each $X_i \in M$, where M is the set of methods in RD represented as controllers, determine the following:
     (a) **RD in-requirements:** $\beta = \{x | \exists (X_j, X_i) \text{ in RD where} \\ (X_j, X_i).label.noun = x\}$.
     (b) **SD in-requirements:** $\beta' = \{x' | \exists X_i \text{ in SD where } x' \text{ is an input} \\ \text{parameter contained in } X_i.label\}$.
     (c) **RD out-requirements:** $\gamma = \{y | \exists (X_i, X_j) \text{ in RD where} \\ (X_i, X_j).label.noun = y\}$.
     (d) **SD out-requirements:** $\gamma' = \{y' | \exists X_i \text{ in SD where } y' \text{ is an output} \\ \text{parameter contained in } X_i.label\}$.
        Note that y' may be found as a parameter in the method definition, however, a pass-by-reference is forcibly enforced upon X_i's method invocation.
     (e) **Ownerships:** Let $O(X_i) = 1$ if X_i is owned by the same class in RD and SD, 0 otherwise.
     (f) **RD class callers:** Let $\tau = \{X_k | \exists (X_j, X_i) \text{ in RD, where } X_k \text{ is the owner of } X_j \text{ and } (X_j, X_i).label \neq nil\}$.
     (g) **SD class callers:** $\tau' = \{X_k | X_i \text{ is an outdegree arc of } X_k \text{ in SD}\}$.
     (h) Compute for the completeness-of-method-representation $I'(X_i)$ of X_i, as follows,
        $$I'(X_i) = \frac{\frac{\beta \cap \beta'}{|\beta|} + \frac{\gamma \cap \gamma'}{|\gamma|} + O(X_i) + \frac{|\tau \cap \tau'|}{|\tau|}}{4}.$$

Therefore,

$$V_m = \frac{\sum\limits_{\forall X_i, X_i \in M} I'(X_i)}{|M|}$$

In these computations, a method will not be considered present in the SD, therefore not satisfying the user requirements indicated in the UCT, when at least 1 required input or output parameters is not included as listed in the RD, when there does not exist one class that owns it, and if not all of its callers, if they exist, are not indicated in the SD.

2. **Temporal Process Capture and Transformation**
   From the execution of Algorithm 1, determine the following quantities,
   (a) $t(X_i)_{RD} = \min\limits_{\forall a \in A} \{a.TE\}$, where A is the set of outgoing arcs from X_i in RD.
   (b) $t(X_i)_{SD}$ is the time of execution of X_i in SD.
   (c) $Cons(X_i, X_j) = \frac{t(X_i)_{RD} - t(X_j)_{RD}}{t(X_i)_{SD} - t(X_j)_{SD}}$.
   (d) If $Cons(X_i, X_j) \geq 0$, then set correct-ordering- representation $I''(X_i, X_j) = 1$, otherwise, 0.
   Therefore,

   $$V_t = \frac{\sum\limits_{i=0}^{n-2} \sum\limits_{j=i+1}^{n-1} I''(X_i, X_j)}{\binom{n}{2}}.$$

3. **Requirements Traceability Values** RTV
   The percentage consistency between Robustness and Sequence Diagram, denoted as $RTV(RD, SD)$, can be computed using the formula:

   $$RTV(RD, SD) = \frac{V_c + V_m + V_t}{3}.$$

2.6   The Class Diagram Verified

It can be easily seen that the concepts and computations in building RDs from UCTs can also be utilized to verify the consistency and completeness of CDs with respect to the UCT in a software's logical design. Since CDs are static representations of the components of modules, the following equivalence can be established with RDs:

- the boundary and entity objects in an RD are the classes in a CD
- a controller in the RD is a methods belonging to exactly one class in a CD
- a controller Co is owned by a entity object Cl when there is an arc with a label of nil in the RD and Co is a method in a class Cl in CD
- a controller Co is owned by a boundary object Cl when there is an arc from Cl to Co, or no arc with a label nil is an indegree of Co in the RD and Co is a method in a class Cl in CD
- the set β and γ defined in Section 2.5 correspond to the inputs and outputs of a controller X_i in RD which would be the in and return parameters of method X_i in CD

Therefore, verifying the completeness and consistency of a CD with its corresponding UCT using RD is a matter of checking the existence of such components and relations in these two diagrams.

2.7 A Note on Lines-of-Code(LOC) Verification via Robustness Diagrams

In order to verify the consistency and completeness of a set of Lines-of-Codes (LOC) with respect to the indicated requirements of its UCT, we will use the Robustness Diagram's set of controllers and the technology for reverse engineering of Doxygen[10]. This can be established by using the following equivalence in RD and LOC:

- The controllers are identified as the methods called upon to execute a functionality in its LOC.
- Controller interaction in the RD established through the arcs will be the function calling in its LOC.
- The set β and γ defined in Section 2.5 correspond to the inputs and outputs of a controller X_i in RD which would be the in and return parameters of method X_i in LOC.
- An outdegree arc of a controller X_i leading to X_j in the RD means X_i invokes X_j in the LOC, and, X_i is executed first before X_j.

To be able to efficiently determine the fourth equivalence to verify the LOC, a call graph[12] is generated from the current LOCs constructed by a software developer. Call graphs are directed graphs that are used to visualize a series of function calls in a computer program. Algorithm 1 generates the ordering and relationships of the methods in RD. As for the call graph, ordering and caller-callee are apparent by a left-right traversal from the graph's starting node.

Therefore, verifying the completeness and consistency of a set of LOCs with its corresponding UCT using RD is just checking the existence of such methods, caller-callee relationships, and order of invocation of methods in RD and the LOC's corresponding call graph.

3 An Example

One of the basic functionalities of Fujitsu Ten's Computer Aided Multi-Analysis System Auto Test Tool (CATT)[13] is to aid users in constructing analog and digital signals in cells within a worksheet. CATT is built mainly as a support tool for creating test specifications to be used by engine control unit developers. Shown below is the UCT for CATT's functionality of creating analog signals. We shall build the RD for this UCT and verify the correctness and completeness of one SD of Fujitsu Ten's Create Analog Signal.

The Create Analog Signal is a functionality in CATT that enables users to create analog signal in a cell, or a group of cells with a row in a worksheet.

This row must be initially set to accept a series composed of analog signals. In creating a signal/s, the user may set different values to vary or set how the signals are projected and how interpolation is done amongst the discrete set of values given by the user. The output expected from the task specifications of this UCT is the signal projected in the specified cells.

A Sample UCT *T*
Title: UCT for Create Analog Signal
Precondition: User adds an analog series

1. User selects single/multiple empty cell(s) in the Chart worksheet
2. User creates an analog signal
   - Pop Up Menu option
     (a) User right clicks on the selected cell(s)
     (b) System displays the pop up menu
     (c) User selects one of the following options:
        - "Fixed Value Input"
        - "Slope Input"
   - Toolbar button
     (a) User clicks one of the ff. toolbar buttons:
        - "Fixed Value Input"
        - "Slope Input"
3. System gets the selected cell(s)
4. System gets the parent Row of one cell in the cell collection
5. System gets the boundary
6. Repeat steps 3-4 until all selected cells are traversed
   (a) System determine Min/Max boundary
7. System displays an input dialog box and asks the input value(s) from the user
   (a) System displays input dialog box
      - If User selects "Fixed Value Input", System asks for a single value from the user
      - Else If User selects "Slope Input", System asks for start and end values from the user
   (b) User inputs values in the dialog box
8. System restricts user's entered values to Max/Min boundary values and Max/Min time values
   - If user input/s is/are out-of-range, System disables OK button.
   - ELSE
     - System enables OK button.
     - User clicks "OK" or "CANCEL"
       * If User Selects "OK", perform steps 9-14
       * Else, skip steps 9-14
9. System gets the user's input
10. System gets the parent row and the parent column of once cell in the collection of cells
11. System gets the time interval of the cells column

12. User creates an analog signal in that cell
    (a) setting Value(s) and Time(s) by calling Add Point
13. System plots an analog signal in that cell by performing createNewPlot() method
14. Repeat steps 10-13 until all selected cells are traversed

Alternate Methods: None

3.1   The Robustness Diagram for T

We build the corresponding RD of T as shown in Figure 1. Below is an illustration of Algorithm 1 performed on the "Creating Analog Signal" UCT of Fujitsu Ten's CATT System. The algorithm yields one of the scenarios embedded in RD.

In Figure 1, the values highlighted in yellow and red which are shown on the arcs correspond to the arc's final TE and backtrack TEs, respectively. Backtrack TEs are the instances of time when a backtrack was done from the arc's incident node a to its source because there exists at least 1 other incident arc of a that is not yet traversed and therefore has not value of TE.

For this example, the DFS strategy yields a cycle with respect to the input to output node since the user specifies the creation of the signal in a selected cell, and outputs the interpolated signal to the same set of cells in the same

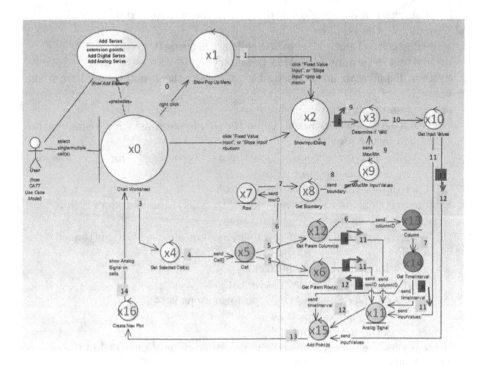

Fig. 1. The RD of T with Algorithm 1 implemented extracting one scenario

worksheet. $x0$ here is the input and output node of the UCT. The sole indegree arc f of $x0$ is the checkpoint of the DFS strategy to mark the end of Algorithm 1, hence, one complete scenario is extracted.

From $x0$, the set of tasks needed to obtain the desired result and the succession of the execution of this tasks can be extracted by listing down the nodes based on the TE's of their outgoing arcs until f has been included in the list. For this example, this list is as follows,

$$x0 \ x1 \ x2 \ x0 \ x4 \ x5 \ [x12 \ x13 \ x14][x6 \ x7 \ x8[x2 \ x9]x3 \ x10]x11 \ x15 \ x16 \ x0.$$

The symbol [ ] enclosing the nodes of the graph corresponds to tasks that may be done in parallel and execution of these tasks do not depend on each other as highlighted by their TEs.

3.2 The Sequence Diagram Verified

Completeness-of-Class/Method-Representation. Shown in Figure 2 is a UCT developed in FTSP corresponding to the UCT of Create Analog Signal. Three classes (2 entity classes and 1 boundary class) and thirteen methods (controllers) were identified in this example. It can be noted that all listed classes and methods in RD are present in the SD in Figure 2. Thus, $V_c = V_m = 1$. This means that SD fully satisfies the user requirements of one scenario of the UCT.

Correct-Ordering-Representation. The time of execution of each controller in the two diagrams are summarized in Table 1.

Table 1. Time-of-Execution of RD and SD using Algorithms 1 and 2, resp

X_i	TEs in RD	TEs in SD	X_i	TEs in RD	TEs in SD	X_i	TEs in RD	TEs in SD
X_0	0	0	X_6	6	4	X_{12}	6	5
X_1	1	1	X_7	7	7	X_{13}	7	6
X_2	9	10	X_8	8	8	X_{14}	11	13
X_3	10	11	X_9	9	9	X_{15}	13	15
X_4	4	2	X_{10}	11	12	X_{16}	14	16
X_5	5	3	X_{11}	12	14			

The Requirements Traceability Value $RTV(RD, SD)$. The individual temporal value consistency of each controller combination are computed. All controller combinations which include instantiation of classes has shown consistent time of executions in the RD and SD. Therefore, we obtain $V_t = 1$. Thus, $RTV(RD, SD)=(1+1+1)/3=1$. This means that the SD is 100% fully representative of the UCT requirements as verified by RD.

62   J.A. Malinao et al.

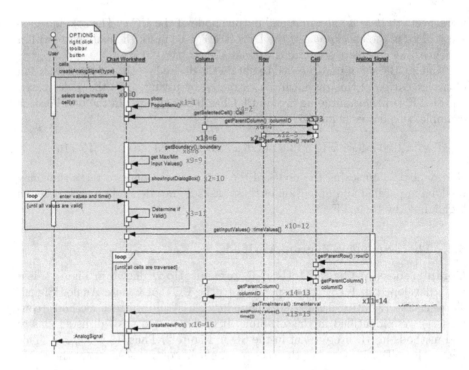

Fig. 2. A Sequence Diagram of T with Algorithm 2 implemented

4   Conclusions

In this paper, we were able to use Robustness Diagrams in the ICONIX paradigm
as a means for User Requirements Traceability in SDs, CDs, and LOCs based on
the user requirements specified in UCTs. This paper also proposed a set of guidelines to effectively map UCTs into specifications in RDs. Additionally, we were
able to develop algorithms to measure the percentage of consistency and completeness of the aforementioned diagrams and LOCs with respect to the UCT's
set of requirements. Finally, we implement these guidelines and algorithms in
verifying Fujitsu Ten's SD for the company's CATT's basic functionality.

Acknowledgments. This work is funded by Fujitsu Ten Solutions Philippines
(FTSP) and Fujitsu Ten Limited (FTL) - Japan. We would like to thank Mr.
Bryan Oroceo of FTSP, Mr. Naoya Kamiyama and Mr. Akira Ikezoe of FTL in
validating the results of this work.

We thank the UP Information Technology Development Center for supporting
this research work.

References

1. Rawat, M.S., Mitaal, A., Dubey, S.K.: Survey on Impact of Software Metrics on Software Quality. International Journal of Advanced Computer Science and Applications 3(1), 137–141 (2012)
2. Heimann, D.: Implementing Software Metrics at a Telecommunications Company a Case Study. Journal of Cases on Information Technology (2004)
3. Marick, B.: Reliable Software Technologies. In: Proceedings of STAREAST 1999 Software Testing Analysis and Review (1999)
4. Whalen, M.W., Rajan, A., Heimdahl, P.E.M., Miller, P.S.: Coverage metrics for requirements-based testing. In: Proc. of the 2006 Intl. Symposium on Software Testing and Analysis, ISSTA, Portland, USA, pp. 25–36 (2006)
5. Stephens, M., Rosenberg, D.: Design Driven Testing: Test Smarter, Not Harder. ISBN-13 (pbk): 978-1-4302-2943-8
6. Dugerdil, P., Belmonte, J., Kony, D.: Using Robustness Diagrams to Help With Software Understanding: An Eclipse Plug-in. Int. J. of Software Engineering, IJSE 2(3), 135–150 (2009)
7. Kanjilal, A., Kanjilal, G., Bhattacharya, S.: Metrics-based Analysis of Requirements for Object-Oriented Systems: Empirical Approach. INFOCOMP Journal of Computer Science 7(2), 26–36 (2008)
8. Rosenberg, D., Stephens, M.: Use Case Driven Object Modeling with UML: Theory and Practice. Apress, Berkley (2007) ISBN-10: 1590597745
9. Honglei, T., Wei, S., Yanan, Z.: The Research on Software Metrics and Software Complexity Metrics. International Forum on Computer Science-Technology and Applications (2009)
10. http://www.stack.nl/dimitri/doxygen/
11. http://iconixprocess.com/iconix-process/
12. Ryder, B.G.: Constructing the Call Graph of a Program. IEEE Transactions on Software Engineering SE-5(3), 216–226 (1979)
13. Fujitsu Ten Limited, CATT Ver. 4.8.0: Reference Manual (2000-2012)

Some Improvements of Parallel Random Projection for Finding Planted (l, d)-Motifs

Jhoirene B. Clemente and Henry N. Adorna

Algorithms & Complexity Lab
Department of Computer Science
University of the Philippines Diliman
Diliman 1101 Quezon City, Philippines
{jbclemente,hnadorna}@up.edu.ph

Abstract. A challenging variant of the motif finding problem (MFP) is called planted (l, d)-motif finding. It is where the pattern has expected number of mismatches on each of its occurrence in the sequence. Difficulty arises when the planted pattern is longer and the expected number of mismatches is high. An algorithm (FMURP) which uses a random projection technique followed by local search algorithms is shown to work better with longer motifs with higher accuracy. A parallel algorithm for FMURP already exist in the literature. However, the parallel algorithm uses an excessive amount of memory. In this paper, we propose some improvements on the existing parallel algorithm. We also prove that the modified parallelization is equivalent to the sequential version of FMURP.

Keywords: Random Projection, Planted (l, d)-motifs.

1 Introduction

Motifs are short recurring patterns in the DNA. These patterns can be *transcription binding sites* which dictate the expression of genes. Finding motifs is considered to be computationally hard. In fact, it is an **NP**-complete problem [7]. Moreover, occurrences of motif in each sequence may vary due to mutations and genome rearrangements. Mutations can occur at random positions of the motif making the problem even more challenging. Let us consider the following *challenge problem*.

Definition 1 (Challenge Problem [8])
Given a set of 20 DNA sequences each with 600 nucleotide bases, identify a consensus pattern of length 15, where each of its occurrence in the set of sequences has exactly 4 mismatches.

Generally, the *challenge problem* presented in Definition 1 can have a variable pattern length l and a variable expected number of mismatches d for any given set of DNA sequences. For simplicity, we will call an l-length string an l-mer all through out the paper. This leads us to a variant called the *planted (l,d)-motif finding*.

S. Nishizaki et al. (Eds.): WCTP 2012, PICT 7, pp. 64–81, 2013.

Definition 2 (Planted (l, d)-Motif Finding Problem [2])
Given a set of t DNA sequences $\mathcal{S} = \{S_1, S_2, \ldots, S_t\}$ where each S_i is of length n_i, a motif length l, and an expected number of mismatches d, identify an l-mer M, such that

$$\forall \, i, \, 0 \leq i \leq (t-1), \, \exists \, j, \, 0 \leq j \leq (n - l + 1)$$

$$d_E(M, S_{i,j}) = d$$

where $S_{i,j}$ is an l-mer in S_i starting at jth position and $d_E(M, S_{i,j})$ is the number of mismatches between two l-mers M and $S_{i,j}$.

Planted (l, d)-motif finding is challenging because even pattern-based approaches for motif finding fail to recover planted (l, d)-motifs [8,2]. Pattern-based approach is shown to work best for finding motifs [10]. The earliest attempt to recover planted motifs is SP-Star which is presented in [8]. A hybrid algorithm (which we will refer to as FMURP all through out the paper) better than SP-STAR is presented in [2]. It is a two step algorithm composed of an initialization and local refinement steps. It also works well with recovering longer planted motifs but not those with higher mutation rate i.e higher l/d. This algorithm will be discussed in detail in Section 2.

A parallel version of FMURP presented in [4] uses an excessive amount of memory. Therefore, in this paper, we present several optimizations to the parallel FMURP as well as proof of its correctness. The parallelization also assumes a massively parallel architecture such as Graphics Processing Units (GPUs) as in [4]. Since memory is limited in the these devices, optimization in memory is necessary. Two improved versions are presented in this paper, the first one attempts to decrease the space complexity of the initial parallel version and the second one attempts to reduce the number of threads being used.

2   Random Projection for Finding Motifs

Finding motifs using random projection (FMURP) [2] is a hybrid algorithm that uses a pattern-based approach called *Projection* to give an initial motif model for local optimization algorithms. In [2], implementation of FMURP used Expectation Maximization and SP-STAR$_\sigma$ [2] to refine the initial motif models identified by *Projection*. We formulate the notation SP-STARσ to avoid confusion between the original SP-STAR presented in [8]. FMURP is presented in Algorithm 1 and is available for download in [3].

2.1   Projection Algorithm

Random Projection is a statistical technique commonly used for dimensionality reduction and visualization. Projection for l-mer $S_{i,j}$ is shown in Definition 3.

INPUT: Set of sequences S, motif length l, expected mismatches d, projection dimension k, and bucket threshold δ
OUTPUT: Best scoring l-mer

1. Generate k random positions for projection.
   Let this be the set $I = \{\hat{i} | \hat{i} \in \{0, \ldots, (l-1)\}\}$ and $|I| = k$.
2. Compute $h_I(S_{i,j})$s for all $S_{i,j}$s in S,
   where $i \in \{0, \ldots, (t-1)\}$, and $\{j \in 0, \ldots, (n-l)\}$.
3. Hash each $S_{i,j}$ with respect to its corresponding $h_I(S_{i,j})$.
4. Identify the set of buckets which contain δ or more l-mers hashed.
5. Perform EM refinement for each enriched bucket.
6. Perform $SP\text{-}STAR\sigma$ for each enriched bucket.
7. Maximize σ score to output best motif.

Algorithm 1. One independent trial of FMURP. This algorithm is performed m times to achieve higher scoring motifs [2].

Definition 3. *Given an l-mer $S_{i,j}$, a set $L = \{0, 1, \ldots, (l-1)\}$, a projection dimension k, and a set $I \subset L$, where $|I| = k$, a k-dimensional projection of $S_{i,j}$ is*

$$h_I(S_{i,j}) = S_{i,j}(I_0)S_{i,j}(I_1) \ldots S_{i,j}(I_{(k-1)}),$$

where I_i denotes the ith element in I when sorted and $S_{i,j}(I_i)$ is the character in $S_{i,j}$ at I_ith position.

The *Projection* algorithm discussed in [2] provides initial motif models (in the form of probability weight matrices) in attempt to find motifs or *planted* (l, d)-*motifs*. Motif models are not limited to probability weight matrices. Generally, these are representations of the possible solution. Examples of which are consensus pattern and a set of starting position of motif occurrences in the set of sequences.

The algorithm starts by generating the set I, i.e. k random positions which is used to compute $h_I(S_{i,j})$s. For each $S_{i,j}$ in the set of sequences S, get the corresponding k-mer $h_I(S_{i,j})$. Maintain $|\Sigma|^k$ buckets, where each bucket is represented by a unique k-mer defined over Σ. The notation Σ pertains to the alphabet of the DNA sequences. Hash each l-mer $S_{i,j}$ using $h_I(S_{i,j})$ to the corresponding bucket. After hashing all l-mers, identify which of the buckets contain a significant amount of l-mers, using the *bucket threshold* δ. The identified buckets which satisfy the threshold are called *enriched buckets*. In the implementation in [3], steps 3 and 4 are performed by sorting the list of $h(S_{i,j})$s and performing a linear search over the sorted list to get the set of buckets. These steps avoid maintaining $|\Sigma|^k$ buckets identified by unique k-mers in Σ.

Projection algorithm serves as an initialization step for local search algorithms. The consensus string or alignment made from the set of hashed l-mers provide a better guess as compared to random starting positions.

2.2   Motif Refinement Using Expectation Maximization (EM)

After identifying enriched bucket(s), the algorithm presented in [2] performed motif refinement using Expectation Maximization (EM) [6]. This algorithm starts by defining an initial motif model θ^0 from δ or more l-mers hashed in a single bucket. The motif model θ^0 is a $(|\Sigma| \times (l+1))$ matrix where each row corresponds to a single symbol in Σ. The order of elements in Σ follows the lexicographic order i. e. for Σ_{DNA} the order is A, C, G, then T, where each symbol in Σ_{DNA} denotes the nucleic acids *Adenine*, *Cytosine*, *Guanine*, and *Thymine* respectively.

$$
\theta_{i,j} = \begin{cases} \text{probability of symbol } \sigma_i \text{ appearing at position } j \text{ of the motif,} & \text{if } 1 \le j \le l \\ \text{probability of symbol } \sigma_i \text{ appearing outside of the motif,} & \text{if } j = 0 \end{cases}
$$

(1)

where $\sigma_i \in \Sigma$ corresponds to the associated symbol in the ith row. The first column represents the background probability distribution of each symbol, i.e. the probability of a symbol to be present in a resulting sequence where the examined motif occurrences are deleted (called *background sequence*).

After getting the initial model θ^0, EM follows an iterative procedure as shown in Algorithm 2. The algorithm constantly updates the motif model θ^j until convergence or until a predefined maximum number of iterations is reached. The first part of the algorithm is the *E-step*. This step computes the expectation of all l-mer w in each sequence S_i using

$$
E(w|\theta^j) = \frac{Pr(w|\theta^j)}{Pr(w|P)},
$$

(2)

where $Pr(w|\theta^j)$ is the probability that motif model θ^j produces an l-mer w and P is the distribution of symbols in the set of sequences. The probability $Pr(w|\theta^j)$ is computed as

$$
Pr(w|\theta^j) = \prod_{i=1}^{l} \theta^j(w(i), i),
$$

(3)

where $w(i)$ is the ith symbol in string w. The probability of w coming from the background sequence denoted by $Pr(w|P)$ is not recomputed for each potential motif instance w. Instead, the background probability of each symbol in the entire input is used. Note that, since all symbols have the same probability of appearing in the background sequence, the probability $Pr(w|P)$ of any l-mer w is constant.

To avoid probability $Pr(w|\theta^j) = 0$, due to some 0 elements in θ^j, *Laplace correction* on each element of θ^j is added [2]. The Laplace correction uses the probability distribution of each symbol in the data, i.e. for a uniform distribution of symbols, the background probability 0.25 is added on each element.

The second step or *M-step* involves maximization of the computed expectation. It identifies the set of l-mers $\{S_{0,a_0}, S_{1,a_1}, \ldots, S_{(t-1),a_{t-1}}\}$ that maximizes the expectation per sequence. From the set of l-mers an updated motif model θ^{j+1} is derived. The likelihood computation of a motif model θ^j is

$$L(\theta^j) = \prod_{i=0}^{(t-1)} max_{a_i} \ E(S_{i,a_i}|\theta^j), \tag{4}$$

where $E(S_{i,a_i}|\theta^j)$ is maximum over all possible starting position a_i in each S_i. The algorithm is expected to converge in linear time.

INPUT: Motif model θ^0 from one enriched bucket, maximum number of iterations, and threshold for convergence δ_{EM}
OUTPUT: Motif model θ^y

1. For j in $\{1,\ldots,y\}$ or until convergence
   (a) **(E-step)** For each l-mer in each sequence S_i,
       compute $E(S_{i,a_i}|\theta^j)$ given the current motif model.
   (b) **(M-step)** For each S_i in $\mathcal{S}$,
       get starting position vector s such that for each $a_i \in s$,
       $E(S_{i,a_i}|\theta^j)$ is maximum $\forall a_i$ in $\{0,\ldots,(n-l)\}$.
   (c) **(Test for Convergence)** Compute $L(\theta^j)$. Compare previous
       likelihood $L(\theta^{j-1})$ to current $L(\theta^j)$.
       If the difference satisfies the threshold δ_{EM}, stop iteration.
   (d) **(Update step)** For the alignment made by starting position vector s
       identified in M-step, get motif model θ^{j+1}.

Algorithm 2. Iterative EM for motif refinement

Although EM is a local optimization algorithm, i.e. does not assure global optimality and is very sensitive to initial configuration, the initial motif model from projection significantly increases the chance of getting the planted motif. Once the planted bucket is refined, EM produces the planted motif [2].

2.3   Further Refinement Using SP-STARσ

To increase the number of correctly identified planted motifs for the set of simulated data, algorithm presented in [2] included another refinement step that is based from SP-STAR originally presented in [8]. Let us define S_b as the set of t number of l-mers identified by EM. Let us define a consensus string C_b from the alignment made by all l-mers in S_b. The algorithm further refines each l-mer in S_b such that the score $\sigma(S_b)$ is maximum. The score $\sigma(S_b)$ is defined as the number of l-mers $S_{i,j}$ in S_b that satisfies $d_E(S_{i,j}, C_b) \leq d$. The maximum value of $\sigma(S_b)$ is t, that is if all l-mers in S_b has number of mismatch less than or equal to d from the consensus string C_b. The algorithm also follows an iterative procedure where the set S_b is updated to get S_b'. The update step is done by getting the best l-mer $S_{i,j}$ in each sequence i with the least distance to the C_b. The algorithm will proceed to the next iteration if the score $\sigma(S_b')$ improves, i.e. the score increases.

3   Parallel Random Projection for Finding Motifs

The parallel algorithms presented in this paper assumes a massively parallel architecture such as Graphics Processing Units (GPU) for the implementation. All throughout the paper, each independent processor will be referred to as *threads* which are running in the *device* or the GPU. Meanwhile, sequential operations will execute in the *host* or the CPU.

3.1   Parallel-FMURP Version 1

A parallel version of Random Projection for finding (l, d)-motifs is shown in Algorithm 3. This algorithm is similar to what is presented in [4] except for some minor revisions. To be able to increase the accuracy of finding planted (l, d)-motifs, an additional refinement step (SP-STAR$_\sigma$) is added after EM. Moreover, computation of expectation and likelihood needed for EM are modified to decrease memory allocations.

The first three steps perform *Projection*, while the last two steps perform *refinements* using Expectation Maximization and SP-STAR$_\sigma$.

INPUT: Set of sequences S, motif length l, expected mismatches d, projection dimension k, and bucket threshold δ
OUTPUT: Best scoring l-mer

1. In CPU, generate k random positions for projection.
   Let this be the set $I = \{\hat{i} | \hat{i} \in \{0, \ldots, (l-1)\}\}$ and $|I| = k$.
2. In GPU, for each thread tid in $\{0, \ldots, (x-1)\}$,
   (a) Using Definition 3, get $h_I(S_{i,j})$ from $S_{i,j}$ in S,
       where i and j are computed using Equations 6 and 7 respectively.
   (b) Convert k-mer $h_I(S_{i,j})$ to its corresponding
       integer representation $k^*_{i,j}$.
   (c) Perform a linear search over all $k^*_{i,j}$s to determine which l-mers
       are 'hashed' in the same bucket.
       The tid of matched $k^*_{i,j}$s are noted instead of the actual l-mer.
3. In CPU, identify the set of enriched buckets,
   and prune unnecessary buckets in preparation for EM refinement. Pruning in this sense is limiting the number of threads to work. Instead of x, a total of e threads will be used for refinements.
4. In GPU, for each tid in $\{0, \ldots, (e-1)\}$,
   (a) Perform EM refinement for each enriched bucket.
   (b) Perform $SP\text{-}STAR\sigma$ for each enriched bucket.
   (c) Maximize σ score to output best motif.

Algorithm 3. One independent trial of Parallel-FMURP version 1

The $tid \in \{0, \ldots, (x-1)\}$ is used to denote the thread ID and it corresponds to an l-mer $S_{i,j}$ in S. Also note that i and j from $S_{i,j}$ identify the sequence number and l-mer's starting position in a sequence respectively. The tid can be computed using i and j, as shown in Equation 5 or vice versa as shown in Equations 6 and 7.

$$tid = i(n - l + 1) + j \tag{5}$$

$$i = \left\lfloor \frac{tid}{(n - l + 1)} \right\rfloor \tag{6}$$

$$j = tid \mod (n - l + 1) \tag{7}$$

Therefore, given a tid or a specific thread that runs in GPU we can identify the position of its corresponding l-mer $S_{i,j}$ and vice versa.

In the actual projection, each thread tid in GPU can access an l-mer $S_{i,j}$ in the set of sequences $\mathcal{S}$ and computes for $h_I(S_{i,j})$ using the random positions in I generated earlier in the host function. The projection is illustrated in Figure 1. The total number of $h_I(S_{i,j})$s generated is similar to the number of $S_{i,j}$s in $\mathcal{S}$.

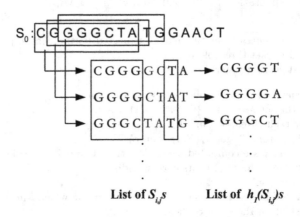

Fig. 1. Illustration of how (represented by an arrow) accesses and projects each l-mer in $\mathcal{S}$ to produce a list of all k-mers. In this example, the set of random position used in projection is $I = \{0, 1, 2, 3, 6\}$ [4].

Given a unique k-mer from projection, a corresponding integer is computed using the following mapping. Let us define

$$
\begin{aligned}
f : \Sigma_{DNA} &\to \{0, 1, 2, 3\}, \\
A &\to 0 \\
C &\to 1 \\
G &\to 2 \\
T &\to 3
\end{aligned}
$$

where each symbol in the DNA alphabet is mapped to a unique integer.

For a string v of length k,

$$f^* : \Sigma^+ \to \mathbb{Z}^+ \cup \{0\}$$

$$v \to \sum_{i=0}^{k-1} f(v_i) 4^i$$

where v_i denotes the symbol at ith position starting from the least significant digit and the integer representation is only defined on the positive integers including $\{0\}$. The computation follows the conversion of base-$|\Sigma|$ to base-10 and can be extended to protein sequences.

For instance, string "$AAAAA$" is equivalent to integer 0 and "$AAACG$" is equal to 6. The integer representation of each k-mer reduces the time complexity of string comparison and also reduces the amount of space used in storing them. A sample conversion of string to integer is illustrated in Figure 2.

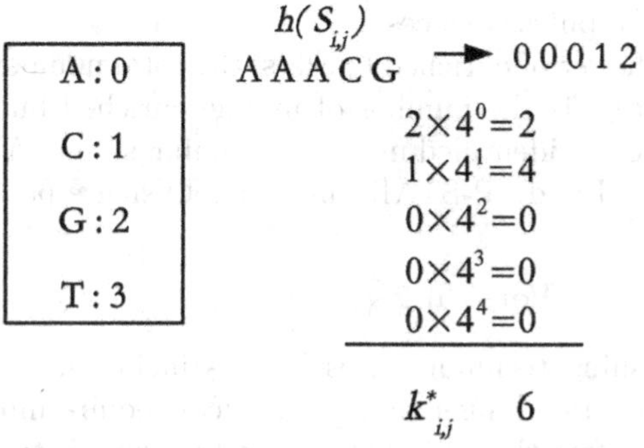

$$
\begin{array}{l}
h(S_{i,j}) \\
A\,A\,A\,C\,G \quad \longrightarrow \quad 0\,0\,0\,1\,2 \\
\end{array}
$$

$$
\begin{array}{|l|}
\hline
A:0 \\
C:1 \\
G:2 \\
T:3 \\
\hline
\end{array}
\qquad
\begin{array}{l}
2\times 4^0 = 2 \\
1\times 4^1 = 4 \\
0\times 4^2 = 0 \\
0\times 4^3 = 0 \\
0\times 4^4 = 0 \\
\hline
\end{array}
$$

$$k^*_{i,j} \quad 6$$

Fig. 2. Conversion of a unique k-mer or $h_I(S_{i,j})$ to a corresponding integer $k^*_{i,j}$ using a symbol-integer mapping displayed on the left side of the illustration. Each $k^*_{i,j}$ will serve as the $S_{i,j}$'s key in hashing, instead of $h_I(S_{i,j})$s.

The uniqueness of the representation we defined using f^* follows from the results below.

Let $\Sigma_k = \{0, 1, 2, \ldots, k-1\}$, and let C_k a regular language such that,

$$C_k = \{\epsilon\} \cup (\Sigma_k - \{0\})\Sigma_k^*.$$

Theorem 1. *Fundamental Theorem of base-k Representation* [1]
Let $k \geq 2$ be an integer. Then every non-negative integer has a unique representation of the form

$$N = \sum_{0 \leq i \leq t} a_i k^i,$$

where $a_t \neq 0$ and $0 \leq a_i < k$ for $0 \leq i \leq t$.

In the case of our representation f^*, we have $k = 4$ and $a_i = f(v_i)$, where $v_i \in \Sigma$. Note that the mapping f is one-to-one and onto by definition. Thus we have the following:

Proposition 1. *The mapping f^* provides a unique integer representation of $h_I(S_{i,j})$, for any given I and $S_{i,j}$.*

For our discussion purposes, let the integer equivalent to a specific $h_I(S_{i,j})$ be $k_{i,j}^*$. After projection, the output is a list of $k_{i,j}^*$s from $S_{i,j}$s in $\mathcal{S}$. The total number of $k_{i,j}^*$ is also equal to the number of $h_I(S_{i,j})$ and $S_{i,j}$ is x, that is $t(n - l + 1)$.

After getting the list of $k_{i,j}^*$s, each thread performs a linear search over all $k_{i,j}^*$s to identify the buckets, which will take $O(n)$ time with x number of threads. Dynamic allocation of space in GPU for instance is not allowed. Therefore, to simulate hash tables, a static 2 dimensional array is needed. It is mentioned in [4] that the dimension of the 2 dimensional array is $(x \times r)$, where r is computed based from the set of input sequences.

After identifying the set of enriched buckets, the total number of threads used is reduced to e, where e is the number of unique enriched buckets. The set of unique enriched buckets is identified using the pruning step in Algorithm 3. Each bucket will perform EM and SP-STAR$_\sigma$ in parallel using e parallel threads.

3.2 Parallel-FMURP Version 2

In version 1, the hashing technique presented is inefficient. Although we can have a faster running time, longer input sequences require unnecessary allocations in hashing. Therefore the aim of the second version, is to remove unnecessary memory allocations in GPU. The second version of the parallel FMURP is summarized in Algorithm 4. Note that in this parallel algorithm, we also used x threads to compute $h_I(S_{i,j})$s and $k_{i,j}^*$s. From Algorithm 3, hashing is performed in GPU, which requires $O(rx)$ amount of space. A way to resolve this issue is to perform hashing in CPU.

The hashing technique uses the *division method*. The item with key $k_{i,j}^*$ will be hashed to a table with x number of items using the function

$$h(k_{i,j}^*) = k_{i,j}^* \mod x \tag{8}$$

where $h(k_{i,j}^*)$ is the position in the hash table (of size x) where we can store the item with key $k_{i,j}^*$.

Since we are interested in the set of items which will be hashed in the same position, rather than avoiding collision or speed up item access, chaining using linked lists is implemented. We do not require that *tids* are sorted in the list, therefore we can insert new items at the beginning of the linked list.

Using division method as the hash function does not guarantee that there are no other items with a different key $k_{i',j'}^*$ that will hash to the same position with $k_{i,j}^*$. This will happen if $h(k_{i,j}^*) = h(k_{i',j'}^*)$. Since we do not want other items with different keys to be included in the list, we look for other positions in the table where we can store them. We used *linear probing* to identify empty locations in the table where we can store $k^*_{i',j'}$. To identify the location, we use the following formula

$$h'(k^*_{i',j'}, i) = h(k^*_{i,j}) + i \mod x \qquad (9)$$

where $h'(k^*_{i',j'}, i)$ is the ith position in the table being probed. The final location of item with key $k^*_{i',j'}$ is the minimum i where hash table at position $h'(k^*_{i',j'}, i)$ is empty.

The hashing technique is proposed because it will only require $O(x)$ space in the hash table, because we are dynamically allocating space as we insert an object. This hashing technique works almost exactly the way sorting does for the purpose of finding elements with the same key.

To retrieve the actual buckets, a linear search $O(x)$ over the hash table is needed. In [5], a theorem stated that the average time for inserting an element with linear probing is $\Theta((1/1 - \alpha))$, where $\alpha = \hat{n}/\hat{m}$ is the *load factor* of the hash table. The load factor is the average number of elements expected to collide when hashing. Notation $\hat{n}$ is the total number of elements to hash and $\hat{m}$ is the total number of slots in the hash table. In our case, $\hat{m}$ is x and $\hat{n}$ is equal to the number of possible $k^*_{i,j}$s, that is 4^k.Therefore, hashing x items will take $O(x(1/(1-\alpha)))$ time. Additionally, pruning is no longer needed before proceeding to EM refinement because resulting buckets will no longer contain duplicates.

Refinement steps for Algorithm 4, i.e. Steps 6 and 7 is the same with Algorithm 3. We use e threads and allocate $O(e(n - l + 1))$ space for refinement.

INPUT: Set of sequences $\mathcal{S}$, motif length l, expected mismatches d, projection dimension k, and bucket threshold δ
OUTPUT: Best scoring l-mer

1. In CPU, generate k random positions for projection.
   Let this be the set $I = \{\hat{i}|\hat{i} \in \{0, \dots, (l-1)\}\}$ and $|I| = k$.
2. In GPU, for each thread tid in $\{0, \dots, (x-1)\}$,
   (a) Using Definition 3, get $h_I(S_{i,j})$ from $S_{i,j}$ in $\mathcal{S}$,
       where i and j are computed using Equations 6 and 7 respectively.
   (b) Convert k-mer $h_I(S_{i,j})$ to its corresponding
       integer representation $k^*_{i,j}$.
3. In CPU, hash the list of $k^*_{i,j}$s .
4. In CPU, identify the set of enriched buckets.
5. In GPU, for each tid in $\{0, \dots, (e-1)\}$,
   (a) Perform EM refinement for each enriched bucket.
   (b) Perform $SP\text{-}STAR\sigma$ for each enriched bucket.
   (c) Maximize σ score to output best motif.

Algorithm 4. One independent trial of Parallel-FMURP version 2

3.3   Parallel-FMURP Version 3

The third version of Parallel-FMURP is summarized in Algorithm 5. The only difference between this algorithm with Algorithm 4 is the number of parallel processors used. In this algorithm, we only used t processors instead of x. Note that x is computed as $t(n - l + 1)$, and $n >> t$, therefore $x >> t$. Compared to Parallel-FMURP in Algorithm 3, we also reduced the space needed for hashing because we followed the hashing scheme made in Algorithm 4. Also, pruning of duplicate buckets required in version 1 is no longer needed.

INPUT: Set of sequences $\mathcal{S}$, motif length l, expected mismatches d, projection dimension k, and bucket threshold δ
OUTPUT: Best scoring l-mer

1. In CPU, generate k random positions for projection.
   Let this be the set $I = \{\hat{i} | \hat{i} \in \{0, \dots, (l-1)\}\}$ and $|I| = k$.
2. In GPU, for each thread tid in $\{0, \dots, (t-1)\}$,
   (a) Using Definition 3, get $h_I(S_{i,j})$ from $S_{i,j}$ in $\mathcal{S}$,
       where i and j are computed using Equations 6 and 7 respectively.
   (b) Convert k-mer $h_I(S_{tid,j})$ to its corresponding
       integer representation $k^*_{tid,j}$.
3. In CPU, hash the list of $k^*_{i,j}$s.
4. In CPU, identify the set of enriched buckets.
5. In GPU, for each tid in $\{0, \dots, (e-1)\}$,
   (a) Perform EM refinement for each enriched bucket.
   (b) Perform $SP\text{-}STAR\sigma$ for each enriched bucket.
   (c) Maximize σ score to output best motif.

Algorithm 5. One independent trial of Parallel-FMURP version 3

4   Proof of Correctness

To formally show that *Projection* (steps 1 to 4 of Algorithm 1) is equivalent to parallel *Projection* (steps 1-3 of Algorithm 3), we need to show that the set of enriched buckets E_B identified by *Projection* is equivalent to enriched bucket $\bar{E}_B$ of parallel *Projection*. Suppose we use input parameters $\mathcal{S}$, pattern length l, projection dimension k, and bucket threshold δ for both algorithms. Let's define notations *Projection*$(\mathcal{S}, l, k, \delta)$ to denote the steps for *Projection* in Algorithm 1 that produces the set of enriched buckets E_B and the notation *Parallel-Projection*$(\mathcal{S}, l, k, \delta)$ to denote the steps for parallel *Projection* in Algorithm 3.

Given the set of sequences $\mathcal{S}$, pattern length l, projection dimension k, and bucket threshold δ, the set of enriched buckets E_B is obtained by performing the steps 1 to 4 in Algorithm 1.

1. Define $I \subset \{0, \dots, (l-1)\}$, where $|I| = k$ and each element in I is randomly chosen from the set $\{0, \dots, (l-1)\}$.
2. Obtain all l-mers $S_{i,j} \in \Sigma^l$ from $\mathcal{S}$, where $i \in \{0, \dots, (t-1)\}$ denotes the sequence number, and $j \in \{0, \dots, (n-l)\}$ denotes the starting position of an l-mer in a sequence.

3. Obtain all k-mers $h_I(S_{i,j}) \in \Sigma_{DNA}^k$ using the Definition 3 and random positions I generated in step 1.
4. Let us define B for each tid p, where $B \subseteq \mathcal{U}_B$ and $B = \{S_{i,j}\}$ The set $\mathcal{U}_B$ is equal to the union of all Bs or simply the set containing all $S_{i,j}$s in $\mathcal{S}$. Note that the intersection of all Bs obtained from this step is equal to the $\emptyset$ set.

Two elements $S_{i,j}$ and $S_{i',j'}$ belong to the same bucket B if they follow the relation R defined below.

Definition 4 $(S_{i,j}$ and $S_{i',j'} \in B)$.

$$(S_{i,j}, S_{i',j'}) \in R \Leftrightarrow h_I(S_{i,j}) = h_I(S_{i',j'})$$

It is easy to prove that the above relation is *reflexive, symmetric,* and *transitive.* Therefore, we have the following proposition.

Proposition 2. R *is an equivalence relation.*

5. Lastly, an enriched bucket from $Projection(\mathcal{S}, l, k, \delta)$ is defined as

$$E_B = \{B| \, |B| \geq \delta\}.$$

Given the same input parameters, the set of enriched buckets $\bar{E}_B$ is obtained from $Parallel\text{-}Projection(\mathcal{S}, l, k, \delta)$ using the the following steps.

The first three steps in obtaining $\bar{E}_B$ and E_B are the same. The differences in their computation starts with the additional step that converts a k-mer to a unique integer representation. Formally we can define the additional conversion step as follows.

From the list of $h_I(S_{i,j})$s derived from step 3 in all parallel algorithms (Algorithm 3, 4, and 5), we compute the corresponding $k_{i,j}^*$s using the mapping f^*.

$$f^*(h_I(S_{i,j})) = k_{i,j}^*$$

From Proposition 1, the set of $h_I(S_{i,j})$s and $k_{i,j}$s are equivalent, that is given a string representation, we can get a unique integer using f^* and vice versa.

From the set of $k_{i,j}^*$s obtained from the previous step, we define a *Parallel-Projection* bucket

$$\bar{B} \subseteq \{0, \dots, (x-1)\}.$$

Let p and q be elements of $\bar{B}$, where $0 \leq p, q \leq (x-1)$. Elements p and q belongs to the same bucket if the following relation holds. Let the relation be denoted by $\bar{R}$.

Definition 5 $(p$ and $q \in \bar{B})$.

$$(p, q) \in \bar{R} \Leftrightarrow k_{i,j}^* = k_{\bar{i},\bar{j}}^*$$

where $i = \lfloor p/(n-l+1) \rfloor$, $j = p \mod (n-l+1)$, $\bar{i} = \lfloor q/(n-l+1) \rfloor$, and $\bar{j} = q \mod (n-l+1)$.

Lemma 1. *The relation R and $\bar{R}$ are equivalent.*

Proof. To prove that relations R and $\bar{R}$ are equivalent, we need to show that

$$(p,q) \in \bar{R} \Leftrightarrow (S_{i,j}, S_{i',j'}) \in R,$$

where $i = \lfloor p/(n-l+1) \rfloor$, $j = p \bmod (n-l+1)$, $i' = \lfloor q/(n-l+1) \rfloor$, and $j' = q \bmod (n-l+1)$. From the definition of relation $\bar{R}$,

$$(p,q) \in \bar{R} \Leftrightarrow k_{i,j}^* = k_{i',j'}^*.$$

Using the mapping shown in Section 3.1, we have

$$k_{i,j}^* = k_{i',j'}^* \Leftrightarrow f^*(h_I(S_{i,j})) = f^*(h_I(S_{i',j'})).$$

Since an equivalence (see Proposition 1) is shown for the set of $k_{i,j}^*$s and $h_I(S_{i,j})$s, it is safe to assume that $h_I(S_{i,j}) = h_I(S_{i',j'})$. From the definition of relation R,

$$h_I(S_{i,j}) = h_I(S_{i',j'}) \Leftrightarrow (S_{i,j}, S_{i',j'}) \in R.$$

Therefore, we have shown that

$$(p,q) \in \bar{R} \Leftrightarrow (S_{i,j}, S_{i',j'}) \in R.$$

$\square$

Lastly from the set of buckets $\bar{B}$s, the set of enriched buckets $\bar{E}_B$ from Algorithm 3 is defined as

$$\bar{E}_B = \{\bar{B} \mid |\bar{B}| \geq \delta\}.$$

We need to show that the set of elements in B is equivalent to the set of elements in $\bar{B}$. Note that elements in B involves $S_{i,j}$s while elements in $\bar{B}$ involve the set of integers $p \in \{0, \ldots, (x-1)\}$. Using Equations 6 and 7 we can retrieve the l-mer $S_{i,j}$ corresponding to p and vice versa. The theorem below follows from the fact that R and $\bar{R}$ are equivalent.

Theorem 2. *The set of enriched buckets produced by Algorithm 1 and Algorithm 3 are equivalent.*

Each bucket identified as enriched in *Projection* and *Parallel-Projection* will proceed to refinement using EM.

Let e be the number of identified enriched buckets in one independent trial, i.e. $|\bar{E}_B| = e$. Since refinement of each enriched bucket is independent from another, *Parallel-Projection* uses e parallel threads to refine each enriched bucket in parallel. However, several modifications are done to EM, to minimize the use of space in GPU. The modified computation of expectation and likelihood is shown below in Equation 10 and 11 respectively.

$$\bar{E}(w|\theta^j) = \sum_{i=0}^{l-1} \theta_j(w(i), i), \tag{10}$$

where $w(i)$ denotes the ith symbol in w.

$$L(\theta^j) = \sum_{i=0}^{(t-1)} E(w_i|\theta^j),$$ (11)

where w_i is the l-mer such that $\bar{E}(w_i|\theta^j)$ is maximum over all l-mer w in S_i. The algorithm is expected to converge linearly.

To show that the modifications do not alter the computation, we need to prove the following claims. We will first show that the two formulas for expectation (2) and (10) are equivalent. Followed by the claim that two formulas for likelihood (4) and (11) are also equivalent.

Lemma 2. *Given $S_{i,j}$ and motif model θ, if $E(S_{i,j}|\theta)$ is maximum, then $\bar{E}(S_{i,j}|\theta)$ is maximum.*

Proof. To get the equivalence of the original expectation and the modified, we just need to show that if we have an l-mer $S_{i,j}$, a motif model θ, and a maximum $E(S_{i,j}|\theta)$, then $\bar{E}(S_{i,j}|\theta)$ is also maximum.

Given that there is a uniform distribution of symbols in the set of input sequences S and we follow the assumption in [2] of using the overall symbol frequency distribution as almost correct representation of the background probability distribution, the probability $Pr(S_{i,j}, P)$ of any l-mer $S_{i,j}$ in S given P is constant.

To illustrate the computation of $Pr(S_{i,j}, P)$, suppose we have an l-mer equal to $aact$, and the background probability vector is equal to $P = (0.25, 0.25, 0.25, 0.25)$ for symbols $\{a, c, g, t\} \in \Sigma_{DNA}$ respectively, the computation of $Pr(S_{i,j}, P)$ is shown as follows.

$$Pr('aact', P) = \prod_{y=0}^{(l-1)} P(S_{i,j}(y)) = (0.25)^4$$

where $P(\sigma)$ is the element in P corresponding to symbol σ.

For any l-mer $S_{i,j}$ and background probability P which has uniform frequency distribution, the probability of $S_{i,j}$ appearing in the background sequence is $Pr(S_{i,j}|P) = (0.25)^l$.

Therefore, the expectation in Equation 2 is directly proportional to $Pr(S_{i,j}|\theta)$, i.e.

$$E(S_{i,j}|\theta) \propto Pr(S_{i,j}|\theta)$$

Maximizing the value of $E(S_{i,j}|\theta)$ is equivalent to maximizing the value of $Pr(S_{i,j}|\theta)$. Given this fact, we only need to show that if $Pr(S_{i,j}|\theta)$ is maximum, then $\bar{E}(S_{i,j}|\theta)$ is also maximum. Based from the definition in Equation 3, $Pr(S_{i,j}|\theta)$ is maximum if every symbol in $S_{i,j}$ corresponds to the maximum element per column of θ.

$$Pr(S_{i,j}|\theta) = \prod_{y=0}^{(l-1)} \theta(S_{i,j}(y), y)$$

where $\theta(S_{i,j}(y), y)$ is max over all $S_{i,j}(y) \in \Sigma_{DNA}$.

Given that $\bar{E}(S_{i,j}|\theta)$ evaluates the same l-mer using the same motif model θ, each symbol in $S_{i,j}$ also corresponds to the maximum element per column in θ, which maximizes

$$\bar{E}(S_{i,j}|\theta) = \sum_{y=0}^{(l-1)} \theta(S_{i,j}(y), y).$$

$\square$

Lemma 3. *If likelihood $L(\theta)$ is maximum, then $\bar{L}(\theta)$ is maximum.*

Proof. The proof for Lemma 3 follows the same proof idea from Lemma 2. The difference between the two likelihood computation is that $L(\theta^i)$ uses products while $\bar{L}(\theta^j)$ uses summations. Given that $L(\theta)$ is maximum, that is

$$L(\theta) = \prod_{i=0}^{(t-1)} E(S_{i,j}|\theta)$$

where $E(S_{i,j}|\theta)$ is the maximum expectation for all $S_{i,j}$ in every sequence S_i.

From Lemma 2, it is shown that if $E(S_{i,j}|\theta)$ is maximum, then $\bar{E}(S_{i,j}|\theta)$ is also maximum which maximizes the computation of likelihood,

$$\bar{L}(\theta) = \sum_{i=0}^{(t-1)} \bar{E}(S_{i,j}|\theta)$$

where $\bar{E}(S_{i,j}|\theta)$ is the maximum expectation for all $S_{i,j}$ per sequence S_i.

$\square$

Theorem 3. *Algorithm EM and Parallel-EM are equivalent.*

Proof. The proof follows from Lemmas 2 and 3.

$\square$

Corollary 1. *FMURP and Parallel-FMURP are equivalent.*

Proof. The proof follows from Theorems 2 and 3.

$\square$

5   Analysis

Given three versions of parallel FMURPs presented in Algorithms 3, 4, and 5, a summary showing their running time and space is presented in Table 2.

Table 1. Total running time and space complexity of the three parallel versions of FMURP in comparison with sequential algorithm

Algorithm	Threads	Time	Space	*Projection* Space
Sequential FMURP	1	$O(x \log x)$	$O(x)$	$O(x)$
Parallel FMURP v1	x	$O(x)$	$O(e(n - l + 1))$	$O(rx)$
Parallel FMURP v2	x	$O(x)$	$O(e(n - l + 1))$	$O(x)$
Parallel FMURP v3	e	$O(x)$	$O(e(n - l + 1))$	$O(x)$

The running time of sequential FMURP is based on the implementation in [3]. To identify the set of buckets, it sorts all l-mers with respect to their keys then performs a linear search to collect l-mers hashed in the same bucket. Assuming that the algorithm in [3] employs an optimal algorithm which uses comparisons, the total running time is $O(x \log x)$ plus the running time of refinements which will take $O(ex)$ for e identified enriched buckets. The first two parallel versions of FMURP use x threads, while the third uses t threads for projection and e threads for refinements. Since e, based from empirical tests is much greater than t, the total number of threads needed is bounded above by e.

Although the total running time and space of all parallel versions of FMURP is the same, space needed in Projection step is minimized from $O(rx)$ to $O(x)$ in the two other versions. The space reduction is due to the hashing technique presented in Section 3.2, which employs linear probing and chaining to collect enriched buckets.

Speedup as the name implies, is the ratio of running time of the parallel implementation with respect to the sequential. To eliminate ambiguity, the speedup computed in this paper is not dependent on the fastest known sequential algorithm that exists. Instead, we compare the running time of the parallel implementations to the sequential implementation in [3].

Meanwhile, *efficiency* in this case is the computed by getting the corresponding speedup divided by the number of processors used. It measures the processors efficiency or the speedup per processor [9]. In the implementation of Parallel FMURP, there are ideally x and e number of threads that run in parallel.

Table 2. Speedup and Efficiency all all parallel versions compared to sequential

Algorithm	Threads	Speedup	Efficiency
Parallel FMURP v1	x	$O(\log x)$	$\frac{\log x}{x}$
Parallel FMURP v2	x	$O(\log x)$	$\frac{\log x}{x}$
Parallel FMURP v3	e	$O(\log x)$	$\frac{\log x}{e}$

The value of efficiency ranges from 0 to 1. Efficiency with value equal to 1 means that the processors used are well utilized. Based from the computation, the efficiency of parallel FMURP with respect to sequential FMURP is equal to 1, while a lower efficiency value is computed with respect to FMURP.

Note however that considering threads as processors highly simplifies the efficiency computation but in actuality threads execute in the device as multiples of the warp size so the actual efficiency will differ from the ideal (or theoretical) efficiency. Also, we assume that the actual speedup is bounded above by the theoretical speedup computed.

Given that the three versions maintain the same speedup of $O((\log x)/x)$, parallel FMURP version 3 has the highest processor efficiency, i.e.

$$E_{FMURP} = E'_{FMURP} < E''_{FMURP}, \tag{12}$$

where E_{FMURP}, E'_{FMURP} , and E''_{FMURP} denotes the efficiency of the three versions respectively.

6 Conclusions

In this paper, we presented some improvements of the parallel algorithm presented in [4].

1. We presented a proof of the parallel algorithm presented in [4].
2. We reduced the space needed in projection by introducing a hashing technique with chaining and linear probing. Instead of performing hashing in parallel, which uses $O(rx)$ space in linear time, hashing is performed in sequential with $O(x)$ space in linear time.
3. In *Projection*, one of the memory optimizations done is the conversion of each unique k-mer to corresponding $k^*_{i,j}$. This optimization does not only saves space needed to store longer k-mers, but also speeds up string comparison by a factor of k.
4. Computation of expectation and likelihood is modified such that it no longer requires higher precision data type in storing them. The formulas used in expectation and likelihood are shown in Equations 10 and 11 respectively.
5. We presented another version (version 3) of the parallel algorithm which uses lesser number of threads with the same asymptotic running and space complexity. This is for parallel architectures that have limited number of processors (e.g. CPUs). On the other hand, version 2 is an advantage for massively parallel architectures such as GPUs.

Acknowledgments. Jhoirene B. Clemente is supported by the DOST-ERDT program. Henry Adorna is funded by a DOST-ERDT research grant and the Alexan professorial chair of the UP Diliman Department of Computer Science.

References

1. Allouche, J.P., Shallit, J.: Automatic Sequences: Theory Applications and Generalization. In: Numeration Systems, ch. 3, pp. 70–73. Cambridge University Press (2003)
2. Buhler, J., Tompa, M.: Finding Motifs Using Random Projections. In: RECOMB 2001 Proceedings of the Fifth Annual International Conference on Computational Biology (2001)

3. Buhler, J.: Projection Genomics Toolkit (2004),
   http://www1.cse.wustl.edu/~jbuhler/pgt/

4. Clemente, J.B., Cabarle, F.G.C., Adorna, H.N.: PROJECTION algorithm for motif finding on gPUs. In: Nishizaki, S.-y., Numao, M., Caro, J., Suarez, M.T. (eds.) WCTP 2011. PICT, vol. 5, pp. 101–115. Springer, Heidelberg (2012)

5. Cormen, T.H., Leiserson, C.E., Rivest, R.L., Stein, C.: Introduction to Algorithms, 2nd edn. MIT Press (2001)

6. Dempster, A.P., Laird, N.M., Rubin, D.B.: Maximum Likelihood from Incomplete Data via the EM Algorithm. Journal of the Royal Statistical Society, Series B (1977)

7. Jones, N., Pevzner, P.: An Introduction to Bioinformatics Algorithms. Massachusetts Institute of Technology Press (2004)

8. Pevzner, P., Sze, S.H.: Combinatorial Approaches to Finding Subtle Signals in DNA Sequences. In: Proceedings of 8th Int. Conf. Intelligent Systems for Molecular Biology (ISMB), pp. 269–278 (2000)

9. Lakshmivarahan, S., Dhall, S.: Analysis and Design of Parallel Algorithms: Arithmetic and Matrix Problems, ch. 1, pp. 18–19. Mcgraw-hill College (1990)

10. Shashidhara, S.H., Joseph, P., Srinivasa, K.G.: Improving Motif Refinement using Hybrid Expectation Maximization and Random Projection. In: Proceedings of the International Symposium on Biocomputing, Calicut, India (2010)

Time after Time: Notes on Delays in Spiking Neural P Systems

Francis George C. Cabarle, Kelvin C. Buño, and Henry N. Adorna

Algorithms & Complexity Lab
Department of Computer Science
University of the Philippines Diliman
Diliman 1101 Quezon City, Philippines
{fccabarle,kcbuno,hnadorna}@up.edu.ph

Abstract. Spiking Neural P systems, SNP systems for short, are bio-
logically inspired computing devices based on how neurons perform computations. SNP systems use only one type of symbol, the spike, in the
computations. Information is encoded in the time differences of spikes or
the multiplicity of spikes produced at certain times. SNP systems with
delays (associated with rules) and those without delays are two of several Turing complete SNP system variants in literature. In this work we
investigate how restricted forms of SNP systems with delays can be simulated by SNP systems without delays. We show the simulations for the
following spike routing constructs: sequential, iteration, join, and split.

Keywords: Membrane Computing, Spiking Neural P systems, delays,
routing, simulations.

1   Introduction

Membrane computing[1] abstracts computational ideas from the way biological
cells perform information processing. The models used in Membrane computing
are known as P systems. Spiking Neural P systems, or SNP systems for short,
are a class of P systems that incorporate the neural-like P systems which only
use indistinguishable electric signals or spikes. SNP systems were introduced in
[8] and were proven in [8] and [4] to be Turing-complete. Because SNP systems
only use one kind of spike symbol a, the output of the computation done by the
system is based on the time interval between the first spike and the second spike
of a designated output neuron. This model represents the fact that spikes in a
biological neural system are almost identical from an electrical point of view.
The time intervals between these signal spikes are crucial to the computations
performed by neurons.

As with the more common results in Membrane computing, SNP systems
have been used to efficiently solve hard problems such as the SAT problem

---

[1] A relatively new field of Unconventional and Natural computing introduced in 1998
by Gheorghe Păun, see e.g. [10].

S. Nishizaki et al. (Eds.): WCTP 2012, PICT 7, pp. 82–92, 2013.

using exponential workspace [9]. Other numerous results also focus on using SNP systems as acceptors, generators, and transducers as in [7] and [1]. SNP systems were also used in order to perform arithmetic operations where the operands are encoded in their binary formats [5]. It has been proven in [6] that delays in applying rules in the neurons of an SNP system are not required for SNP systems to be Turing complete. However SNP systems with delays might still prove to be useful for modeling purposes. In [13] SNP systems without delays were represented as matrices and the computations starting from an initial configuration can be performed using linear algebra operations. By simulating SNP systems with delays using SNP systems without delays, the work in [13] can be used to further study SNP systems.[2]

In this work we investigate how SNP systems with delays can be simulated by SNP systems without delays. In particular we focus on a special type of SNP systems that consist of a source neuron and a sink neuron. The initial configuration consists of a spike in the source neuron only and a final configuration where only the sink neuron has a spike. The spike from the source neuron to the sink neuron is routed through other intermediary neurons using four routing primitives: *sequential, iteration, joins,* and *splits.*

The contents of this paper are organized as follows: Section 2 provides definitions and notations that will be used in this work. Section 3 presents our results. We end this paper in Section 4 with some final notes, conjectures and open problems.

2   Preliminaries

It is assumed that the readers are familiar with the basics of Membrane Computing[3] and formal language theory. We only briefly mention notions and notations which will be useful throughout the paper. Let V be an alphabet, V^* is the free monoid over V with respect to concatenation and the identity element λ (the empty string). The set of all non-empty strings over V is denoted as V^+ so $V^+ = V^* - \{\lambda\}$. We call V a *singleton* if $V = \{a\}$ and simply write a^* and a^+ instead of $\{a^*\}$ and $\{a^+\}$. The length of a string $w \in V^*$ is denoted by $|w|$. If a is a symbol in V, $a^0 = \lambda$. A language $L \subseteq V^*$ is regular if there is a regular expression E over V such that $L(E) = L$. A regular expression over an alphabet V is constructed starting from λ and the symbols of V using the operations union, concatenation, and $+$, using parentheses when necessary to specify the order of operations. Specifically, (*i*) λ and each $a \in V$ are regular expressions, (*ii*) if E_1 and E_2 are regular expressions over V then $(E_1 \cup E_2)$, $E_1 E_2$, and E_1^+ are regular expressions over V, and (*iii*) nothing else is a regular expression over V. With each expression E we associate a language $L(E)$ defined in the following way: (i) $L(\lambda) = \{\lambda\}$ and $L(a) = \{a\}$ for all $a \in V$, (ii)

---

[2] A massively parallel SNP system simulator in [2,3] is based on the matrix representation in order to perform simulations.

[3] A good introduction is [10] with recent results and information in the P systems webpage at `http://ppage.psystems.eu/` and a recent handbook in [12].

$L(E_1 \cup E_2) = L(E_1) \cup L(E_2)$, $L(E_1 E_2) = L(E_1)L(E_2)$, and $L(E_1^+) = L(E_1)^+$, for all regular expressions E_1, E_2 over V. Unnecessary parentheses are omitted when writing regular expressions, and $E^+ \cup \{\lambda\}$ is written as E^*.

We define an SNP system of a finite degree $m \geq 1$ as follows:

Definition 1.

$$\Pi = (O, \sigma_1, \ldots, \sigma_m, syn),$$

where:

1. $O = \{a\}$ is the singleton alphabet (a is called spike).
2. $\sigma_1, \ldots, \sigma_m$ are neurons of the form $\sigma_i = (\alpha_i, R_i), 1 \leq i \leq m$, where:
   (a) $\alpha_i \geq 0$ is an integer representing the initial number of spikes in σ_i
   (b) R_i is a finite set of rules of the general form

$$E/a^c \to a^b; d$$

   where E is a regular expression over O, $c \geq 1$, $b \geq 0$, with $c \geq b$.
3. $syn \subseteq \{1, 2, \ldots, m\} \times \{1, 2, \ldots, m\}$, $(i, i) \notin syn$ for $1 \leq i \leq m$, are synapses between neurons.

A *spiking rule* is where $b \geq 1$. A *forgetting rule* is a rule where $b = 0$ is written as $E/a^c \to \lambda$. If $L(E) = \{a^c\}$ then spiking and forgetting rules are simply written as $a^c \to a^b$ and $a^c \to \lambda$, respectively. Applications of rules are as follows: if neuron σ_i contains k spikes, $a^k \in L(E)$ and $k \geq c$, then the rule $E/a^c \to a^b \in R_i$ is enabled and the rule can be fired or applied. If $b \geq 1$, the application of this rule removes c spikes from σ_i, so that only $k - c$ spikes remain in σ_i. The neuron fires b number of spikes to every σ_j such that $(i, j) \in syn$. If the delay $d = 0$, the b number of spikes are sent immediately i.e. in the same time step as the application of the rule. If $d \geq 1$ and the rule with delay was applied at time t, then the spikes are sent at time $t + d$. From time t to $t + d - 1$ the neuron is said to be *closed*[4] and cannot receive spikes. Any spikes sent to the neuron when the neuron is closed are *lost* or removed from the system. At time $t + d$ the neuron becomes *open* and can then receive spikes again. The neuron can then apply another rule at time $t + d + 1$. If $b = 0$ then no spikes are produced. SNP systems assume a global clock, so the application of rules and the sending of spikes by neurons are all synchronized.

A configuration of the system can be denoted as $\langle n_1/t_1, \ldots, n_m/t_m \rangle$, where each element of the vector is the configuration of a neuron σ_i, with n_i spikes and is open after $t_i \geq 0$ steps. Since no rules at time step 0 with or without delay are yet to be applied, $t_i = 0, \forall i \in \{1, ..., m\}$. t_i becomes d_i where d_i is the delay of the rule applied in σ_i.

As an illustration, let us have an SNP system with three neurons, σ_1, σ_2, and σ_3 with synapses $(1, 2)$, $(2, 3)$. Only σ_1 has some number of spikes, say x. σ_1 has a rule $a^+/a \to a; 2$, σ_2 has a rule $a \to a$, and σ_3 is a sink neuron. At a time step k where no rules are yet to be applied, the configuration of this system would

---

[4] This corresponds to the *refractory period* of the neuron in biology.

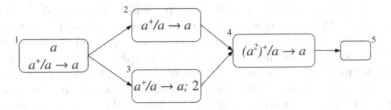

Fig. 1. An example of an SNP system where the source neuron is σ_1 and sink neuron is σ_5. Neuron σ_4 will only spike once it accumulates two spikes, one each from σ_2 and σ_3.

be $\langle x/0,\ 0/0,\ 0/0 \rangle$. At time step $k+1$, we apply the rule in σ_1; the configuration would then become $\langle x-1/2,\ 0/0,\ 0/0 \rangle$. At time step $k+2$, $\langle x-1/1,\ 0/0,\ 0/0 \rangle$. At time step $k+3$, σ_1 is open again and will now fire one spike to σ_2, the configuration would now be $\langle x-1/0,\ 1/0,\ 0/0 \rangle$. At time step $k+4$, σ_1 can again use its rule, and now σ_2 can also apply its rule; the configuration would then become $\langle x-2/2,\ 0/0,\ 1/0 \rangle$.

The initial number of spikes of σ_i is denoted as α_i, $\alpha_i \geq 0$. The initial configuration therefore is $\langle \alpha_1/0, \ldots, \alpha_m/0 \rangle$. A *computation* is a sequence of transitions from an initial configuration. A computation may halt (no more rules can be applied for a given configuration) or not.

In this work[5] and unless otherwise stated we deal with the following SNP systems and assumptions: we have only one *source* neuron (a neuron without any incoming synapses to it) and one *sink* neuron (a neuron without any outgoing synapses from it). The initial configuration will always be a single spike found only at the source neuron. The objective is to route this spike from the source to the sink neuron. The routes will involve paths in the system, where a *path* consists of at least two neurons σ_i, σ_j such that $(i,j) \in syn$. Using this idea of a path, we can have four basic routing constructs: (1) sequential, where σ_{i+1} spikes after σ_i and $(i, i+1) \in syn$, (2) iteration, where at least two neurons spike multiple (possibly an infinite) number of times and a loop is formed e.g. synapses (i,j) and (j,i) exist between neurons σ_i and σ_j, (3) join, where spikes from at least two input neurons σ_m, σ_n are sent to a neuron σ_i, where $(m,i), (n,i) \in syn$, so that σ_i produces a spike to at least one output neuron σ_j, and $(i,j) \in syn$, (4) split, where a spike from σ_j is sent to at least two output neurons σ_k and σ_l and $(j,k), (j,l) \in syn$. An example of an SNP system that we deal with in this work is shown in Fig. 1.

SNP systems considered in this work are those with neurons having exactly one rule only. For SNP systems with delays, notice that if there exists a sequential path from σ_i (with delay $d1$) to σ_j (with delay $d2$) if $d1 < d2$ and σ_i spikes more than once, it is possible for the spikes from σ_i to be lost. This is due to the

---

[5] More information on SNP systems having input and output neurons as well as non-deterministic SNP systems (we focus on deterministic systems in this work) can be found in [11] and [12] for example.

possibility that σ_j may still be closed when spikes from σ_i arrive. We therefore avoid lost spikes by assuming $d1 \geq d2$ whenever a sequential path from σ_i to σ_j exists, and leave the cases where $d1 < d2$ as an open problem. Given an SNP system with delays Π and an SNP system without delay $\overline{\Pi}$, by simulation in this work we mean the following: (a) the arrival time t of spikes arrive at sink neurons of Π is the the same time t for the arrival of spikes at the sink neurons of $\overline{\Pi}$ or offset by some constant integer k, (b) the number of spikes that arrive in the sink neurons of Π is the same for the sink neurons in $\overline{\Pi}$ or a factor of the delay of Π. We define the *total runtime* of Π and $\overline{\Pi}$ as the total time needed for a spike to arrive to the sink neurons from the source neurons.

3   Main Results

Now we present our results in simulating SNP systems with delays using SNP systems without delays. Once again we denote SNP systems with delays as Π and those without delays as $\overline{\Pi}$, subject to the restrictions and assumptions mentioned in the previous section.

Definition 2. *Given an SNP system $\Pi = (O, \sigma_1, \ldots, \sigma_m, syn)$, a routing Π' of Π is defined as*

$$\Pi' = (O, \Sigma', syn'),$$

where:

1. *$syn' \subseteq syn$*
2. *Σ is a subset of the neurons of Π so that if $\sigma_i, \sigma_j \in \Sigma$, then $(\sigma_i, \sigma_j) \in syn'$*

Lemma 1. *Given an SNP system with delay Π that contains a sequential routing, there exists an SNP system without delay $\overline{\Pi}$ that contains a routing that can simulate the sequential routing of Π.*

Proof. We refer to Fig. 2 for illustrations and designate empty neurons as sink neurons. Let Π be the system in Fig. 2(a). Total runtime for Π is d steps, where final spike count for σ_{12} is $\alpha_{12} = 1$. Now let $\overline{\Pi}$ be Fig. 2(b) that simulates Π. The initial spike count for σ_{21} is $\alpha_{21} = 1 + d$. Total runtime for $\overline{\Pi}$ is $1 + d$ steps, and σ_{22} receives $1 + d$ number of spikes. Final spike count for σ_{23} is $\alpha_{23} = 1$.

Now let Π be the system in Fig. 2(c) and $\overline{\Pi}$ be the system in Fig. 2(d). Π now has two delays, $d1$ and $d2$, and the total runtime for Π is $1 + d1 + d2$. For $\overline{\Pi}$ (with initial spikes $\alpha = (1 + d1)(1 + d2)$), the total runtime is $(1 + d1)(1 + d2) + 3 - 1$. The always open neurons of $\overline{\Pi}$ except for σ_{44} each introduce one time delay i.e. 3, minus one time delay since σ_{44} does not spike. The final spike count for σ_{33} is $\alpha = 1$, which is also the final spike count for σ_{44}. It can be easily shown that addition of neurons without delayed rules in either Π or $\overline{\Pi}$ contribute to a delay of one time step (per neuron) being added to the total runtimes for both systems (see Fig. 3). □

The total runtime for $\overline{\Pi}$ is one additional time step from the total runtime of Π. The additional time step for $\overline{\Pi}$ comes from delay due to the first spiking of σ_{21}.

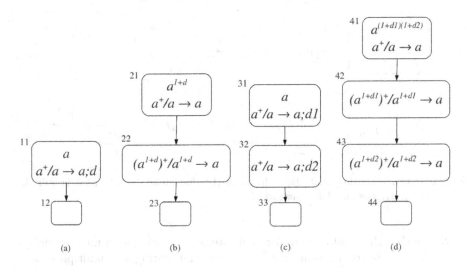

Fig. 2. Sequential routing, with single and multiple delays

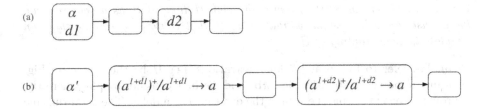

Fig. 3. Sequential routing with additional neurons without delays

Lemma 2. *Given an SNP system with delay Π that contains an iterative routing, there exists an SNP system without delay $\overline{\Pi}$ that contains a routing that can simulate the iterative routing of Π.*

Proof. We refer to Fig. 4 and 5 for illustrations. Let Π be the system in Fig. 4(a) and $\overline{\Pi}$ be the system in Fig. 4(b). Notice that σ_{11} and σ_{12} continuously replenish the spike of one another, forming an infinite loop so that σ_{13} will keep on accumulating spikes. The total runtime is $1 + d$ steps. Neurons σ_{21} and σ_{22} similarly replenish the spike of one another in an infinite loop. The total runtime for $\overline{\Pi}$ is $2 + d$, with one additional delay due to σ_{23}. Due to the operation of σ_{23}, both σ_{13} and σ_{24} obtain one final spike each. It can be easily shown that if the delay of Π is at σ_{11} instead of σ_{12}, the $\overline{\Pi}$ in Fig. 4(b) can still simulate Π.

For iterations with more than one delay, let Π be the system in Fig. 5(a) and $\overline{\Pi}$ be the system in Fig. 5(b). The total runtime of Π is $d1 + d2$ and $(1 + d1)(1 + d2) + 3 - 1$ for $\overline{\Pi}$. The final spike count for Π is $\alpha_{13} = 1$ which is the same as the final spike count for $\overline{\Pi}$ (due to the operation of σ_{24}). □

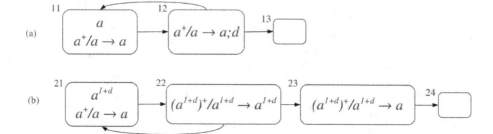

Fig. 4. Iterative routing. Note that the $\overline{\Pi}$ in (b) that simulates Π in (a) is still the same even if σ_{11} has the delay instead of σ_{12}.

Notice that the total runtime for $\overline{\Pi}$ for iterative routing with multiple delays is the same as the total runtime for $\overline{\Pi}$ for sequential routing with multiple delays. Again, every additional neuron without delay adds one time delay to the total runtime of Π and $\overline{\Pi}$.

Lemma 3. *Given an SNP system with delay Π that contains a split routing, there exists an SNP system without delay $\overline{\Pi}$ that contains a routing that can simulate the split routing of Π.*

Proof. We refer to Fig. 6 and 7 for illustrations. Let Π be the system in Fig. 6(a) and $\overline{\Pi}$ be the system in Fig. 6(b) and we first consider a split where the neuron σ_{12} that performs the split is the one that has a delay. The total runtime for Π is $1 + d$, with a final spike count of $\alpha_{13} = \alpha_{14} = 1$. The total runtime for $\overline{\Pi}$ is also $1 + d$ with a similar final spike count of $\alpha_{23} = \alpha_{24} = 1$.

Now let Π be the system in Fig. 7(a) and $\overline{\Pi}$ be the system in Fig. 7(b) and we consider a split where the delay is in one of the receiving "child" neurons (σ_{13}). In this case the sink neurons do not receive a spike at the same time. For Π, if σ_{11} spikes at time t, σ_{14} receives a spike at time $t+1$ while σ_{15} receives a spike at time $t + 1 + d$. The total runtime therefore (time when both sink neurons obtain one spike) is $1 + d$. For $\overline{\Pi}$ if σ_{21} first spikes at time t then σ_{24} first receives a spike at $t + 1$. The total runtime is $1 + d$ where $\alpha_{25} = 1$ and $\alpha_{24} = 1 + d$. □

Note that since σ_{24} accumulates $1 + d$ spikes (from σ_{21}), if the split SNP system $\overline{\Pi}$ is a subsystem of a larger SNP system, an additional neuron σ_i with a rule $a^{1+d} \rightarrow a$ needs to be added in order to return the spike number back to one[6]

Lemma 4. *Given an SNP system with delay Π that contains a join routing, there exists an SNP system without delay $\overline{\Pi}$ that contains a routing that can simulate the join routing of Π.*

Proof. We refer to Fig. 8 and 9 for illustrations. Let Π be the system in Fig. 8(a) and $\overline{\Pi}$ be the system in Fig. 8(b) so that we first we consider a join where the

---

[6] This additional neuron once again adds one time step of delay.

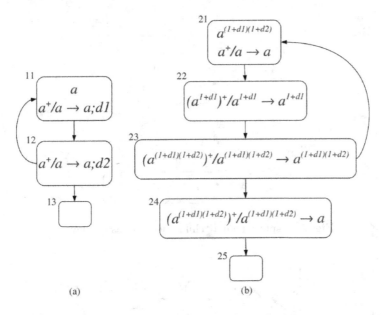

(a)   (b)

Fig. 5. Iteration with more than one delay

delay is in a neuron σ_{12} before the neuron σ_{13} that performs the join. The total runtime for Π is $1 + d$ since σ_{13} will wait for the spike from sending "parent" neuron σ_{12} to arrive before sending a spike to σ_{14}. The total runtime for $\overline{\Pi}$ is $2 + d$ because of the additional source neuron σ_{21}, since σ_{24} will wait for the spike from σ_{23}. For Π and $\overline{\Pi}$ the final spike count is 1.

Now let Π be the system in Fig. 9(a) and $\overline{\Pi}$ be the system in Fig. 9(b) and we consider next a join where the delay is in the neuron σ_{13} that performs the join. Total runtime for Π is $1 + d$ which is also the total runtime for $\overline{\Pi}$. The final spike count for both systems is 1.   □

Following our assumptions from Section 2 we have the following theorem.

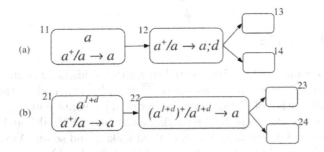

Fig. 6. A split where the neuron performing the split has a delay

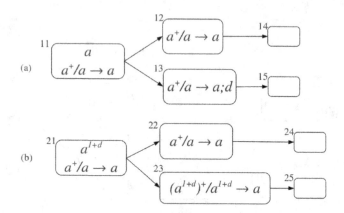

Fig. 7. A split where a "child" neuron has a delay

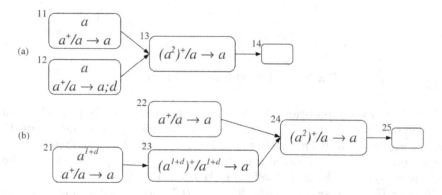

Fig. 8. A join where a parent neuron has a delay

Theorem 1. *Let Π be an SNP system with delays containing the following routings: sequential, iterative, split, join.*

Then there exists an SNP system without delays $\overline{\Pi}$, that simulates Π.

Proof. Proof follows from Lemma 1, 2, 3, and 4.   □

4 Final Remarks

Since $\overline{\Pi}$ has to simulate Π, it is expected that either additional neurons, regular expressions, or spikes are to be added to $\overline{\Pi}$ in order to simulate the delay(s) and final configuration of Π (with some offset or spike count multiple of Π). We have shown in this work how to the routing constructs of Π, although we could perhaps do better i.e. less neurons, less initial spikes, and so on. As part of our future work we intend to investigate further the following ideas along the lines of this work:

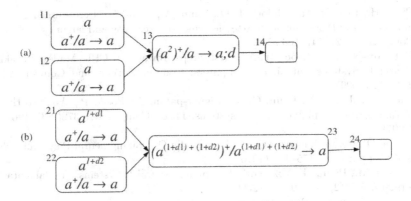

Fig. 9. A join where the neuron performing the join has a delay

1. For the case $\alpha = a^k$ where $k > 1$ is an integer, SNP systems without delays can still simulate SNP systems with delays.
2. It is possible to make the time of spiking of SNP systems with delays exactly coincide with the time of spiking of SNP systems without delays (or at least to lessen the neurons or time difference as presented in this work).

We did not consider a split where the delays of the child neurons are not equal, since Lemma 3 only considers a split where only one of the child neurons have a delay (also, it is easy to show that if both child neurons of a split have the same delay d then the $\overline{\Pi}$ in Lemma 3 still holds). As mentioned earlier, we also leave as an open problem on how to simulate SNP systems with delays where $d1 < d2$ given a sequential routing.

Acknowledgments. F.G.C. Cabarle is supported by the DOST-ERDT program and the UP Information Technology Dev't Center. K.C. Buño is supported by the UP Diliman Department of Computer Science (UPD DCS). H.N. Adorna is funded by a DOST-ERDT research grant and the Alexan professorial chair of the UPD DCS.

References

1. Alhazov, A., Freund, R., Oswald, M., Slavkovik, M.: Extended Spiking Neural P Systems. In: Hoogeboom, H.J., Păun, G., Rozenberg, G., Salomaa, A. (eds.) WMC 2006. LNCS, vol. 4361, pp. 123–134. Springer, Heidelberg (2006)
2. Cabarle, F.G.C., Adorna, H., Martínez, M.A.: A Spiking Neural P system simulator based on CUDA. In: Gheorghe, M., Păun, G., Rozenberg, G., Salomaa, A., Verlan, S. (eds.) CMC 2011. LNCS, vol. 7184, pp. 87–103. Springer, Heidelberg (2012)
3. Cabarle, F.G.C., Adorna, H.N., Martínez-del-Amor, M.A., Pérez-Jiménez, M.J.: Improving GPU Simulations of Spiking Neural P Systems. Romanian Journal of Information Science and Technology 15(1) (2012)

4. Chen, H., Ionescu, M., Ishdorj, T.-O., Păun, A., Păun, G., Pérez-Jiménez, M.J.: Spiking neural P systems with extended rules: universality and languages. Natural Computing 7(2), 147–166 (2008)
5. Gutiérrez-Naranjo, Leporati, A.: First Steps Towards a CPU Made of Spiking Neural P Systems. Int. J. of Computers, Communications and Control IV(3), 244–252 (2009)
6. Ibarra, O., Păun, A., Păun, G., Rodríguez-patón, A., Sosik, P., Woodworth, S.: Normal forms for spiking neural P systems. Theor. Comput. Sci. 372(2-3), 196–217 (2007)
7. Ibarra, O., Pérez-Jiménez, M.J., Yokomori, T.: On spiking neural P systems. Natural Computing 9, 475–491 (2010)
8. Ionescu, M., Păun, G., Yokomori, T.: Spiking Neural P Systems. Fundamenta Informaticae 71(2, 3), 279–308 (2006)
9. Pan, L., Păun, G., Pérez-Jiménez, M.J.: Spiking neural P systems with neuron division and budding. In: Proc. of the 7th Brainstorming Week on Membrane Computing, RGNC, Sevilla, Spain, pp. 151–168 (2009)
10. Păun, G.: Membrane Computing: An Introduction. Springer (2002)
11. Păun, G., Pérez-Jiménez, M.J.: Spiking Neural P Systems. Recent Results, Research Topics. In: Condon, A., et al. (eds.) Algorithmic Bioprocesses. Springer (2009)
12. Păun, G., Rozenberg, G., Salomaa, A.: The Oxford Handbook of Membrane Computing. Oxford University Press (2010)
13. Zeng, X., Adorna, H., Martínez-del-Amor, M.Á., Pan, L., Pérez-Jiménez, M.J.: Matrix Representation of Spiking Neural P Systems. In: Gheorghe, M., Hinze, T., Păun, G., Rozenberg, G., Salomaa, A. (eds.) CMC 2010. LNCS, vol. 6501, pp. 377–391. Springer, Heidelberg (2010)

A Grammar for Detecting Well-Handled 2-split, 2-join Workflow Nets without Cycles

Nestine Hope S. Hernandez, Richelle Ann B. Juayong, and Henry N. Adorna

Algorithms & Complexity Lab
Department of Computer Science
University of the Philippines Diliman
Diliman 1101 Quezon City, Philippines
{nshernandez,rbjuayong,hnadorna}@up.edu.ph

Abstract. In this paper, we propose a grammar-based technique to analyze the property of well-handledness in a class of Petri net used for workflows called Workflow net (WF-net). In particular, a set of rewriting rules is presented for a restricted type of WF-net. Using this grammar, we determine the conditions needed to assure such property and show that these conditions can be utilized to detect a well-handled 2-split, 2-join WF-net without cycles.

Keywords: Petri net, Workflow net, well-handledness.

1 Introduction

Petri nets are abstract devices used to model concurrent systems. The study of Petri nets is initiated in [3] and since then has been extensively analyzed as mathematical representation for complex systems (as in biological systems) and workflow management. Recently, Petri nets have also been used to analyze properties of relatively new computing models [1]. For an in-depth discussion of Petri net, refer to [2,4].

In this paper, we focus on a class of Petri nets employed for modeling workflows. These nets are called Workflow nets (WF-nets); its properties and applications have been investigated in [5,6]. Through WF-nets, a formal model of complex workflows allows for computerized support in managing business processes. Moreover, such representation can be used to determine, so-called "correctness" of a workflow. In WF-nets, this property is called soundness and is formally defined in [6]. Also in [6], van der Aalst defined well-handledness as a requirement to assure efficient verification of soundness of a WF-net.

In this study, we propose a technique for detecting a specific type of WF-nets with no cycle using grammars. However, our proposed grammar is not used to define languages generated using workflows, instead, functions as a device for determining possible paths. We therefore start with defining a grammar for an input WF-net, we also specify how the grammar computes; afterwards, we enumerate a set of criteria that needs to be satisfied for well-handledness. Finally,

S. Nishizaki et al. (Eds.): WCTP 2012, PICT 7, pp. 93–106, 2013.

we present the set of cases that triggers non-well-handledness and show that at least one criterion will be unsatisfied for each case.

This paper is organized as follows: Section 2 discusses preliminaries and notations for Petri nets and WF-nets, including our proposed grammar for WF-nets. The main contribution of this paper is discussed in Section 3. Finally, we conclude with a possible extension of our proposed grammar in Section 4.

2   Definitions

A Petri net is a directed bipartite graph having two types of node: *place* and *transition*. A pair of nodes can be connected via *arcs*. However, an arc can only connect two nodes of differing type. The formal definition of Petri net shown below is adapted from [5].

Definition 1 *(Petri Net). A Petri net PN is a triple $PN = (P, T, F)$ where*

- ◦ *P is a finite set of places,*
- ◦ *T is a finite set of transitions $(P \cap T = \emptyset)$*
- ◦ *$F \subseteq (P \times T) \cup (T \times P)$ is a set of arcs (flow relation)*

Figure 1 is an example of a Petri net (P, T, F) where $P = \{p_1, p_2, p_3, p_4\}$, $T = \{t_1, t_2, t_3\}$, $F = \{(p_1, t_1), (p_1, t_2), (t_1, p_2), (t_2, p_3), (p_2, t_3), (p_3, t_3), (t_3, p_4)\}$. In our graphical illustrations, a place is represented as a circle while a square represents a transition.

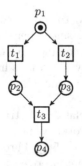

p_1

Fig. 1. An example of a Petri net

Additionally, any node in the resulting Petri net is regarded as *element* of the Petri net. Given a transition t, the set $\bullet t$ of *input places of t* consists of places having a directed arc to t; the set $t\bullet$ of *output places of t* consists of places having a directed arc from t. Formally, a place $p \in \bullet t$ implies that there exists $(p, t) \in F$, a place $p \in t\bullet$ implies that there exists $(t, p) \in F$. Similarly, given a place p, the set $\bullet p$ of *input transitions of p* consists of transitions having a directed arc to p; the set $p\bullet$ of *output transitions of p* consists of transitions having a directed arc from p. Note that it is assumed that there is no place p (or transition t) such that $\bullet p = p \bullet = \emptyset$ (or $\bullet t = t \bullet = \emptyset$). Moreover, the use of shaded dots ($\bullet$) to denote input and output places and transitions have been used to conform with the well-established notations in Petri net theory.

Petri nets compute by initially providing a set of *tokens* to a set $P' \subseteq P$ of places. A transition t is said to be *enabled* if and only if every input place $p \in \bullet t$ has at least one token. An enabled transition t is said to be *fired* when a token is removed from each $p \in \bullet t$ and a token is added to each place $p' \in t \bullet$. When a place p has multiple output transitions, i.e. $|p \bullet| > 1$, a token in p can be used to nondeterministically enable exactly one transition $t \in p \bullet$. We shall also use $\bullet$'s to denote tokens in our graphical illustration.

In [5], a set of building blocks is used to determine several ways of routing tokens. We shall use these building blocks to label connections from a (set of) places to a (set of) transitions and vice versa. Specifically, listed below are place-transition connections with their corresponding labels:

AND-split. Connection where a transition t has a directed arc to more than one place ($|t \bullet| > 1$)

OR-split. Connection where a place p has a directed arc to more than one transition ($|p \bullet| > 1$)

AND-join. Connection where a transition t has a directed arc coming from more than one place ($|\bullet t| > 1$)

OR-join. Connection where a place p has a directed arc coming from more than one transition ($|\bullet p| > 1$)

Shown in Figure 2 are the graphical illustrations of these building blocks.

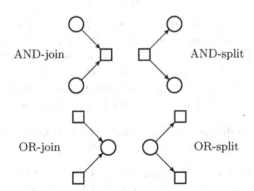

Fig. 2. The building blocks for Petri nets

We now introduce the concept of a k-split, k-join Petri net where splits and joins are limited to at most k. In this paper, our study focuses on analysis regarding a restricted type of 2-split, 2-join Petri net.

Definition 2 (k-split, k-join Petri Net). *A Petri net $W = (P, T, F)$ is a k-split, k-join Petri net, $k \geq 2$, if and only if for every transition $t \in T$, $|t \bullet|$, $|\bullet t| \leq k$ and for every place $p \in P$, $|p \bullet|, |\bullet p| \leq k$.*

For the suceeding sections, we will only be dealing with a type of Petri net used for modeling workflows. Our definition of Workflow net (WF-net) is adapted from [5].

Definition 3 *(WF-net)* *A Petri net $W = (P, T, F)$ is a WF-net (Workflow net) if and only if:*

- *W has two special places: i, a source place and o, a sink place. Note that $\bullet i = \emptyset = o \bullet$.*
- *If we add a transition t^* to PN which connects place o with i, (i.e. $\bullet t^* = \{o\}$ and $t^* \bullet = \{i\}$), then the resulting Petri net, called extended WF-net, is strongly connected.*

Unlike other Petri nets, WF-net start its computation by assigning a single token to source i. The goal is for the assigned token to reach sink o. In [5], van der Aalst specifies several conditions to characterize correctness of such computation. When these conditions are satisfied, the WF-net is called a *sound* WF-net. In order to determine soundness of a WF-net in polynomial time, the corresponding extended WF-net of the given net must be well-handled.

Before we introduce well-handledness in the context of Petri net, we first define the concepts of a *path*, an elementary path and a WF-net without cycles.

Definition 4 *(Path, Elementary, Cycle, Sequential).* *Given a WF-net $W = (P, T, F)$, a path C_{xy}, where x and y are elements in W, $x \neq y$, is a sequence consisting of the set of places and transitions that connects p to t. Specifically,*

$$C_{xy} = z_1 z_2 \ldots z_n$$

where $x = z_1$ and $y = z_n$. The node $z_i \in \bullet z_{i+1}$ for $1 \leq i \leq n-1$ and $z_i \in z_{i-1} \bullet$ for $2 \leq i \leq n$. We introduce the use of alphabet of C_{xy}, $\alpha(C_{xy})$, where for all k, $1 \leq k \leq n$, $z_k \in \alpha(C_{xy})$. A path is elementary when all symbols in this path are distinct. A WF-net without cycles is a WF-net with all associated paths being elementary. We say that a path C_{xy} is composed of sequential connections when element $\{z_i\} = \bullet z_{i+1}$ for $1 \leq i \leq n-1$ and $\{z_i\} = z_{i-1} \bullet$ for $2 \leq i \leq n$.

We are now ready to define a well-handled Petri net. Our definition is a slight modification of the one used in [5].

Definition 5 *(Well-handled Petri Net).* *A Petri net PN is well-handled if and only if for any pair of nodes x and y such that one of the nodes is a place and the other a transition and for any pair of elementary paths C_{xy} and C'_{xy}, $\alpha(C_{xy}) \cap \alpha(C'_{xy}) = \{x, y\}$ implies $C_{xy} = C'_{xy}$.*

Figure 3 shows both a well-handled WF-net and a not well-handled net.

We now define a grammar for a given WF-net. This grammar manipulates a multiset of symbols, therefore no order is imposed. We use string representation to denote a multiset, therefore there can be multiple strings that represent the same multiset (i.e. $p'p'p'' = p''p'p'$ among other representations). The rules used

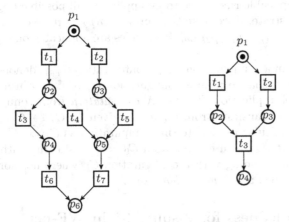

Fig. 3. On the left is an example of a well-handled WF-net. However, the WF-net on the right is not well-handled due to existence of paths $p_1t_1p_2t_3$ and $p_1t_2p_3t_3$.

in the grammar are multiset-rewriting rules similar to context-free grammars. However, there can be more than one symbol in the left-hand side of a rule and the manner of how rules are applied is slightly modified.

Definition 6 (Grammar for WF-net). *Given a WF-net $W = (P, T, F)$ with source i and sink o, a grammar $G(W)$ is a tuple $G(W) = (O, R, s)$ where*

- $O = P$ *is the alphabet of symbols used,*
- $s = i$ *is the start symbol,*
- R *is a set of multiset-rewriting rules, each rule r having the form $q \to q'$ where q and q' are non-empty multiset of distinct symbols from O.*

In order to define how a rule is applied, we use the concept of a configuration. A configuration C_j is a string representing a multiset of symbols from O at a certain time step j. The initial configuration consists of only one symbol, $C_0 = i$. We say that rule r is applicable on a configuration if the symbols in q are present in the configuration. In the next time step, rule r is then applied; the multiset q will be replaced with the multiset q'. To illustrate, if at a time step we have a rule $r : p_1p_3 \to p_4$ and a configuration $C_j = p_1p_2p_3$, rule r is applicable since both p_1 and p_3 are in C_j. In the next time step, the configuration will be p_2p_4.

Rules are applied in a nondeterministic and maximally parallel manner. Nondeterminism, in this case, has the following meaning: when there are multiple rules applicable on a multiset, the grammar nondeterministically chooses only a single rule. For example, we have two rules $r : p_1p_2 \to p_4$ and $r' : p_2p_3 \to p_5$ and a configuration $C_j = p_1p_2p_3$. Although both rules are applicable on C_j, we only choose one rule on C_j so that C_{j+1} can be one of p_3p_4 and p_1p_5. Maximal parallelism, on the other hand, imposes that all applicable rules must be applied.

This means all applicable rules have to be applied to all possible objects at the same time. To illustrate, if we have two rules $r : p_1p_2 \to p_4$ and $r' : p_2p_3 \to p_5$ and a configuration $C_j = p_1p_2p_2p_3p_6$. Both rules are applicable on C_j, therefore $C_{j+1} = p_5p_4p_6$.

A *transition* from a configuration C_j to configuration C_{j+1}, denoted by $C_j \Rightarrow C_{j+1}$ is defined as a change from a configuration C_j to C_{j+1} through a set of maximally parallel application of rules. A *computation* from configuration C_j to C_k is a series of transition from C_j to C_k given by $C_j \to C_{j+1} \to \cdots \to C_{k-1} \to C_k$. In particular we denote this computation as $C_j \Rightarrow^* C_k$. We define a *successful computation* as a computation $C_0 \Rightarrow^* C_h$ starting with the initial configuration C_0 and ending with a configuration C_h where no more rules are applicable, C_h is called a *halting configuration*.

3   Well-Handledness for 2-split, 2-join WF-net without Cycles

In this section, we present a method for detecting well-handled 2-split, 2-join WF-net without cycles using grammar for WF-net. First, we shall show how we construct rewriting rules from a given 2-split, 2-join WF-net. Afterward, we evaluate how this grammar determines if the input WF-net is well-handled.

Grammar for 2-split, 2-join WF-net

Given a 2-split, 2-join WF-net $W = (P, T, F)$ with source i and sink o, we can define a grammar $G(W) = (O, R, s)$ where $O = P$, $s = i$ and

- $\forall t \in T$, if $|\bullet t| = 1$ where $\bullet t = \{p\}$ and
  - (a) if $|t \bullet| = 1$, $t\bullet = \{p_1\}$, then rule $(p \to p_1) \in R$
  - (b) if $|t \bullet| = 2$, $t\bullet = \{p_1, p_2\}$, then rule $(p \to p_1p_2) \in R$
- $\forall t \in T$, if $|\bullet t| = 2$ where $\bullet t = \{p, p'\}$ and
  - (a) if $|t \bullet| = 1$, $t\bullet = \{p_1\}$, then rule $(pp' \to p_1) \in R$
  - (b) if $|t \bullet| = 2$, $t\bullet = \{p_1, p_2\}$, then rule $(pp' \to p_1p_2) \in R$
- $\forall p \in P$, if $|p \bullet| = 2$ where $p\bullet = \{t_1, t_2\}$, $\forall t_j \in p\bullet$, and
  - (a) if $|t_j \bullet| = 1$, $t_j\bullet = \{p_1\}$, then rule $(p \to p_1) \in R$
  - (b) if $|t_j \bullet| = 2$, $t_j\bullet = \{p_1, p_2\}$, then rule $(p \to p_1p_2) \in R$

From the grammar $G(W)$, we say that a 2-split, 2-join WF-net W without cycles is well handled if and only if all the following requirements are satisfied:

R1. All successful computation $C_0 \Rightarrow^* C_h$ ends with $C_h = o$.
R2. All rules must be applied in at least one successful computation.

To show that the grammar defined above determines whether a given WF-net is well-handled, we first illustrate that a well-handled net satisfies all requirements listed above. Next, we enumerate all possibilities that may cause a 2-split 2-join WF-net without cycles to be not well-handled. Afterwards, we show that in all these cases, there exists at least one requirement for well-handledness that will not be satisfied.

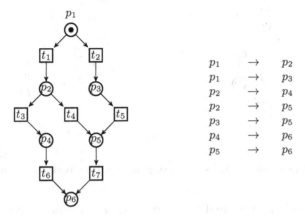

$$
\begin{aligned}
p_1 &\rightarrow p_2 \\
p_1 &\rightarrow p_3 \\
p_2 &\rightarrow p_4 \\
p_2 &\rightarrow p_5 \\
p_3 &\rightarrow p_5 \\
p_4 &\rightarrow p_6 \\
p_5 &\rightarrow p_6
\end{aligned}
$$

Fig. 4. A WF-net and its set of rewriting rules

An example of a WF-net along with its set of rewriting rules is shown in Figure 4. Its computations from p_1 are as follows:

$$
\begin{aligned}
p_1 &\rightarrow p_2 \rightarrow p_4 \rightarrow p_6 \\
p_1 &\rightarrow p_2 \rightarrow p_5 \rightarrow p_6 \\
p_1 &\rightarrow p_3 \rightarrow p_5 \rightarrow p_6
\end{aligned}
$$

As can be observed, all computations end in p_6, thus satisfying R1. Furthermore, all rules were used in at least one computation, hence, R2 is satisfied. We therefore conclude that the WF-net is well-handled.

Next, we examine different cases triggering a not well-handled WF-net:

Proposition 1 (OR-split Closed by AND-join). *If a part of a WF-net contains an OR-split in which all paths coming from that split are joined by an AND-join, requirement R2 will not be satisfied.*

Proof. Suppose a given WF-net have an OR-split at place p, an AND-join at transition t_q and a set of elements connecting p and t_q such that there exists only two paths $C'_{pt_q} = p t'_0 p'_0 \ldots t'_m p'_m t_q$ and $C''_{pt_q} = p t''_0 p''_0 \ldots t''_n p''_n t_q$ connecting p and t_q. Furthermore, suppose both $C'_{t'_0 p'_m}$ and $C''_{t''_0, p''_n}$ are composed of sequential connections and except for p and t_q, no elements in C'_{pt_q} are present in C''_{pt_q} (and vice versa).

The resulting grammar of the given WF-net will contain the following rules:
(a) $p \rightarrow p'_0$ and $p \rightarrow p''_0$ (for the OR-split at p)
(b) $p'_m p''_n \rightarrow p_q$ (for the AND-join at t_q where $t_q \bullet = \{p_q\}$)
(c) $p'_0 \rightarrow p'_1; p'_1 \rightarrow p'_2; \cdots; p'_{m-1} \rightarrow p'_m$ (to represent connections from p'_0 to p'_m)
(d) $p''_0 \rightarrow p''_1; p''_1 \rightarrow p''_2; \cdots; p''_{n-1} \rightarrow p''_n$ (to represent connections from p''_0 to p''_n)

Thus computations from p will be as follows: $p \rightarrow p'_0 \rightarrow^* p'_m$ and $p \rightarrow p''_0 \rightarrow^* p''_m$. The rule $p'_m p''_n \rightarrow p_q$ will not be applied in any of the computations thus failing to satisfy R2.

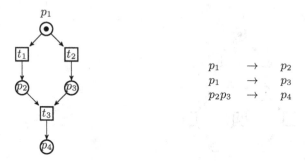

Fig. 5. A not well-handled WF-net containing an OR-split closed by an AND-join

An example of the case handled in Proposition 1 is shown in Figure 5. Notice that in the figure, for paths $C_{p_1 t_3} = p_1 t_1 p_2 t_3$ and $C'_{p_1 t_3} = p_1 t_2 p_3 t_3$, $\alpha(C_{p_1 t_3}) \cap \alpha(C'_{p_1 t_3}) = \{p_1, t_3\}$ but $C_{p_1 t_3} \neq C'_{p_1 t_3}$. From Definition 5, the given WF-net is not well-handled. Shown below are the computations for its corresponding grammar from p_1:

$$
\begin{array}{ccc}
p_1 & \rightarrow & p_2 \\
p_1 & \rightarrow & p_3
\end{array}
$$

The rule $p_2 p_3 \rightarrow p_4$ will not be applied in any of the computations thus failing to satisfy R2. This simulates the case where transition t_3 is not enabled since only one of places p_2 and p_3 will contain a token at a particular time step. From the resulting computation, it can be concluded that the WF-net is not well-handled.

Proposition 2 (AND-split Closed by OR-join). *If a part of a WF-net contains an AND-split in which all paths coming from that split are joined by an OR-join, R1 will not be satisfied.*

Proof. Suppose a given WF-net have an AND-split at transition t where $\bullet t = \{p\}$ and $t\bullet = \{p'_0, p''_0\}$, an OR-join at place p_q and a set of elements connecting t and p_q such that there exists only two paths $C'_{tp_q} = tp'_0 t'_0 \ldots p'_m t'_m p_q$ and $C''_{tp_q} = tp''_0 t''_0 \ldots p''_n t''_n p_q$ connecting t and p_q. Furthermore, suppose both $C'_{p'_0 t'_m}$ and $C''_{p''_0 t''_n}$ are composed of sequential connections and except for t and p_q, no elements in C'_{tp_q} are present in C''_{tp_q} (and vice versa).

Then our grammar will contain the following rules:
(a) $p \rightarrow p'_0 p''_0$ (for the AND-split at t where $\bullet t = \{p\}$ and $t\bullet = \{p'_0, p''_0\}$)
(b) $p'_m \rightarrow p_q$ and $p''_n \rightarrow p_q$ (for the OR-join at p_q)
(c) $p'_0 \rightarrow p'_1; p'_1 \rightarrow p'_2; \cdots; p'_{m-1} \rightarrow p'_m$ (to represent connections from p'_0 to p'_m)
(d) $p''_0 \rightarrow p''_1; p''_1 \rightarrow p''_2; \cdots; p''_{n-1} \rightarrow p''_n$ (to represent connections from p''_0 to p''_n)

Thus computations from p will be as follows: $p \rightarrow p'_0 p''_0 \rightarrow^* p'_m p''_n \rightarrow p_q p_q$. Any path chosen from p_q to the sink o will result to a computation that ends in oo causing violation of condition R1.

In Figure 6, the case tackled in Proposition 2 is depicted. If we consider paths $C_{t_1 p_4} = t_1 p_2 t_2 p_4$ and $C'_{t_1 p_4} = t_1 p_3 t_3 p_4$, $\alpha(C_{t_1 p_4}) \cap \alpha(C'_{t_1 p_4}) = \{t_1, p_4\}$ but

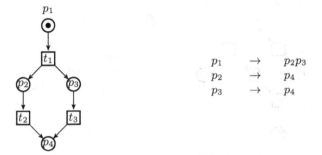

Fig. 6. A not well-handled WF-net containing an AND-split closed by an OR-join

$C_{t_1 p_4} \neq C'_{t_1 p_4}$ causing the WF-net to fail the well-handledness property. The only computation from p_1 for the grammar of the WF-net in this figure is as follows:

$$p_1 \quad \rightarrow \quad p_2 p_3 \quad \rightarrow \quad p_4 p_4$$

Notice that the computation ends in $p_4 p_4$ and not in p_4. Thus any path chosen from p_4 to the sink o will result to a computation that ends in oo causing violation of condition R1. This simulates the case where place p_4 receives two tokens at a particular time step. Again, we say that the WF-net is not well-handled.

Proposition 3 (OR-split and AND-split Partially Closed by AND-join). *If a part of a WF-net contains an OR-split and an AND-split wherein an AND-join connects a branch from the OR-split with a branch from the AND-split, R2 will not be satisfied.*

Proof. Suppose a given WF-net have an OR-split at place p, an AND-split at transition t''_{n+1} where $\bullet t''_{n+1} = \{p''_n\}$ and $t''_{n+1}\bullet = \{p''_{n_1}, p''_{n_2}\}$, an AND-join at transition t_q where $t_q\bullet = \{p_q\}$ and a set of elements connecting p and t_q such that there only exist two paths $C'_{pt_q} = pt'_0 p'_0 \ldots t'_m p'_m t_q$ and $C''_{pt_q} = pt''_0 p''_0 \ldots t''_n p''_n t''_{n+1} p''_{n_1} \ldots p''_r t_q$ connecting p and t_q. Furthermore, suppose both $C'_{t'_0 p'_m}$, $C''_{t''_0 p''}$ and $C''_{p''_{n_1} p''_r}$ are composed of sequential connections and except for p and t_q, no elements in C'_{pt_q} are present in C''_{pt_q} (and vice versa). Moreover, $p''_{n_2} \notin \alpha(C'_{pt_q})$.

Then our grammar will contain the ff rules:

(a) $p \rightarrow p'_0$ and $p \rightarrow p''_0$ (for the OR-split at p)
(b) $p'_0 \rightarrow p'_1; p'_1 \rightarrow p'_2; \cdots ; p'_{m-1} \rightarrow p'_m$ (to represent connections from p'_0 to p'_m)
(c) $p''_0 \rightarrow p''_1; p''_1 \rightarrow p''_2; \cdots ; p''_{n-1} \rightarrow p''_n$ (to represent connections from p''_0 to p''_n)
(d) $p''_n \rightarrow p''_{n_1} p''_{n_2}$ (for the AND-split at t''_{n+1} where $\bullet t''_{n+1} = \{p''_n\}$)
(e) $p''_{n_1} \rightarrow^* p''_r$ (to represent connections from p''_{n_1} to p''_r)
(f) $p'_m p''_r \rightarrow p_q$ (for the AND-join at t_q where $t_q\bullet = \{p_q\}$)

Thus computations from p will be as follows: $p \rightarrow p'_0 \rightarrow^* p'_m$ and $p \rightarrow p''_0 \rightarrow^* p''_n \rightarrow p''_{n_1} p''_{n_2} \rightarrow^* p''_r p''_{n_2}$. The rule $p'_m p''_r \rightarrow p_q$ will not be applied in any of the computations thus violating R2.

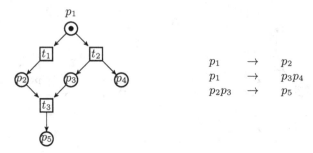

$$
\begin{aligned}
p_1 &\rightarrow p_2 \\
p_1 &\rightarrow p_3 p_4 \\
p_2 p_3 &\rightarrow p_5
\end{aligned}
$$

Fig. 7. Part of a not well-handled WF-net containing an OR-split and AND-split partially closed by AND-join

Figure 7 gives an illustration of the case for Proposition 3. Notice that in the figure, for paths $C_{p_1 t_3} = p_1 t_1 p_2 t_3$ and $C'_{p_1 t_3} = p_1 t_2 p_3 t_3$, $\alpha(C_{p_1 t_3}) \cap \alpha(C'_{p_1 t_3}) = \{p_1, t_3\}$ but $C_{p_1 t_3} \neq C'_{p_1 t_3}$. This suggests that the WF-net in Figure 7 is not well-handled. To show how a condition will be violated by computations in its corresponding grammar, we first enumerate all computations starting from p_1:

$$
\begin{aligned}
p_1 &\rightarrow p_2 \\
p_1 &\rightarrow p_3 p_4
\end{aligned}
$$

As shown, the two possible transitions from p_1 proceed with one containing only a p_2 and another containing a p_3, respectively, but not with both p_2 and p_3. Therefore, however computation proceeds, rule $p_2 p_3 \rightarrow p_5$ will not be applied in any computation, thus violating R2. Regardless of the resulting computation, we can conclude that this WF-net is not well-handled.

Proposition 4 (AND-split and OR-split Partially Closed by OR-join).
If a part of a WF-net contains an AND-split and an OR-split wherein an OR-join connects a branch from the AND-split with a branch from the OR-split , R1 will not be satisfied.

Proof. Suppose a given WF-net have an AND-split at transition t where $\bullet t = \{s\}$, an OR-split at place s''_n, an OR-join at place s_q and a set of elements connecting t and s_q such that there exists only two paths $C'_{ts_q} = t s'_0 t'_0 \dots s'_m t'_m s_q$ and $C''_{ts_q} = t s''_0 t''_0 \dots s''_n t''_n s''_{n_1} \dots s''_r t''_r s_q$. The OR-split is manifested in a place s''_n where $s''_n \bullet = \{t''_n, t''_{n'}\}$, $t''_n \bullet = \{s''_{n_1}\}$, $t''_{n'} \bullet = \{s''_{n_2}\}$. Furthermore, suppose both $C'_{s'_0 t'_m}$, $C''_{s''_0 t''_{n-1}}$ and $C'''_{s_{n_1} t''_r}$ are composed of sequential connections and except for t and s_q, no elements in C'_{ts_q} are present in C''_{ts_q} (and vice versa). Moreover, $s''_{n_2} \notin \alpha(C'_{ts_q})$.

Then our grammar will contain the ff rules:
(a) $s \rightarrow s'_0 s''_0$ (for the AND-split at t where $\bullet t = \{s\}$)
(b) $s'_0 \rightarrow s'_1; s'_1 \rightarrow s'_2; \dots; s'_{m-1} \rightarrow s'_m$ (to represent connections from s'_0 to s'_m)
(c) $s''_0 \rightarrow s''_1; s''_1 \rightarrow s''_2; \dots; s''_{n-1} \rightarrow s''_n$ (to represent connections from s''_0 to s''_n)
(d) $s''_n \rightarrow s''_{n_1}$ and $s''_n \rightarrow s''_{n_2}$ (for the OR-split at s''_n)

(e) $s''_{n_1} \to^* s''_r$ (to represent connections from s''_{n_1} to s''_r)
(f) $s'_m \to s_q$ and $s''_r \to s_q$ (for the OR-join at s_q)

Thus computations from s will be as follows: $s \to s'_0 s''_0 \to^* s'_m s''_n \to s'_m s''_{n_1} \to s_q s''_r \to s_q s_q$ and $s \to s'_0 s''_0 \to^* s'_m s''_n \to s'_m s''_{n_2} \to s_q s''_{n_2}$.

Note that all paths from s''_{n_2} and s_q going to sink o will eventually meet at some place or transition, that is,

(a) there exists $t_1, \ldots, t_k \in T$ and $p'_0, \ldots, p'_k \in P$ such that $\forall \tau \in s_q \bullet \cup s''_{n_2} \bullet$,

$$C_{\tau, p'_k} = \tau \ldots p'_0 t_1 p'_1 \ldots t_k p'_k$$

where $p'_k = o$ and $k \geq 0$, or
(b) there exists $t_0, \ldots, t_k \in T$ and $p'_0, \ldots, p'_k \in P$ such that $\forall \tau \in s_q \bullet \cup s''_{n_2} \bullet$,

$$C_{\tau, p'_k} = \tau \ldots t_0 p'_0 t_1 p'_1 \ldots t_k p'_k$$

where $p'_k = o$ and $k \geq 0$.

We now explore how configurations $s_q s_q$ and $s_q s''_{n_2}$ proceed in each of the cases above.

(a') If paths from s_q and s''_{n_2} meet at a place p'_0, then there is a computation that proceeds as follows: $s_q s''_{n_2} \Rightarrow^* p'_0 p'_0 \to \ldots \to p'_k p'_k = oo$ which is not a desirable successful computation. This violates R1.
(b') If paths from s_q and s''_{n_2} meet at a transition t_0 and $|\bullet t_0| = 1$, configurations will proceed as in (a'). Otherwise, there is a place $p' \in \bullet t_0$ such that there is a computation $s_q s_q \Rightarrow^* p' p'$ in which R1 is violated.

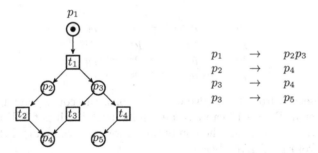

Fig. 8. Part of a not well-handled WF-net containing an AND-split and OR-split partially closed by OR-join

An example of the case discussed in Proposition 4 is shown in Figure 8. The WF-net in the figure is not well-handled since if we consider paths $C_{t_1 p_4} = t_1 p_2 t_2 p_4$ and $C'_{t_1 p_4} = t_1 p_3 t_3 p_4$, $\alpha(C_{t_1 p_4}) \cap \alpha(C'_{t_1 p_4}) = \{t_1, p_4\}$ but $C_{t_1 p_4} \neq C'_{t_1 p_4}$. The computations for the grammar of the WF-net in Figure 8 from p_1 are as shown below:

$$p_1 \quad \rightarrow \quad p_2p_3 \quad \rightarrow \quad p_4p_4$$
$$p_1 \quad \rightarrow \quad p_2p_3 \quad \rightarrow \quad p_4p_5$$

Note that all paths from p_4 and p_5 going to sink o will eventually meet at some place or transition, as mentioned in Proposition 4. We now explore how configurations p_4p_4 and p_4p_5 proceed in each of the enumerated scenarios in the proposition.

(a′) If paths from p_4 and p_5 meet at a place p'_0, then there is a computation that proceeds as follows: $p_4p_4 \Rightarrow^* p'_0p'_0 \rightarrow \dots \rightarrow p'_kp'_k = oo$ which is not a desirable successful computation. This violates R1.

(b′) If paths from p_4 and p_5 meet at a transition t_0 and $|\bullet t_0| = 1$, configurations will proceed as in (a′). Otherwise, there is a place $p' \in \bullet t_0$ such that there is a computation $p_4p_4 \Rightarrow^* p'p'$ in which R1 is violated.

4   Conclusion

We have presented a method for detecting well-handledness of 2-split, 2-join WF-net without cycles using our proposed grammar for WF-nets. We conjecture that a 2-split, 2-join WF-net without cycles is well-handled if and only if all computation in its corresponding grammar satisfies requirement R1 and R2.

The reader is encouraged to explore the possibility of extending the proposed grammar for other classes of Petri nets. Our future works include a formal proof for our conjecture and extending our method to determining well-handledness of WF-nets with cycles and k-split, k-join WF-nets, $k > 2$. We conclude our work by giving an illustration for 3-split 3-join WF-net shown in Figures 9 and 10.

The computation for the grammar of the WF-net in Figure 9 from p_1 is as follows:

$$p_1 \quad \rightarrow \quad p_2p_3 \quad \rightarrow \quad p_6 \quad \rightarrow \quad p_9$$
$$p_1 \quad \rightarrow \quad p_4 \quad\quad \rightarrow \quad p_7 \quad \rightarrow \quad p_9$$
$$p_1 \quad \rightarrow \quad p_5 \quad\quad \rightarrow \quad p_7 \quad \rightarrow \quad p_9$$
$$p_1 \quad \rightarrow \quad p_5 \quad\quad \rightarrow \quad p_8 \quad \rightarrow \quad p_9$$

As can be observed from the above computations, the corresponding grammar for the WF-net in Figure 10 can also determine that the input WF-net is well-handled. On the other hand, the computation for the grammar of the WF-net in Figure 10 from p_1 is given below:

$$p_1 \quad \rightarrow \quad p_2p_3 \quad \rightarrow \quad p_5$$
$$p_1 \quad \rightarrow \quad p_2p_3 \quad \rightarrow \quad p_2p_5$$
$$p_1 \quad \rightarrow \quad p_3 \quad\quad \rightarrow \quad p_5$$
$$p_1 \quad \rightarrow \quad p_4 \quad\quad \rightarrow \quad p_5$$

In the resulting computations, R1 is violated, leading to the conclusion that the input WF-net is not well-handled. These illustrations express the possibility of using the proposed method to generalized WF-nets.

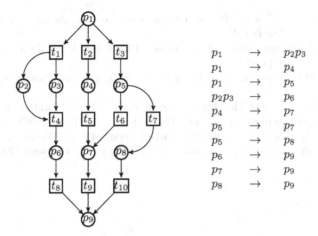

Fig. 9. A well-handled 3-split 3-join WF-net

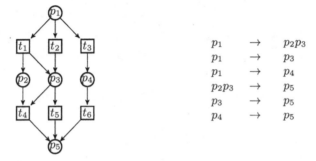

Fig. 10. A 3-split 3-join WF-net that is not well-handled

Acknowledgments. N.H.S.Hernandez would like to thank the UP Diliman College of Engineering through the Cesar Nuguid Professorial Chair for the financial support. R.A.B. Juayong is supported by the Engineering Research and Development (ERDT) Scholarship Program. H.N. Adorna is funded by a DOST-ERDT research grant and the Alexan professorial chair of the UP Diliman College of Engineering.

The authors would also like to acknowledge the help of people from the Algorithms and Complexity Laboratory, especially F.G.C. Cabarle whose comments and suggestions have been useful in the discussions of the results and editing of this work.

References

1. Cabarle, F.G.C., Adorna, H.N.: On Structures and Behaviors of Spiking Neural P systems and Petri nets. In: Csuhaj-Varjú, E., Gheorghe, M., Rozenberg, G., Salomaa, A., Vaszil, G. (eds.) CMC 2012. LNCS, vol. 7762, pp. 145–160. Springer, Heidelberg (2013)

2. Murata, T.: Petri Nets: Properties, analysis and application. Proc. of the IEEE 77(4), 541–580 (1989)
3. Petri, C.A.: Kommunikation mit Automaten. PhD thesis, Institut für instrumentelle Mathematik, Bonn (1962)
4. Reisig, W., Rozenberg, G. (eds.): APN 1998. LNCS, vol. 1491. Springer, Heidelberg (1998)
5. van der Aalst, W.M.P.: The Application of Petri Nets to Workflow Management. The Journal of Circuits, Systems and Computers 8(1), 21–66 (1998)
6. van der Aalst, W.M.P.: Structural Characterizations of Sound Workflow Nets. Computing Science Reports 96/23. Eindhoven University of Technology (1996)

A Process Algebra Model of Interleukin-2 Trafficking in Hematopoeitic Cells

John Justine S. Villar[1] and Adrian Roy L. Valdez[2]

[1] Institute of Mathematics, University of the Philippines Diliman
john_justine.villar@up.edu.ph
[2] Department of Computer Science, University of the Philippines Diliman
alvaldez@dcs.upd.edu.ph

Abstract. Interleukin-2 (IL-2) is a signaling molecule involved in the development of cellular immunity, and has also been identified as an important regulator in cell proliferation and differentiation in hematopoeitic cells. Furthermore, this has been explored as means of treatment in several disorders, such as melanoma and kidney cancer, among others. In this paper, a model of IL-2 trafficking is presented using stochastic π-calculus. π-calculus is a process algebra that allows the components of a biological system to be modeled independently, rather than modeling the individual reactions. The model simulation is carried out using Stochastic Pi Machine (SPiM), an implementation of stochastic π-calculus using the Gillespie's algorithm. The proposed model will illustrate the behavior of the ligand system and identify various processes that influence its dynamics.

Keywords: stochastic π-calculus, Interleukin-2, process algebra model.

1 Introduction

In the recent years, modeling biological processes using process algebra has been recognized due to its rigorous compositional description of the different interactions and processes of the components. As opposed to traditional mathematical techniques, which employs set of differential equations to describe the behavior of the system, process-algebraic approaches are modular and make the interactions and/or constraints explicit among elements in the state space [4]. This provides us with deeper understanding of the behavior of the biological system being studied as the system processes and interactions are imitated to produce similar characteristics. Furthermore, the results obtained from using process algebra techniques have been observed that it is consistent with the results obtained from *in vivo* experiments and with the ODE methods.

Various stochastic process algebras have been used to model biological systems, including Bio-PEPA [2], stochastic π-calculus [11], BioAmbients calculus [10], and others. As vast majority of biological systems, such as the cellular systems, involve the exchange of information between different compartments,

S. Nishizaki et al. (Eds.): WCTP 2012, PICT 7, pp. 107–115, 2013.

and modelers are inclined to use process algebras that correctly handle compartments. However, stochastic π-calculus models [12] boast its minimality in abstracting biological systems, while maintaining its robustness and flexibility. Also, the main benefits of the representation is its ability to clearly highlight the existence of cycles, which are a key aspect of many biological systems, and its ability to animate interactions between biological system components, in order to clarify the overall system function [8].

In this paper, the authors attempt to adapt the π-calculus formalism in modeling the trafficking dynamics of Interleukin-2 (IL-2) [3], a signaling molecule that attracts lymphocytes that is involved in the development of the immune system. It is produced by T-helper cells, a kind of white blood cell, when they are stimulated by an infection. IL-2 increases the growth and activity of other T lymphocytes and B lymphocytes, which are also responsible for cellular immunity. This protein has also been identified as an important regulator in cell proliferation and differentiation in hematopoeitic cells. It was originally called the T cell growth factor because of its role as a proliferative signal for CD4$^+$ T-lymphocytes and has been involved in numerous clinical studies in the treatment of AIDS [1]. It has also been utilized as an enhancer for cancer therapy on the basis of its stimulation of cytotoxic T cells.

This paper intends to develop a simple process algebra abstraction for relating molecular-level binding and trafficking events to cell-level function as part of an integrated dynamic system. This will help the researchers to give us a better understanding of the different processes that evoke particular cellular responses. In particular, the paper looks into the sorting of the receptors at different cellular compartments (endosome, lysosome and recycling compartment), and the effects of cell surface and endosomal ligand/receptor binding to receptor availability and cell proliferation.

Furthermore, the authors would like to test if a process algebra model of receptor trafficking, which includes different compartments, can be developed using basic π-calculus.

The results of the simulation are obtained by using Stochastic Pi Machine (SPiM) [7], a tool developed by Andrew Philipps to design and simulate computer models of biological processes through the Gillespie's algorithm. Furthermore, the results are compared to the experimental results and to the computational analysis of [3].

2   The Stochastic π-Calculus

The π-calculus was designed to express, run, and reason about concurrent systems. These are abstract systems composed of processes, i.e., autonomous, independent processing units that run in parallel and eventually communicate, by exchanging messages through channels. This was first presented by Milner, Parrow and Walker in 1992 [6] as a calculus of communicating systems in which it describes concurrent computations whose network configuration may change during the computation.

Regev and Shapiro [11] proposed the stochastic π-calculus as a modeling language for systems biology. As a refinement of Priami's [9] synchronous π-calculus by a notion of time, stochastic π-calculus processes are seen as chemical reactions, in which molecules interact concurrently. Moreover, the reaction speeds are governed by stochastic parameters, defining exponential distribution of waiting times, and the process execution is done based on the Gillespie's algorithm [5], generating a stochastic simulation.

In this section, the syntax, operational semantics and structural congruence rules used in the stochastic π-calculus are presented below, as implemented in SPiM [7]. It is assumed that the existence of a set of names or channels $\mathcal{N}$, and let $\{n, m, z, s, \ldots\} \in \mathcal{N}$.

Definition 1. *Syntax of the stochastic π-Calculus: Processes range over $P,Q,\ldots$ Below, fn(P) denotes the set of names that are free in P.*

P,Q	::=	M	*Choice*
		| X(n)	*Instance*
		| P | Q	*Parallel*
		| new x P	*Restriction*
E	::=	{}	*Empty*
		| E,X(m)=P	*Definition, fn(P) $\subseteq$ m*
M	::=	()	*Null*
		| π;P	*Action*
π	::=	?x(m)*r	*Input*
		| !x(n)*r	*Output*
		| delay@r	*Delay*

Expressions above are considered equivalent up to the least congruent relation given by the equivalence relation $\equiv$ defined as follows.

$$
\begin{array}{rcl}
P \mid () & \equiv & P \\
P \mid Q & \equiv & Q \mid P \\
P \mid (Q \mid R) & \equiv & (P \mid Q) \mid R \\
X(m)=P \qquad X(n) & \equiv & P\{m:=n\} \\
\text{new } x \; () & \equiv & () \\
\text{new } x \text{ new } y \; P & \equiv & \text{new } y \text{ new } x \; P \\
x \in fn(P) \qquad \text{new } x \; (P \mid Q) & \equiv & P \mid \text{new } x \; Q
\end{array}
$$

The reduction rules of the calculus are given below. Each rule is labeled with a corresponding rate that denotes the rate of a single reaction, which can be ether a communication or a delay. The rules are standard except for the second communication rule, where the rate of the communication is given by the weights of the input and output actions.

Definition 2. *Structural Congruence on Stochastic π-Calculus [7]*

$$\text{do delay@r; P or } \ldots \quad \xrightarrow{r} \quad P$$

$$(\text{do !x(n)*r1;; P1 or } \ldots)$$

$$| \ (\text{do ?x(m)*r2;; P2 or } \ldots) \xrightarrow{\rho(x)\cdot r1\cdot r2} P1 \ | \ P2\{m:=n\}$$

$$P \xrightarrow{r} P' \qquad \text{new x P} \quad \xrightarrow{r} \quad \text{new x P'}$$

$$P \xrightarrow{r} P' \qquad P \ | \ Q \quad \xrightarrow{r} \quad P' \ | \ Q$$

$$Q \equiv P \xrightarrow{r} P' \equiv Q' \quad \xrightarrow{r} \quad Q \xrightarrow{r} Q'$$

A process can send a value n on channel x with weight r_1 and then do P_1, written !x(n)*r1;P1, or it can receive a value m on channel x with weight r_2 and then do P_2, written ?x(m)*r2;P2. With respect to the reduction semantics above, if these complementary send and receive actions are running in parallel, they can synchronize on the common channel x and evolve to P1 | P2{m:=n}, where m is replaced by n in process P_2. This allows messages to be exchanged from one process to another. Each channel name x is associated with an underlying rate $\rho(x)$. The resulting rate of the interaction is given by $\rho(x)$ times the rates r_1 and r_2. These weights decouple the ability of two processes to interact on a given channel x from the rate of the interaction, which can change over time depending on the evolution of the processes. If no weight is given then a default of 1 is used.

The variant of the stochastic pi-calculus presented above has been used as the basis for the current version of the SPiM, developed by Phillips and Cardelli in 2005 [7].

3   Mechanisms of Interleukin-2 Trafficking Dynamics

The computational ligand trafficking model, adapted from [3] and shown in Fig. 1, shows the path of IL-2 from ligand-receptor binding to internalization, ligand recycling and receptor/complex degradation. As detailed in [1,3,13], the regulation of IL-2 on T-lymphocytes starts with the binding of a ligand to the heterotrimeric IL-2 receptor, forming a ligand-receptor complex with association rate k_{on} and dissociation rate k_{off}. Then, the complexes are internalized with rate k_{eC}, while consequently inducing the production of the receptors at the rate of k_{syn}. Also, the receptors are also constitutively internalized at the rate of k_{eR}. After entering the endosome, the internalized ligands and receptors bind at the rate of k_{fe} and unbind at the rate of k_{re}. From these internalized complex and receptors, these are either sent to the lysosome for degradation, with rate k_{deg}, or the ligands are recycled back to the plasma membrane at the rate of k_{rec}. Lastly, the receptors are synthesize at the rate V_R.

The parameters that will be used in the construction of the process algebra model and subsequent simulations are cited from [3] and are presented in Table 1.

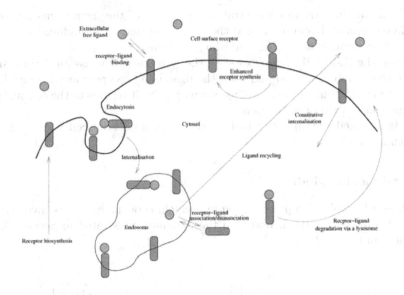

Fig. 1. Kinetic Model Representation of Interleukin-2 Trafficking Dynamics [3]

Table 1. Model Parameters

Parameter	Definition	Base Value(s)
k_{on}	dissociation rate (cell surface)	$0.0138\ \text{min}^{-1}$
k_{off}	association rate (cell surface)	$k_r/11.1\ \text{min}^{-1}$
k_{re}	dissociation rate (endosome)	$8\ k_r\ \text{pM}^{-1}\text{min}^{-1}$
k_{fe}	association rate (endosome)	$k_{re}/1000\ \text{pM}^{-1}\text{min}^{-1}$
k_{eC}	internalization rate	$0.04\ \text{min}^{-1}$
k_{rec}	recycling rate	$0.15\ \text{min}^{-1}$
k_{deg}	degradation rate	$0.035\ \text{min}^{-1}$

4   Construction of the Process Algebra Model

In this section, the construction of the process algebra model of IL-2 trafficking is presented. Specifically, the assumptions in building the model and the schematic representation of the trafficking system is presented.

4.1   Assumptions

In constructing the stochastic π-calculus model of IL-2 trafficking, the following assumptions have been used:

1. It is assumed that one IL-2 protein will bind to the IL-2 receptor (IL2R).
2. When the IL-2:IL2R complex enters the early endosome, it is assumed that they remain bound.

3. The model intends to concentrate on the flow of the ligand into different compartments. Because of this, the authors excluded the induced receptor synthesis process in the IL2 trafficking pathway.
4. When the IL-2:IL2R complex enters the lysosomes, it is assumed that they are completely degraded and both the ligand and receptor are destroyed.
5. When the IL2 enters the recycling compartment, it returns to the cell surface immediately, ready to be used.
6. It is assumed that initially had no receptors inside the cell at any of the compartments.

4.2 π-Calculus Model

The directed graphical representation in π-calculus of the processes involved in the IL-2 trafficking system, based on the dynamics illustrated in Section 3, is presented in Figure 2.

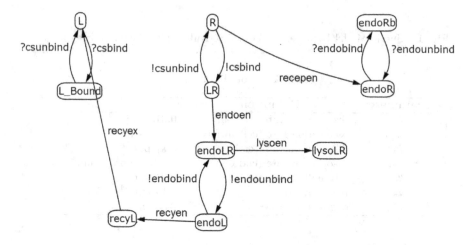

Fig. 2. Model Schematic of Interleukin-2 Trafficking Dynamics

The sketch of the code of IL-2 trafficking system in SPiM is presented in Appendix A.

5   Results and Discussion

In this section, the results of the simulation runs of the π-calculus model and comparisons to the computational analysis of [3] are discussed. Moreover, the results of the simulation is presented in Figure 3.

From the figure above, one can infer that 30% of the complexes entered the endosome, in which less than 10% of the total internalized complexes entered the

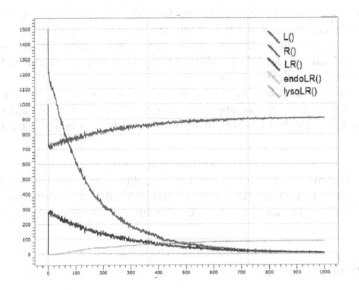

Fig. 3. Simulation Results (Intracellular Sorting) [3]

lysosome. Also, checking the time series data from the simulation results show the first complex enters the lysosome after approximately 15 min. This means that there is a significant delay in the degradation process, in which it needs further sensitivity analysis. The results obtained from the *in silico* experiment is similar to the results presented from [3].

Furthermore, it is prominent in the graph that the ligand in the cell surface increases through time because of the high cell surface dissociation rate and the enhanced ligand recycling due to endosomal dissociation rate greater than the association rate by the thousand-fold. This shows us that binding affinity significantly affect the ligand routing to the different cellular compartments.

6   Conclusion and Future Directions

In this paper, a process algebra model in stochastic π-calculus is presented for illustrating the availability and trafficking dynamics of Interleukin-2 (IL-2) in hematopoeitic cells. The model attempts to abstract the ligand trafficking though the different compartments to stochastic π-calculus, while keeping the model simple to analyze. Moreover, the simulations are done using SPiM [7], and it is observed from the *in silico* experiment that the ligand/receptor binding kinetics directly affect the intracellular sorting of the ligands and receptors in the model. The results of the *in silico* experiment agree with those in the literature.

However, the stochastic π-calculus model is clearly not complete, but it describes the main dynamics of intracellular sorting of the IL-2 proteins. It can be made more accurate in the future by compositionally adding new components, and extending the model using variants of the stochastic π-calculus, which involves compartmentalization.

Also, as a limitation of this study, this paper only discussed the simulation results arising from the constructed process algebra model. The deductions from the formalism is not included in this study, and this is intended for future research.

Acknowledgement. John Justine S. Villar would like to thank the Department of Science and Technology, through the Accelerated Science and Technology Human Resource Development Program, for funding his graduate studies and research.

References

1. Cantrell, D.A., Smith, K.A.: The interleukin-2 t-cell system: a new cell growth model. Science 224 (1984)
2. Ciocchetta, F., Hillston, J.: Bio-pepa: a framework for the modelling and analysis of biochemical networks. Theoret. Comp. Sci. 410 (2009)
3. Fallon, E.M., Lauffenburger, D.A.: Computational model for effects of ligand/receptor binding properties on interleukin-2 trafficking dynamics and t-cell proliferation response. Biotechnol. Prog. 16 (1982)
4. Fokkink, W.J.: Introduction to Process Algebra. Springer-Verlag New York, Inc., New York (2000)
5. Gillespie, D.T.: Exact stochastic simulation of coupled chemical reactions. J. Phys. Chem. 81 (1977)
6. Milner, R., Parrow, J., Walker, D.: A calculus of mobile processes part 1. Information and Communication 100(1) (1992)
7. Philipps, A.: The Stochastic Pi-Machine (2006),
   http://research.microsoft.com/~aphillip/spim/
8. Philipps, A., Cardelli, L.: A graphical representation for the stochastic pi-calculus. Theoret. Comp. Sci. 410 (2009)
9. Priami, C.: Stochastic π-calculus. Computer J. 6 (1995)
10. Regev, A., Panina, E.M., Silverman, W., Cardelli, E., Shapiro, E.: Bioambients: An abstraction for biological compartments. Theoret. Comp. Sci. 325 (2004)
11. Regev, A., Shapiro, E.: Cells as computation. Nature 419 (2002)
12. Regev, A., Silverman, W., Shapiro, E.: Representation and simulation of biochemical processes using the π-calculus process algebra. In: Altman, R.B., Dunker, A.K., Hunter, L., Klein, T.E. (eds.) Pacific Symposium on Biocomputing, pp. 459–470 (2001)
13. Robb, R.J.: Human t-cell growth factor: purification, biochemical characterization, and interaction with a cellular receptor. Immunobiol. 161 (1982)

Appendix A. Representation of the IL-2 Trafficking System in SPiM

```
L ::= ?csbind; L_Bound
L_Bound ::= ?csunbind; L
R ::= do !csbind; LR or delay@recepen; endoR
LR ::= do !csunbind; R or delay@endoen; endoLR
endoRb ::= ?endounbind; endoR
endoR ::= ?endobind; endoRb
endoLR ::= do !endounbind; endoL or delay@lysoen; lysoLR
endoL ::= do !endobind; endoLR or delay@recyen; recyL
recyL ::= delay@recyex; L
lysoLR ::= ()

SYSTEM ::= 1000 L | 1500 R
```

Membrane Computing with Genetic Algorithm for the Travelling Salesman Problem

Pablo Manalastas[1,2]

[1] Department of Information Systems and Computer Science,
Ateneo de Manila, Quezon City, Philippines
pmanalastas@ateneo.edu
[2] Department of Computer Science, University of the Philippines,
Quezon City, Philippines
pmanalastas@gmail.com

Abstract. We propose a heuristic solution to the travelling salesman problem that uses membrane computing to allow for distributed asynchronous parallel computation, and genetic algorithm to select the Hamiltonian cycles that are to be included in the computation in each membrane. We applied this heuristic solution to several asymmetric problems in the TSP Lib website, and obtained solutions that are more costly than the known optimal solutions. We propose modifications to improve this heuristic solution.

Keywords: Travelling salesman problem, heuristic solution, membrane computing, genetic algorithm, TSPLib.

1 The Travelling Salesman Problem

1.1 Statement of the Problem

Given a complete directed graph G on d vertices, in which each arc (x, y) has a weight $w(x, y)$ associated with it, where x and y are vertices of G. Given a path P in G, the cost $C(P)$ of path P is the sum of the weights of all arcs in P. A path that passes through each vertex of G exactly once and returns to the starting vertex is called a Hamiltonian cycle. The travelling salesman problem looks for a Hamiltonian cycle with minimum cost among all Hamiltonian cycles of G.

1.2 Computational Complexity

An exhaustive brute force enumeration of all Hamiltonian cycles that starts with a selected starting vertex x_0 will be a list of $(d - 1)!$ Hamiltonian cycles. Thus an algorithm that solves the travelling salesman problem based on brute force enumeration will have a time complexity of $O((d - 1)!)$ and so takes too much time. We are looking for a 'travelling salesman algorithm' (TSA) that executes in less time, even if the solution arrived at is not a minimum cost Hamiltonian cycle, as long as we can have an idea of how 'near enough' the TSA solution is to a minimum cost.

S. Nishizaki et al. (Eds.): WCTP 2012, PICT 7, pp. 116–123, 2013.

2 Membrane-Computing-Genetic-Algorithm-(MCGA)-Based Heuristic Solution

2.1 General Considerations for MCGA-Based Heuristic

Unlike Gheorghe Paun's model of membrane computing [PA1], we shall use membranes only as structures that hold programs and data, such that a program in a membrane runs independently of a program in another membrane. Occasionally, these programs communicate to exchange data.

We shall assume that graph G is a complete directed graph on d vertices, the weights $w()$ of all arcs are given, and $w(x,y)$ need not be equal to $w(y,x)$, for any pair of vertices x and y. Further, we shall assume that the vertices are labelled x_0, x_1, x_2, ..., x_{d-1}. The TSA that we are proposing uses membrane computing to allow for distributed parallel computation, and genetic algorithm to select the Hamiltonian cycles that are to be included in the computation in each membrane. We start with d membranes, namely M_0, M_1, M_2, ..., M_{d-1}, as shown in Figure 1.

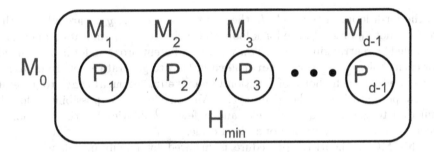

Fig. 1. The membranes of MCGA

Membrane M_0 is a storage membrane that holds, at any given time, a Hamiltonian cycle H_{min}, that currently has least cost. The membranes M_1, M_2, ..., M_{d-1} are computing membranes that can be described as follows. All the $(d-1)$ membranes M_1, M_2, ..., M_{d-1} compute asynchronously, not waiting for each other, but each proceeds in its own computation independently of other membranes. Membrane M_j includes in its computations only Hamiltonian cycles that have x_0 and x_j as the first two vertices in the cycle. It generates Hamiltonian cycles H starting with x_0 and x_j as the first two vertices, using a suitable GenerateNewCycle() function. A number Q of these cycles are generated. Then it generates permutations H' of cycle H, using some suitable function GenerateNewPermutation() to obtain these permutations. The function GenerateNewPermutation() is a genetic algorithm [KGR] that computes from cycle H a new cycle H' that preserves the Hamiltonian property. For each cycle H, a number P of these permutations are computed. If any of these permutations H' has cost less than or equal to the original cycle H, then H' replaces H, and becomes the new interim minimum cost cycle for the membrane M_j. The minimum cost cycle for M_j is hoped to be computed when all Q cycles are tried. When membrane M_j discovers a new interim minimum cost cycle H, if $C(H)$ is less than or equal to $C(H_{min})$, then H replaces H_{min} in the storage membrane M_0. The program P_j of M_j is as follows:

```
for(h = 0; h < Q; ++h) {
  GenerateNewCycle(): H = x_0 x_j y_2 y_3 ... y_{d-1} x_0 ;
  for(k = 0; k < P; ++k) {
    GenerateNewPermutation(): H' = x_0 x_j y_2' y_3' ... y_{d-1}' x_0 ;
    if(C(H') <= C(H) ) {
      Replace H by H' in membrane M_j ;
      if(C(H') <= C(H_{min}) ) {
        Replace H_{min} in membrane M_0 by H' ;
      }
    }
  }
}
```

In the for-h loop, in the cycle H the vertices $y_2, y_3, \ldots, y_{d-1}$ are all the other vertices not x_0 or x_j. In the for-k loop, the cycle H' is a permutation of cycle H obtained by rearranging the ys. Now we get a different program for membrane M_j for each choice of function `GenerateNewCycle()` to generate a new Hamiltonian cycle, and for each choice of function `GenerateNewPermutation()` to generate a new permutation of the given cycle. We consider a few possibilities for the function to generate a new cycle, and a few possibilities for the function to generate a new permutation of a given cycle.

This MCGA heuristic procedure is inspired by membrane algorithms described in [NIS], except that the best solution will appear in the outermost membrane M_0 as the Hamiltonian cycle H_{min}.

2.2 Brute Force Complete Enumeration (CompEnum) of New Cycles

CompEnum MCGA. The function `GenerateNewCycle()` can be defined to completely enumerate the $(d-2)!$ permutations of $y_2, y_3, \ldots, y_{d-1}$. One can use inversion vectors and factoradic numbering as described in [PE1] to generate all the cycles. The for-h loop takes $Q = (d-2)!$ time, and the inner for-k loop takes $O(1)$ time, since we can choose $P = 1$ and we can choose the `GenerateNewPermutation()` as the identity function that does nothing. In this case, the computational complexity is $O((d-2)!)$ which is not really much of an improvement over $O((d-1)!)$. Any improvement is due to the fact that we have now $(d-1)$ membranes to do the complete enumeration. We call this method CompEnum MCGA.

2.3   Randomly Generating (RanGen) a Fixed Number of Cycles

RanGen MCGA. Another method uses a pseudo-random number generator. Each time through the for-h loop the function `GenerateNewCycle()` will generate a new cycle H by using a pseudo-random number generator to obtain the values of the vertices y_2, y_3, ..., y_{d-1}. Let us say that we do the for-h loop a fixed number of times, say $Q = 100 * d$. We omit the for-k loop. Instead, for each cycle H randomly generated in the for-h loop, we select two vertices y_j and y_k from among y_2, y_3, ..., y_{d-1} , such that when y_j and y_k are swapped, we obtain the biggest reduction in cost of cycle H, from among all possible choices of y_j and y_k. The swap is illustrated in Figure 2.

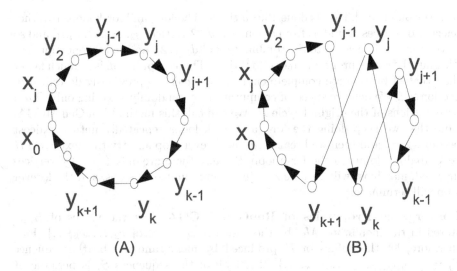

(A)   (B)

Fig. 2. Hamiltonian cycle H, before the swap(A), and H', after the swap(B)

In this case, the program P_j of membrane M_j is as follows:

```
for(h = 0; h < 100*d; ++h) {
GenerateNewCycleUsingRandomNumberGen():
    H = x_0 x_j y_2 y_3 ... y_{d-1} x_0 ;
    for(;;) { // forever loop
       for(j = 2, maxdcost = 0; j <= d-2; ++j) {
          for(k = j+1; k <= d-1; ++k) {
             Let H' be obtained from H by swapping y_j and y_k ;
             dcost = C(H) - C(H') ;
             if(dcost > maxdcost) {
                maxdcost = dcost;
```

```
      }
    }
  }
  // if no decrease in cost is found, break out of forever loop
  if(maxdcost == 0) break;
  // Here H' has lower cost than H and has the lowest such cost
  Replace H by H' in membrane M_j ;
  if(C(H') <= C(H_{min}) ) {
    Replace H_{min} in membrane M_0 by H' ;
  }
  } // repeat forever loop
} // repeat for-h loop
```

Now the outer for-h loop is done $100*d$ times. The for-j and for-k loops together select two vertices y_j and y_k from among the $d-2$ vertices $y_2, y_3, \ldots, y_{d-1}$, and so has execution time equal to the combination $C(d-2, 2)$, and so is proportional to d^2. Thus the entire program takes $O(d^3)$ time. This computation is so much faster than that of brute force complete enumeration method previously described in Section 2.2. We achieve speed of computation by randomly selecting only $O(d^2)$ permutations of the original cycle H. We can call this method RanGen MCGA. Note that we keep doing the for-j and for-k loops repeatedly until maxdcost becomes zero, and then we break from the forever loop and try the next cycle H generated randomly in the for-h loop. Because there are only $(d-2)$ y-vertices involved, maxdcost will become zero in a finite number of steps, and the forever loop will terminate.

Convergence Properties of RanGen MCGA. Let the values of H_{min} stored in the membrane M_0, be the sequence $S = \{c_0, c_1, c_2, \ldots, c_k, \ldots\}$. Furthermore, let the values of H produced by membrane M_j be the sequence $S_j = \{c_{j,0}, c_{j,1}, c_{j,2}, \ldots, c_{j,k}, \ldots\}$. If not all of the sequences S_j is periodic, it is clear that S is a decreasing sequence that should converge to a local minimum, at worst, and to a global minimum, at best.

RanGen MCGA Implementation Issues. We actually did a computer implementation of RanGen MCGA. There were several implementation issues that we had to consider. Do we use d computers in a grid, each computer representing a membrane? Or do we use d threads of computation on a single multi-core computer, each thread representing a membrane? If we used a grid of d computers, we might not have available a grid with enough computers when d is large. Furthermore, network communication between computers will be a bottleneck, since each of the $(d-1)$ computers running the programs of the membranes M_1, $M_2, \ldots, M_{d-1}$ have to communicate regularly with the computer representing M_0. So we decided to use d threads in a single multi-core computer.

Using a single multi-core computer to run d threads solves the problem of communication between membranes, since these threads (membranes) can communicate using global variables. However, when d is large, and our computer has only four cores, then a load-balancing Linux will need to run an average of $d/4$ threads per core. When d is large, computation will be very slow. Since our objective is to give a proof of concept to show that RanGen MCGA can work, then speed really does not matter too much.

RanGen MCGA Implementation Details. Our implementation used $100*d$ randomly generated cycles in each of the $d-1$ membranes M_1, M_2, ..., M_{d-1}. The parent process was used to store the global variables of the membrane M_0. The thread attributes were set to default values using the pthread_attr_init() function. Then we selected a round-robin schedule policy for thread execution, by calling the function pthread_attr_setschedpolicy() with argument of SCHED_RR. Then we created $(d-1)$ threads using the function pthread_create() to run the programs P_1, P_2, ..., P_{d-1}, for the membranes M_1, M_2, ..., M_{d-1}. Whenever one of the programs P_j needs to access the global variables of membrane M_0, then the program uses the function pthread_mutex_lock() to gain exclusive access to the global variables, then uses the function pthread_mutex_unlock() to release exclusive access to the global variables. Moreover, from time to time within the for(;;) forever loop, P_j calls the sleep() function, that is, P_j blocks, in order to give other threads a chance to execute.

Table 1. Results from Running RanGen MCGA

Problem Name	MinCost from TSPLib	MinCost found by RanGen	Hamiltonian Cycle found by RanGen
br17	39	39	0 11 13 2 12 10 9 1 15 6 5 14 4 3 8 16 7
ftv33	1286	1431	0 12 15 14 16 24 23 17 18 19 20 21 31 11 8 10 9 32 7 4 6 5 30 26 22 27 28 29 25 1 33 2 3 13
ft53	6905	8805	0 17 16 23 21 20 11 10 14 13 12 8 6 9 7 5 51 52 50 49 48 26 25 27 29 28 24 22 47 46 37 19 18 15 39 36 35 40 38 4 2 1 41 42 43 45 44 34 32 31 30 33 3
ft70	38673	42387	0 25 4 31 30 24 56 9 7 13 8 3 60 33 51 46 45 62 61 59 58 23 26 6 20 18 19 47 55 53 52 5 11 40 29 21 28 54 57 44 49 35 38 42 41 36 39 37 69 65 68 67 63 66 64 48 43 34 32 27 22 12 10 14 50 17 16 15 2 1
kro124p	36230	48849	0 5 62 90 44 22 31 20 58 35 73 16 14 10 71 9 48 91 7 41 66 88 79 30 55 96 89 27 92 67 38 95 77 47 13 82 54 61 59 19 85 26 11 6 56 50 80 68 39 63 53 1 43 49 81 94 12 75 36 32 4 51 29 33 8 86 24 57 60 72 84 40 99 70 2 42 28 45 34 76 97 46 52 25 64 65 3 74 18 15 21 93 69 87 78 17 37 23 98 83

Some Experimental Results. The data that we used are from the TSPLib website [TSP], namely asymmetric TSP data of the problems br17, ftv33, ft53, ft70, and kro124p. We wrote the RanGen MCGA program in C, based on the pseudocode in Section 2.3, and ran it on a Linux laptop with Intel Core i3 Duo with two processors and 4MB of ram. In Table 1 below, we give the problem name, the minimum cost published in TSPLib, the minimum cost found by the RanGen MCGA program running for 30 minutes to two hours each, and the actual Hamiltonian cycle with that cost.

3   Conclusion

Except for br17, the costs of the Hamiltonian cycles found by our program for RanGen MCGA are higher than the costs published at the TSP Lib website. The nice thing about our code is that the longer we run our program, or the higher the value of Q that we use in the for-h loop, the smaller the costs of the Hamiltonian cycles that the program discovers. Therefore it seems evident that if we had used a computer with more processors, or if we had run the program for a longer period of time, using higher value of Q, on each data set, we could have obtained lower costing cycles than those that we found at present. However, we have to take into consideration the total amount of actual memory installed, the amount of memory used up by each thread, and the maximum number of threads allowed under Linux in determining if the Linux system can actually handle that many threads required to solve each problem in the TSP Lib website. For example, StackOverflow reports in [CHE] that each Linux process can allow a number of threads not exceeding one-fourth the number of memory pages available for running processes.

4   Recommendations for Modifying RanGen MCGA

4.1   Trying a Different GenerateNewCycle() Function

Instead of randomly generating the initial cycle H in each step of the for-h loop, we can select that cycle that takes, at each step, the local minimum. That is, from vertex x_0 we go to vertex x_j, and from there select the least costly arc not yet taken to go to a next vertex y_2, and from y_2, we choose the least costly arc not yet taken to go to a vertex y_3, and so on.

4.2   Trying a Different GenerateNewPermutation() Function

Given a starting cycle $H = x_0 x_j y_2 y_3 \ldots y_{d-1} x_0$, we select two vertices y_j and y_k such that the new cycle H' is the cycle that goes from x_0 to x_j, then to all the vertices until y_j, then skip to y_k, then back to y_{j+1}, then y_{j+2}, and so on to all the vertices until y_{k-1} then skip to y_{k+1}, and then on to all the vertices until y_{d-1} then back to x_0. This is diffferent from just swapping y_j and y_k. It is more like removing y_k, joining up y_{k-1} to y_{k+1}, then inserting y_k, between y_j and y_{j+1}. We let j and k go through all the y-vertices until we find the new cycle H' with the least cost.

4.3   Running MCGA on a GPU

A GPU allows execution of massively parallel instructions. The MCGA algorithm can be redesigned to run the programs P_j in parallel, and also to do parallel generation of new cycles, and parallel computation of permutations of these cycles.

4.4   Running MCGA on a Grid

The problem of communication between the computing membranes M_1, M_2, M_3, ..., M_{d-1}, and the skin membrane M_0 can be minimized by transfering only the final results of the search. Thus, MCGA may be implemented on a grid, by assigning one computer to each of the programs P_1, P_2, P_3, ..., P_{d-1}, and another computer to run the program for the skin membrane M_0.

References

[CHE] Cox, A., Hosi, E., et al.: Maximum number of threads per process in Linux (2011), http://stackoverflow.com/questions/5635362/max-thread-per-process-in-linux

[KGR] Kjellstrm, G., Gong, Y., Renier, M., et al.: Genetic Algorithm (2012), http://en.wikipedia.org/wiki/Genetic_algorithm

[NIS] Nishida, T.: Membrane Algorithms: Approximate Algorithms for NP-Complete Optimization Problems (2006), http://www.springerlink.com/content/pw578u2456342074/fulltext.pdf

[PA1] Paun, G.: Introduction to Membrane Computing, http://psystems.disco.unimib.it/download/MembIntro2004.pdf

[PE1] Pesko, S.: Enumerating permutations using inversion vectors and factoradic numbering (2007), http://lin-ear-th-inking.blogspot.com/2012/11/enumerating-permutations-using.html

[TSP] TSPLIB: Library of travelling salesman problems, http://comopt.ifi.uni-heidelberg.de/software/TSPLIB95/

Solving the Exact Pattern Matching Problem Constrained to Single Occurrence of Pattern P in String S Using Grover's Quantum Search Algorithm

Brian Kenneth A. de Jesus, Jeffrey A. Aborot, and Henry N. Adorna

Algorithms and Complexity Laboratory
Department of Computer Science
College of Engineering
University of the Philippines Diliman
Diliman 1101 Quezon City, Philippines
{badejesus,jeffrey.aborot}@up.edu.ph, hnadorna@dcs.upd.edu.ph

Abstract. In this paper we present an application of Grover's Quantum Search Algorithm in solving the Exact Pattern Matching Problem. We discuss the details of each step of the algorithm, from the preparation, initialization, evolution and measurement of the final state of the quantum memory registers and the verification of the measurement result. We also discuss the inner workings of the *Oracle* in identifying the solution state $|i_x\rangle$ from among the other superpositioned states $|i\rangle$s. Analysis of the running time of the algorithm will lead us to the worst-case running time of $O(\sqrt{N})$ (such that N is the length of the string) proportional to the length of the pattern we are searching for within the string. Some basic concepts in quantum mechanics necessary for understanding the topic are also presented.

Keywords: Quantum Computing, Grover's Search Algorithm, Exact Pattern Matching Problem.

1 Introduction

Quantum Mechanics is a radically different model from the Newtonian model of the physical world. The premise is, at the subatomic level, particles behave differently as how we observe it in the macro world. There are a few basic principles that differentiate Quantum Mechanics and Newtonian Mechanics that will be relevant to our study.

Basics of Quantum Mechanics

Crucial to understanding Quantum Mechanics is the mathematical concept of a vector space called Hilbert Space. A finite-dimensional Hilbert Space is a vector space defined over the field of complex numbers equipped with an inner product

S. Nishizaki et al. (Eds.): WCTP 2012, PICT 7, pp. 124–142, 2013.

$H \times H \to \mathbb{C}, (x, y) \to \langle x \,|\, y \rangle$. Each Hilbert Space H has a set of basis vectors which can linearly represent all other vectors within the Hilbert Space. A vector x in the Hilbert Space H with a basis set $(v_1, v_2, v_3, \dots, v_n)$ can be represented as the linear combination of basis vectors as

$$x = c_1 v_1 + c_2 v_2 + c_2 v_2 + \dots + c_n v_n$$

such that $c_1, c_2, c_3, \dots, c_n$ are complex numbers. The Dirac notation is used in denoting quantum states in quantum mechanics. We can then represent the above vector as

$$|x\rangle = c_1 \,|v_1\rangle + c_2 \,|v_2\rangle + c_3 \,|v_3\rangle + \dots + c_n \,|v_n\rangle$$

using the Dirac notation, where the notation $|\ \rangle$ is called a ket. A Hilbert Space H which has a set of basis vectors of size n is said to be n-dimensional. Operations in composite quantum systems are transformations on individual quantum systems which compose a composite quantum system represented by vectors in an n-dimensional complex Hilbert Space.

Superposition. In Newtonian Physics, if one throws a projectile with given initial velocity and angle, you can predict the location, velocity, and acceleration at any given time. Subatomic particles such as photons differ from a Newtonian projectile in two basic concepts that will be relevant to our study: the first is that while a quantum particle is unobserved, it may exist as a superposition of states. It can, for example, be in two places at the same time, or exhibit multiple values for properties such as spin or momentum at the same time. When the particle is observed or measured, the particle collapses into one of all the possible states it is superpositioned into.

The current state of a quantum particle can be described mathematically using the Dirac notation, as shown below:

$$|\psi\rangle = p_0 \,|a_0\rangle + p_1 \,|a_1\rangle + p_2 \,|a_2\rangle \dots. p_{n-1} \,|a_{n-1}\rangle$$

where we call the notation $|\ \rangle$ a ket and the p_i as amplitudes of the states $|a_i\rangle$.

In the formula, p_1 describes the probability that the state $|a_1\rangle$ will be observed. More precisely, the probability that the state $|a_1\rangle$ will be observed is $|p_1|^2$. By the properties of probabilities we add a restriction to the above formula, which is:

$$|p_0|^2 + |p_1|^2 + |p_2|^2 + \dots + |p_{n-1}|^2 = 1$$

or simply,

$$\sum_{i=0}^{n-1} |p_i|^2 = 1$$

Entanglement. The second concept relevant to this study is that particles exhibiting quantum behavior also follow the entanglement property. Two particles are said to be entangled if, by manipulating or observing one, there will be an instantaneous effect on the other. Expanding on the formula above, entangled states are mathematically represented as the following:

$$
\begin{aligned}
|\psi\rangle = |ab\rangle = \; & p_{00}\,|a_0b_0\rangle + p_{10}\,|a_1b_0\rangle + \dots + p_{n-1\,0}\,|a_{n-1}b_0\rangle + \\
& p_{01}\,|a_0b_1\rangle + p_{11}\,|a_1b_1\rangle + \dots + p_{n-1\,1}\,|a_{n-1}b_1\rangle + \\
& p_{02}\,|a_0b_2\rangle + p_{12}\,|a_1b_2\rangle + \dots + p_{n-1\,2}\,|a_{n-1}b_2\rangle + \\
& \dots + \\
& p_{0\,m-1}\,|a_0b_{m-1}\rangle + p_{1\,m-1}\,|a_0b_{m-1}\rangle + \dots + p_{n-1\,m-1}\,|a_{n-1}b_{m-1}\rangle
\end{aligned}
$$

With the restriction that:

$$
\sum_{i=0}^{n-1}\sum_{j=0}^{m-1} |p_{ij}|^2 = 1
$$

As described above, the Dirac notation enumerates all possible combinations of states of particle a and states of particle b. Entanglement is apparent once either particle is observed. If, for example, particle b is measured, we get a single one of the m possible states of b. What happens is that the state of particle a will be altered simply by measuring b. For example, if we measure b and observe the state b_2, we get a new description of the particle a:

$$
|ab\rangle = p_0\,|a_0b_2\rangle + p_1\,|a_1b_2\rangle + p_2\,|a_2b_2\rangle + \dots + p_{n-1}\,|a_{n-1}b_2\rangle, \qquad \sum_{i=0}^{n-1} |p_i|^2 = 1
$$

We will get a new set of values for the probabilities. Since b_2 is constant, we can factor it out from the equation due to distributive property.

$$
|ab\rangle = (p_0\,|a_0\rangle + p_1\,|a_1\rangle + p_2\,|a_2\rangle + \dots + p_{n-1}\,|a_{n-1}\rangle)\,|b_2\rangle
$$

or simply,

$$
|a\rangle = p_0\,|a_0\rangle + p_1\,|a_1\rangle + p_2\,|a_2\rangle + \dots + p_{n-1}\,|a_{n-1}\rangle
$$

It is worth noting that some of the resulting probabilities will result to zero. In fact, some quantum algorithms take advantage of this fact. By observing certain qubits it is possible to alter other entangled qubits without directly manipulating them.

Basics of Quantum Computing

Computation is a physical process. In classical computing we consider systems with two-level state such as the voltage level on wires. In this case, the two levels of state (the amount of voltage on a wire) that define the system in classical computing are the state in which the voltage is low and the state in which the voltage reaches a certain threshold. This two-level state can be encoded by a classical bit in which we could assign the classical value of 0 to the state in which the voltage is low and the classical value 1 to the state in which the voltage reaches the threshold. A composite system made up of individual two-level state systems can then encode information as series of 0s and 1s. In quantum computing we commonly are also concerned with systems with two-level state. Since quantum mechanics deal with subatomic particles, we are presented with various choices for the type of state of the a quantum system to consider. Some types of state, like the location of a particle, have infinite distinct levels as a particle can possibly be in infinite number of locations in a region of space. Practicality wise, this choice of type of state of a system would not be good given finite computing resource. We would also have hard time differentiating very close distinct locations in space. For these reasons we would like to limit our attention to types of state of systems with only finite distinct levels, for which the state of the system can be described by a unit vector on a finite-dimensional complex Hilbert space H. Since we are focusing on systems with two-level state, the state of this system is described by a unit vector on a 2-dimensional complex Hilbert space. The state of a composite system made up of individual of these two-level systems is then described by an 2^n-dimensional complex Hilbert space H.

Qubit. Quantum Computing uses the properties of Quantum Mechanics as a means for computation. For example, the classical bit implemented by on and off transistors has its quantum equivalent known as a quantum bit, or a qubit. A qubit is physically represented by the state of a subatomic particle. Since we focus our attention only to two-level states, we will be concerned only with a specific property of the particle with only two possible stable states. Examples of this property would be an electron's spin, two different polarizations of an atom, and the nuclear spin in a uniform magnetic field [13]. Stability is required so that the states do not jump from one to the other and vice versa uncontrollably. The two possible qubit states are arbitrarily represented as $|0\rangle$ and $|1\rangle$, analogous to the states 0 and 1 of a classical bit. Deriving from the Dirac notation of a quantum state, a qubit, which we denote here as $|\psi\rangle$, therefore will have a general form of

$$|\psi\rangle = p_0 |0\rangle + p_1 |1\rangle, \quad |p_0|^2 + |p_1|^2 = 1$$

which is a superposition of the states $|0\rangle$ and $|1\rangle$ (representing the basis vectors of the 2-dimensional Hilbert space H) with respective amplitudes p_0 and p_1 which are complex numbers. Since Quantum Computing is governed by the principles of quantum mechanics, a qubit can be in either of the two states or both at the same time. As a consequence, a qubit need not have a definite value until it is observed, or in other words its state is unknown until it is observed. This is in contrast with what is assumed in classical bits having a predefined definite value even prior to measurement. Also, observing a qubit in a superposition of states causes the qubit to irreversibly collapse into a definite state and the other information about its previous state is lost. Another difference between classical bits and qubits is that qubits in an unknown state cannot be cloned or copied exactly without causing change to its current state. This is known as the No Cloning Theorem. Also, due to entanglement of particles in quantum mechanics, observing a qubit entangled with another qubit affects the other. This is not the case with classical bits which are defined separately from each other. [12]

Quantum Gates. Similar to Boolean gates, Quantum gates act like operators on qubits and will act as the components of quantum devices. Quantum gates also have NOT, AND , OR gates similar to their classical counterparts. In addition, quantum gates offer something more. Quantum gates operate on a superposition of values and can adjust the probabilities of each state.

A basic example of a quantum gate is the Walsh-Hadamard gate. A Walsh-Hadamard gate creates a balanced superposition of quantum states. When applied to the state $|0\rangle$, it produces

$$H|0\rangle = \frac{1}{\sqrt{2}}|0\rangle + \frac{1}{\sqrt{2}}|1\rangle$$

which is a balanced superposition of the states $|0\rangle$ and $|1\rangle$ (since states $|0\rangle$ and $|1\rangle$ both have probabilities $\left|\frac{1}{\sqrt{2}}\right|^2 = \frac{1}{2}$ and therefore both have equal probability of occurring).

The special property being that the Walsh-Hadamard gate can create an equal superposition of the two possible states, the emphasis being that the probability for the two states are equal. If Walsh-Hadamard gates are applied in parallel to entangled states (or product states), we get similar behavior. For two qubits both initialized to the state $|0\rangle$, it will be

$$(H \otimes H)|00\rangle = \frac{1}{2}|00\rangle + \frac{1}{2}|01\rangle + \frac{1}{2}|10\rangle + \frac{1}{2}|11\rangle$$

which is a superposition of the four possible states of the two qubits with equal probability. The general case for n qubits is described below as [2]

$$(H^{\otimes n})|0\rangle^{\otimes n} = \frac{1}{\sqrt{2^n}} \sum_{i \in \{0,1\}^n} |i\rangle$$

This means then that if a Walsh-Hadamard gate is applied to each of n qubits all initialized to $|0\rangle$, then the result is an equal superposition of all the possible states of the n qubits. If the n qubits represent numbers in binary format, then the resulting state after application of Walsh-Hadamard gate to the n qubits can represent a superposition of the numbers 0 to $2^n - 1$. This is particularly useful when initializing qubits before use in quantum algorithms, and will be used in the algorithm presented in this paper.

Quantum Operators. Qubits and gates can be mathematically represented by matrices and can be useful both for theoretical study as well as simulation on classical computers. The states $|0\rangle$ and $|1\rangle$, for example can be represented as such:

$$|0\rangle = \begin{bmatrix} 1 \\ 0 \end{bmatrix} \qquad |1\rangle = \begin{bmatrix} 0 \\ 1 \end{bmatrix}$$

These form the basis for two dimensional Hilbert space. Operators also exist as matrices and the quantum gate output can be computed by matrix multiplication of the quantum operator and the quantum bit. For example, the Walsh-Hadamard Gate as explained before has a corresponding operator:

$$H = \frac{1}{\sqrt{2}} \begin{bmatrix} 1 & 1 \\ 1 & -1 \end{bmatrix}$$

Below we demonstrate that when a Walsh-Hadamard gate is applied to the pure state $|0\rangle$, the result is an equal superposition of the two pure states.

$$H\,|0\rangle = \frac{1}{\sqrt{2}} \begin{bmatrix} 1 & 1 \\ 1 & -1 \end{bmatrix} \begin{bmatrix} 1 \\ 0 \end{bmatrix} = \frac{1}{\sqrt{2}} \begin{bmatrix} 1 \\ 0 \end{bmatrix} + \frac{1}{\sqrt{2}} \begin{bmatrix} 0 \\ 1 \end{bmatrix} = \frac{1}{\sqrt{2}}\,|0\rangle + \frac{1}{\sqrt{2}}\,|1\rangle$$

Entanglement is demonstrated by applying tensor products to matrices:

$$|01\rangle = |0\rangle \otimes |1\rangle = \begin{bmatrix} 1 \\ 0 \end{bmatrix} \otimes \begin{bmatrix} 0 \\ 1 \end{bmatrix} = \begin{bmatrix} 1\begin{bmatrix} 0 \\ 1 \end{bmatrix} \\ 0\begin{bmatrix} 0 \\ 1 \end{bmatrix} \end{bmatrix} = \begin{bmatrix} 0 \\ 1 \\ 0 \\ 0 \end{bmatrix}$$

These vectors form an n-dimensional Hilbert Space in which all vectors can be expressed as a linear combination of the basis states $|0\rangle$ and $|1\rangle$. The space is closed under addition and scalar multiplication.[1]. Matrix notation will be used in the demonstration of Grover's Algorithm in this paper.

Quantum Algorithms. Quantum algorithms are being developed to take advantage of the properties of quantum computation to solve computationally difficult problems.

Deutch's Algorithm first demonstrated the power of quantum parallelism. The algorithm creates a superposition of all possible inputs to a simple one bit function and then is able to test if it is a balanced function or a constant function[1]. It is also the first algorithm to have a physical implementation using Nuclear Magnetic Resonance (NMR) technology[2].

---

[1] A balanced function is a function whose outputs are opposite for half the inputs. A constant function is a function whose output is constant.

Shor's Algorithm uses a combination of classical and quantum algorithms to factor a large integer in polynomial time. Prime Factorization is previously believed to be a difficult problem, and it is being used for encryption algorithms such as RSA. Implementing Shor's algorithm will potentially break some current encryption systems [14].

Grover's Search algorithm again demonstrated the power of parallel computation but uses amplitude amplification to be used for database search [1]. The core of the quantum database search algorithm is the amplitude amplification iteration and that it can been used for any general function in which we are looking for the input given the desired answer. It has also been formulated as a decision, search, and exact counting algorithm [15]. Although these functions are not related to database search, the use of the crucial amplification process is generally referred to as Grover's Algorithm or Grover Iterate. Grover's Algorithm can be applied to a wide variety of difficult problems.

Grover's Search Algorithm. Quantum algorithms take advantage of the special properties of qubits and quantum gates to solve problems. For this paper, Grover's Quantum Search Algorithm will be utilized to solve the Exact Pattern Matching Problem. In summary, Grover's Quantum Search Algorithm's outline is as follows:

Problem: Given an indexed table T with size m, whose ith element is t_i such that $0 \leq i \leq m - 1$, and a unique element K, find the index i such that $t_i = K$.

1. Represent all m possible values of table indices of T using n qubits ($n = \lceil ln\, m \rceil$) initialized to the state $|0\rangle$ and then applied with Walsh-Hadamard gate each. This will create an equal superposition of the m possible quantum states of the n qubits, representing all the m table indices of T.
2. Use the equally superpositioned m states to check all elements of T if there is any element t_i which matches with K. If a t_i is found, the probability of state i will be slightly increased, while that of the other states will be proportionally decreased.
3. Repeat step 2 an optimal number of times to maximize the probability of the found state i as output when we observe the final state of the n qubits.
4. Finally, observe the n qubits and check if the resulting state i is correct by verifying that $t_i = K$. If not, go back to step 1 and repeat the whole process again.

Further details of the algorithm is described in the Discussion section of the paper.

For a more in-depth introduction to Quantum Computing, the reader may read on [7,1,8,2].

2   Preliminaries

String Matching Problem

A common problem with multiple applications is the String Matching Problem and its variants. The problem is stated as follows [4]:

We define a string S of length l, identified as the Text; and another string P of length t, identified as the Pattern.

Both T and P uses the same alphabet Σ; that is $T \in \Sigma^*$ and $P \in \Sigma^*$.

A shift is defined with an integer i. The pattern P occurs in text T with a valid shift i if the pattern P matches the text T starting at position $i + 1$, or more precisely, for all integer j, $0 \leq j \leq t - 1$, $T[i + j + 1] = P[j]$.

The pattern matching problem finds all possible *valid shifts* of the pattern P over the text S.

Shift of Occurrence of Pattern P in String S

For the Exact Pattern Matching Problem we are mainly concerned in finding the point of occurrence of a pattern in a string of symbols. The string and pattern are defined over a finite set of symbols called an alphabet, which we denote as $\Sigma = \{\sigma_0, \sigma_1, \sigma_2, \ldots, \sigma_m\}$. We denote the string S with length l as $S[0\ldots l\text{-}1]$ and the pattern P with length t as $P[0\ldots t\text{-}1]$ and the length of S and P are related such that $l \gg t$. From these we define the *shift* i with which P occurs in S as the index in string S from which the occurrence of P starts, that is, if P occurs with *shift* i in S, then $0 \leq i \leq l - t$ and $S[i+1\ldots i+t]=P[0\ldots t\text{-}1]$. In this case in which P occurs in S starting from index $i+1$ (or P occurs in S with *shift* i), index i is called a *valid shift*. Otherwise, index i is an *invalid shift* [4].

From these we then formally define a constrained variant of the Exact Pattern Matching Problem.

Exact Pattern Matching Problem Constrained to Single Occurrence of Pattern P in String S

Given string S and pattern P, find a *valid shift* i with which pattern P occurs in string S. It is assumed that only a single occurrence of P exists in S (which implies that we may find only a single *valid shift* i in S).

3 Finding an Exact Pattern P in String S

Given the Exact Pattern Matching Problem, we present the following quantum algorithm based on Grover's Quantum Search Algorithm for solving the problem and discuss later the details of each step. This approach makes use of a "sliding-window" mechanism in finding P over S. In this approach though, we exploit the property of superposition in quantum mechanics in which a quantum particle may exist in a superposition of several quantum states at the same time. This gives the advantage of evaluating the content of the "sliding window" in just one step. We now present the algorithm as follows:

1. Prepare the quantum memory registers and initialize their qubits. We assume for simplicity that for the length l of string S, $l = 2^n$.

   (a) Prepare an n-qubit quantum memory register, which we will call register A, and initialize each qubit into the computational basis state $|0\rangle$.

(b) Prepare a 1-qubit quantum memory register, which we will call register B, whose sole qubit we call as the *ancilla* qubit, and initialize this qubit into the computational basis state $|1\rangle$.

2. Create a superposition of the qubits in register A and B.

(a) Prepare a superposition of the n qubits of register A by acting on each qubit a Walsh-Hadamard operator. This operation will result to

$$|\psi\rangle = \frac{1}{\sqrt{2^n}} \sum_{i \in \{0,1\}^n} |i\rangle$$

such that i is a string of length n that is composed of 0's and 1's; and is the binary representation of the indices of S to represent each of the n qubits of the register A.

(b) Act on the *ancilla* qubit of register B a Walsh-Hadamard operator. This operation will result to

$$|-\rangle = \frac{|0\rangle - |1\rangle}{\sqrt{2}}$$

3. We define a function f which can identify a solution i_x from a set of candidate solutions $(i_0, i_1, i_2, \ldots, i_{2^n-1})$, defined as

$$f : (0, ..., 2^n - 1) \to \{0, 1\}$$

$$f(i) : \begin{cases} 0 & if \ S\,[i+1\ldots i+t] \neq P\,[1\ldots t] \\ 1 & if \ S\,[i+1\ldots i+t] = P\,[1\ldots t] \end{cases}, \quad 0 \leq i \leq l-t$$

such that $S\,[i+1\ldots i+t]$ is the substring of S starting at index $i+1$ and ending at index $i+t$.

4. We define a linear unitary operator which is dependent on the previously defined function f. We denote this operator as U_f and we call it the *Oracle*, and is defined as

$$U_f\,(|i\rangle\,|j\rangle) = |i\rangle\,|j \oplus f(i)\rangle$$

where the operator $\oplus$ is the logical XOR operator. For any state $|i\rangle$ of the register A and the state $|-\rangle$ of the register B, we can see that

$$U_f\,(|i\rangle\,|-\rangle) = U_f\left(|i\rangle\left(\frac{|0\rangle - |1\rangle}{\sqrt{2}}\right)\right)$$

$$= \frac{U_f\,(|i\rangle\,|0\rangle) - U_f\,(|i\rangle\,|1\rangle)}{\sqrt{2}}$$

$$= \frac{|i\rangle\,|0 \oplus f(i)\rangle - (|i\rangle\,|1 \oplus f(i)\rangle)}{\sqrt{2}}$$

since

$$1 \oplus f(i) = \begin{cases} 0 & if \; f(i) = 1, \; S\,[i+1\ldots i+t] = P\,[1\ldots t], \; which \; denotes \; a \; match \\ 1 & if \; f(i) = 0, \; S\,[i+1\ldots i+t] \neq P\,[1\ldots t], \; which \; denotes \; a \; non-match \end{cases}$$

5. Identifying the solution basis $|i_x\rangle$. We let the *Oracle* act on the superposition of basis states $|\psi\rangle$ of the qubits of register A together with that of register B, $|-\rangle$, and this will result to the state $|\psi_1\rangle |-\rangle$

$$U_f\,(|\psi\rangle\,|-\rangle) = \frac{1}{\sqrt{2^n}} \sum_{i \in \{0,1\}^n} U_f\,(|i\rangle\,|-\rangle)$$

$$|\psi_1\rangle\,|-\rangle = \left(\frac{1}{\sqrt{2^n}} \sum_{i \in \{0,1\}^n} (-1)^{f(i)} |i\rangle \right) |-\rangle$$

The resulting state $|\psi_1\rangle |-\rangle$ will be a superposition of all the basis states of the qubits of register A while the state of the ancilla qubit of register B will be kept unchanged, with the amplitude of the identified (by the *Oracle*) solution basis $|i_x\rangle$ set to negative while the rest of the candidate basis $|i\rangle$s remain positive.

6. Amplitude amplification. We define an operator denoted by $2\,|\psi\rangle\,\langle\psi| - I$ called the *"inversion about the mean"* operator and we let it act on the resulting state $|\psi_1\rangle$ of the register A to amplify the amplitude of the solution basis $|i_x\rangle$ previously marked by the *Oracle*. We denote the resulting state from the operation as

$$|\psi_G\rangle = (2\,|\psi\rangle\,\langle\psi| - I)\,|\psi_1\rangle$$

Note that the state of register B remains to be $|-\rangle$. We call the operator $(2\,|\psi\rangle\,\langle\psi| - I)\,U_f$ as the Grover operator G [2], or the G operator.

7. Iterative application of the G operator. We let the G operator act on the succeeding states of registers A and B for $\lceil \frac{\pi}{4}\sqrt{2^n} \rceil - 1$ more times to maximize the amplification of the amplitude of the solution basis $|i_x\rangle$ marked by the *Oracle* and to minimize the amplitude of the other basis states $|i\rangle$s.

8. After $\lceil \frac{\pi}{4}\sqrt{2^n} \rceil$ [2] number of iterations of the application of the G operator on the qubits of registers A and B, we measure the state of the qubits of register A and convert the resulting solution i_x to its decimal representation. We let this be the index i for which we shall verify if it is a *valid shift* of P in S.

9. Given index i, we verify if $S[i+1\ldots i+t] = P[1\ldots t]$ or if *shift* i is a *valid shift*. If *shift* i is indeed a *valid shift*, we end the computation. We can say then that pattern P occurs with *shift* i in string S. Else, if *shift* i is an *invalid shift*, we then repeat the computation starting from step 1.

4   Discussion

We now present the details of each step of the presented quantum algorithm for
solving the Exact Pattern Matching Problem.

Initialization

First, we prepare two quantum memory registers which we shall call registers
A and B. Register A contains n qubits while register B contains only a single
qubit, which we call the *ancilla* qubit. Each of the n qubits of register A we
will initialize into the computational basis state $|0\rangle$ while the *ancilla* qubit of
register B we initialize into the computational basis state $|1\rangle$. Each index i in S
will be represented by a bit string configuration denoted as a quantum state $|i\rangle$
(such that $|i\rangle$ is an element of the computational basis set of the 2^n-dimensional
complex space $\mathbb{C}^{2^n}$) of the n qubits of register A. All the possible quantum states
of the n qubits of register A will then range from $|000\ldots0\rangle$ to $|111\ldots1\rangle$ (which
are all the elements of the computational basis set of the n-dimensional complex
space $\mathbb{C}^{2^n}$), which are 2^n in count.

We introduce into the quantum system a superposition of all the possible
quantum states of the n qubits of register A by applying a Walsh-Hadamard operation denoted by H on each of the n qubits. We then initialize all the n qubits
of register A into the computational basis state $|0\rangle$. Applying the H operation
to a single qubit of register A will result to:

$$H \otimes |0\rangle = H |0\rangle = \tfrac{1}{\sqrt{2}} \begin{bmatrix} 1 & 1 \\ 1 & -1 \end{bmatrix} \begin{bmatrix} 1 \\ 0 \end{bmatrix}$$

$$= \tfrac{1}{\sqrt{2}} \left(\begin{bmatrix} 1 \\ 0 \end{bmatrix} + \begin{bmatrix} 0 \\ 1 \end{bmatrix} \right)$$

$$= \tfrac{1}{\sqrt{2}} (|0\rangle + |1\rangle)$$

$$= \tfrac{|0\rangle + |1\rangle}{\sqrt{2}}$$

Applying the H operation to each qubit of register A creates an equal superposition of all the possible quantum states of all the qubits of register A as follows
and we denote this superposition as the quantum state $|\psi\rangle$ of register A:

$$(H \otimes H \otimes \ldots \otimes H)(|0\rangle \otimes |0\rangle \otimes \ldots \otimes |0\rangle) = H^{\otimes n} |0\rangle^{\otimes n}$$

$$= \left(\tfrac{1}{\sqrt{2}} \begin{bmatrix} 1 & 1 \\ 1 & -1 \end{bmatrix} \right)^{\otimes n} \begin{bmatrix} 1 \\ 0 \end{bmatrix}^{\otimes n}$$

$$= \tfrac{1}{\sqrt{2^n}} (|000\ldots00\rangle + |000\ldots01\rangle + \ldots + |111\ldots11\rangle)$$

$$|\psi\rangle = \tfrac{1}{\sqrt{2^n}} \sum_{i=0, i \in \{0,1\}^n}^{2^n - 1} |i\rangle$$

We can see that now all the possible quantum states $|i\rangle$ have equal probabilities $(\left|\frac{1}{\sqrt{2^n}}\right|^2)$ of being the result when the state of register A is measured at this time. This is what we call as an equal superposition of quantum states. We then can't do the measurement of the state of the register A as early as this stage because it is very likely that we will get a result which is not the answer we are looking for. We now want to have a high certainty that when we do a measurement of the state of register A we will get, with high probability, the answer we are looking for. In order to do that, the algorithm must first be able to identify which among all the possible quantum states $|i\rangle$ (representing all the indices i of S such that $0 \leq i \leq l - t$) of the n qubits of register A is the solution, $|i_x\rangle$, and then make the probability of this quantum state approach 1 while that of the other states approach 0. After this will we only have a high certainty that the result which we will get after measuring the state of register A is the solution we are looking for, the *valid shift* i of which pattern P occurs in string S.

On the other hand, the result of acting the H operation on the *ancilla* qubit of register B will result to the state $|-\rangle$ as follows

$$H \otimes |1\rangle = H |1\rangle$$

$$= \frac{1}{\sqrt{2}} \begin{bmatrix} 1 & 1 \\ 1 & -1 \end{bmatrix} \begin{bmatrix} 0 \\ 1 \end{bmatrix}$$

$$= \frac{1}{\sqrt{2}} \left(\begin{bmatrix} 1 \\ 0 \end{bmatrix} - \begin{bmatrix} 0 \\ 1 \end{bmatrix} \right)$$

$$= \frac{1}{\sqrt{2}} (|0\rangle - |1\rangle)$$

$$|-\rangle = \frac{|0\rangle - |1\rangle}{\sqrt{2}}$$

After step 2, register A and register B will then be in the state

$$|\psi\rangle |-\rangle = \left(\frac{1}{\sqrt{2^n}} \sum_{i \in \{0,1\}^n} |i\rangle \right) |-\rangle$$

$$= \frac{1}{\sqrt{2^n}} \sum_{i \in \{0,1\}^n} |i\rangle |-\rangle$$

Marking the Solution

To identify the solution state $|i_x\rangle$ from among all the other states $|i\rangle$ in the superposition state of $|\psi\rangle$, we introduce a unitary operator U_f which we call as the *Oracle* and is defined as

$$U_f (|i\rangle |j\rangle) = |i\rangle |j \oplus f(i)\rangle$$

We can see that the Oracle U_f is dependent on a function f defined as

$$f : (0, \ldots, 2^n - 1) \rightarrow \{0, 1\}$$

$$f(i) : \begin{cases} 0 & if \ S[i+1 \ldots i+t] \neq P[0 \ldots t-1] \\ 1 & if \ S[i+1 \ldots i+t] = P[0 \ldots t-1] \end{cases}$$

For any state $|i\rangle$ of the superposition state of register A and the state $|-\rangle$ of the superposition state of register B, the *Oracle* works as follows:

$$U_f(|i\rangle |-\rangle) = U_f\left(|i\rangle \left(\frac{|0\rangle - |1\rangle}{\sqrt{2}}\right)\right)$$

$$= U_f\left(\frac{|i\rangle|0\rangle - |i\rangle|1\rangle}{\sqrt{2}}\right)$$

$$= \frac{U_f(|i\rangle|0\rangle) - U_f(|i\rangle|1\rangle)}{\sqrt{2}}$$

$$= \frac{(|i\rangle|0 \oplus f(i)\rangle) - (|i\rangle|1 \oplus f(i)\rangle)}{\sqrt{2}}$$

$$= (-1)^{f(i)} |i\rangle |-\rangle$$

When we consider all of the superpositioned states of register A, $|\psi\rangle$, and the superpositioned states of register B, $|-\rangle$, and apply the *Oracle* U_f, we get a new superposition state for register A, $|\psi_1\rangle$, but the superpositioned states of register B are kept.

$$U_f(|\psi\rangle |-\rangle) = U_f\left(\left(\frac{1}{\sqrt{2^n}} \sum_{i \in \{0,1\}^n} |i\rangle\right) |-\rangle\right)$$

$$= U_f\left(\frac{1}{\sqrt{2^n}} \sum_{i=0, i \in \{0,1\}^n} |i\rangle |-\rangle\right)$$

$$= \frac{1}{\sqrt{2^n}} \sum_{i \in \{0,1\}^n} U_f(|i\rangle |-\rangle)$$

$$= |\psi_1\rangle |-\rangle$$

such that

$$|\psi_1\rangle = \frac{1}{\sqrt{2^n}} \sum_{i \in \{0,1\}^n} (-1)^{f(i)} |i\rangle$$

In this superpositioned states, $|\psi_1\rangle$, of the qubits of register A, the state $|i_x\rangle$ identified by the *Oracle* as the solution from among the rest of the superpositioned states $|i\rangle$ is marked by a negative sign $(-)$ while the other states remain to have a positive sign $(+)$.

$$|\psi_1\rangle = \frac{1}{\sqrt{2^n}} |000 \ldots 00\rangle + \frac{1}{\sqrt{2^n}} |000 \ldots 01\rangle + \ldots - \frac{1}{\sqrt{2^n}} |i_x\rangle + \ldots + \frac{1}{\sqrt{2^n}} |111 \ldots 11\rangle$$

Since the Oracle U_f's part in the algorithm in determining the solution $|i_x\rangle$ is crucial, we also present the construction of the *Oracle*.

Oracle

The *Oracle* will be tasked to query the entire string S for the pattern P. Given an index i, it will return 1 if the substring of S starting at index $i + 1$ matches the pattern P. Without loss of generality we assume an alphabet $\{0, 1\}$. We can say that the jth bit matches if $S[i + j + 1] = P[j]$ for $0 \leq j \leq t - 1$. In terms of Boolean algebra, the jth bit matches if $1 =\sim (S[i + j + 1] \oplus P[j])$ where $\oplus$ is the logical XOR operation and $\sim$ is the NOT operation. Below is the truth table to prove the equivalence of the formulas.

Table 1. Oracle Truth Table

$|a\rangle$	$|b\rangle$	$|a\rangle = |b\rangle$?	$|a\rangle \oplus |b\rangle$	$\sim (|a\rangle \oplus |b\rangle)$
$|1\rangle$	$|1\rangle$	$|1\rangle$	$|0\rangle$	$|1\rangle$
$|1\rangle$	$|0\rangle$	$|0\rangle$	$|1\rangle$	$|0\rangle$
$|0\rangle$	$|1\rangle$	$|0\rangle$	$|1\rangle$	$|0\rangle$
$|0\rangle$	$|0\rangle$	$|1\rangle$	$|0\rangle$	$|1\rangle$

For the entire substring to match the pattern P, all of the bits should match, and the equality formula must be all 1 for all j. The formula for the *Oracle* will then be below ($\bigwedge$ is the logical AND operator):

$$f(i) = \bigwedge_{j=1}^{t} \sim (S[i + j] \oplus P[j])$$

By this design, the *Oracle* will run in a single step, but the circuit complexity of this will require $t - 1$ Toffoli gates (logical AND), and $2t$ CNOT gates, (t gates act as logical XOR, and t acts as logical NOT) gates. The *Oracle* circuit does not depend on the length of the string S, but it will have a linear circuit complexity with the length of the pattern P. This will be practical for patterns with short length, much shorter than the length of S, that is, $|P| \ll |S|$.

For alphabets with size greater than two, we will represent the characters as binary strings, and thus the number of gates will be multiplied by a factor of $\lceil ln(|\Sigma|) \rceil$, and the circuit complexity of the Oracle will then be $O\left(ln(|\Sigma|) \times t\right)$.

Inversion about the Mean

Our first aim of identifying (marking) the solution is now done. Next, we need to increase the probability that when we measure the state of register A we will get the marked solution $|i_x\rangle$. We make use of an operator called *"inversion about the mean"* [3] which is defined as

$$2 |\psi\rangle \langle\psi| - I$$

Letting the *"inversion about the mean"* operator act on the previous resulting state $|\psi_1\rangle$ of register A, we get a new resulting state for the qubits of register A and we denote this as $|\psi_G\rangle$.

$$|\psi_G\rangle = (2\,|\psi\rangle\,\langle\psi| - I)\,|\psi_1\rangle$$

We can represent $|\psi_1\rangle$ as

$$|\psi_1\rangle = |\psi\rangle - \frac{2}{\sqrt{2^n}}\,|i_0\rangle$$

and we can see that[2]

$$\langle\psi\,|i_x\rangle = \left(\frac{1}{\sqrt{2^n}}\,\langle i_1| + \frac{1}{\sqrt{2^n}}\,\langle i_2| + \ldots + \frac{1}{\sqrt{2^n}}\,\langle i_x| + \ldots + \frac{1}{\sqrt{2^n}}\,\langle i_{2^n-1}|\right)|i_x\rangle$$

$$= \frac{1}{\sqrt{2^n}}\,\langle i_1\,|i_x\rangle + \frac{1}{\sqrt{2^n}}\,\langle i_2\,|i_x\rangle + \ldots + \frac{1}{\sqrt{2^n}}\,\langle i_x\,|i_x\rangle + \ldots + \frac{1}{\sqrt{2^n}}\,\langle i_{2^n-1}\,|i_x\rangle$$

$$= \frac{1}{\sqrt{2^n}}\cdot 0 + \frac{1}{\sqrt{2^n}}\cdot 0 + \ldots + \frac{1}{\sqrt{2^n}}\cdot 1 + \ldots + \frac{1}{\sqrt{2^n}}\cdot 0$$

$$= 0 + 0 + \ldots + \frac{1}{\sqrt{2^n}} + \ldots + 0$$

$$= \frac{1}{\sqrt{2^n}}$$

We can then evaluate $|\psi_G\rangle$ as follows:

$$|\psi_G\rangle = (2\,|\psi\rangle\,\langle\psi| - I)\,|\psi_1\rangle$$

$$= (2\,|\psi\rangle\,\langle\psi| - I)\left(|\psi\rangle - \frac{2}{\sqrt{2^n}}\,|i_x\rangle\right)$$

$$= 2\,|\psi\rangle - |\psi\rangle - \frac{2^2}{2^n}\,|\psi\rangle + \frac{2}{\sqrt{2^n}}\,|i_x\rangle$$

$$= \frac{2^{n-2}-1}{2^{n-2}}\,|\psi\rangle + \frac{2}{\sqrt{2^n}}\,|i_x\rangle$$

The operation of the *"inversion about the mean"* operator increases the amplitude of the marked solution state while decreasing that of the other states in the superposition of states of the qubits of register A. We can see that from the first few iterations of the G operator, the amplitude of the marked solution state $|i_x\rangle$ increases by approximately $\frac{2}{\sqrt{2^n}}$. The G operator is then just

$$G = (2\,|\psi\rangle\,\langle\psi| - I)\,U_f$$

and the state $|\psi_G\rangle$ is the resulting state of the qubits of register A after the first application of the G operator to registers A and B.

---

[2] Most of the formulations are lifted from [3] with greater details of the computation added for a clearer presentation of the evaluation.

Iteration

After the application of the *"inversion about the mean"* operator on the super-position of states of the qubits of register A, our second aim of increasing the probability of getting the identified solution $|i_x\rangle$ (as identified by the *Oracle*) is achieved. With the iterative application of the G operator we are therefore, first, able to identify a solution state from among the superpositioned states of the qubits of register A, and secondly, able to increase the certainty that we will get the identified solution when we measure the state of the qubits of register A. The bound on how many times we will apply the G operator on the qubits of register A and B will be $\left\lceil \frac{\pi}{4}\sqrt{2^n} \right\rceil$ as detailed in [2]. If the number of applications of the G operator exceeds this value the amplitude of the identified solution will start to decrease, while any number of applications of the G operator lesser than this value will not give us the optimal probability of getting the identified solution $|i_x\rangle$ as result when we measure the state of the qubits of register A.

Measurement

After optimal number of applications of the G operator, we are then ready to measure the state of the qubits of register A. With high probability we will get as result of measurement the identified solution $|i_x\rangle$. We then convert it to its decimal representation i. We then verify if indeed *shift i* is a *valid shift* for the occurrence of P in S, that is if

$$S[i+1\ldots i+t] = P[0\ldots t-1]$$

If *shift i* is indeed a *valid shift*, we end the computation and we have already determined the position of occurrence of P in S. Else, we repeat the computation starting from the preparation of the registers A and B and the initialization of qubits of each respective register, which is step 1. Since we are guaranteed that the amplitude of the identified solution state $|i_x\rangle$ is near to 1 and the amplitude of the other states $|i\rangle$ is near to 0, it will be very unlikely that we will get an *invalid shift* from the computation.

Aside from operator format, we can also represent Grover's Quantum Search Algorithm in quantum circuit format.

Figure 1 is the quantum circuit diagram for Grover's Quantum Search Algorithm. Register A is the top register with n qubits while register B is the bottom register with only the *ancilla* qubit. The iterative application of the G operator is presented as a series of G gates in the circuit acting on the qubits of register A and B. At the end of the circuit is a measurement operator for each of the qubits of register A. The state of the qubits which will result from the measurement operation will be, with high probability, the identified solution state $|i_x\rangle$.

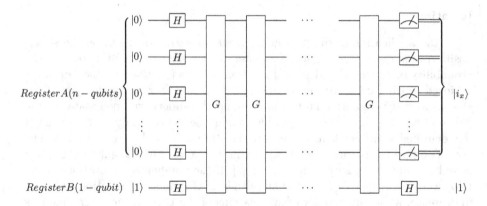

Fig. 1. Grover's Search Algorithm

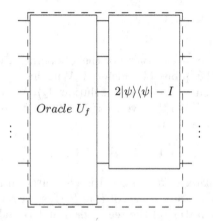

Fig. 2. Details of the G operator

In Figure 2 the details of the G operator is presented. It consists of the Oracle for identification of the solution state $|i_0\rangle$ and the *"inversion about the mean"* operator for amplifying the amplitude of the identified solution state $|i_0\rangle$.

For a geometric representation of the computation of the algorithm (specifically Grover's Quantum Search Algorithm) represented in a Bloch Sphere, the reader may read on [3]. Also, for a more detailed discussion on Grover's Quantum Search Algorithm applied to database search and spatial search, the reader may further read on [6,3,1]. The quantum circuit diagrams were drawn using [9].

Performance. The algorithm running time will depend on the number of times the Grover iterate is repeated, which is $\lceil \frac{\pi}{4}\sqrt{2^n} \rceil$. This running time of $O(\sqrt{2^n})$ can also be represented by $O(\sqrt{l})$. Classical algorithms such as Rabin-Karp algorithm runs in $O((l - t + 1)t)$ worst case but in many applications it runs in $O(l + t)$. Knuth-Morris-Pratt algorithm runs in $\Theta(l + t)$ time.[4]

The complexity of the problem remains exponential in n (linear in l). The complexity class for the classical algorithms is in P, while for the Quantum algorithm, since it is a probabilistic quantum algorithm, falls under the class BQP.[1]

5   Conclusion

In this paper we were able to present an application of Grover's Quantum Search Algorithm to the Exact Pattern Matching Problem. Using the property of superposition of states of qubits, the computation (identification if a state $|i\rangle$ is a solution and amplification of the amplitude of the identified solution state $|i_x\rangle$) on all the states $|i\rangle$ is done in a single step for $\lceil \frac{\pi}{4}\sqrt{2^n} \rceil$ number of iterations of the G operation. This can mainly be attributed to the single step evaluation of all the superpositioned states (indirectly representing all the possible positions of the "sliding window" in the string S) of the qubits of the registers in each of the $\lceil \frac{\pi}{4}\sqrt{2^n} \rceil$ iterations of the G operation. The *Oracle* defined will have a $O(1)$ running time and thus no effect on the overall performance of the algorithm. However, its circuit complexity is $O(ln(|\Sigma|) \times t)$, linearly dependent on the length of the pattern and logarithmically dependent on the alphabet size.

Acknowledgments. We are much grateful to H. N. Adorna, N. H. S. Hernandez and J. B. Clemente for dedicating time discussing with us the details of the problem tackled and for the suggested improvements on the presented algorithm. We are also thankful to F. G. Cabarle for lending us a hand in formatting the document.

References

1. Gruska, J.: Quantum Computing. McGraw-Hill Publishing Company (1999)
2. McMahon, D.: Quantum Computing Explained. John Wiley & Sons, Inc. (2008)
3. Lavor, C., Manssur, L.R.U., Portugal, R.: Grover's Algorithm: Quantum Database Search, arXiv:quant-ph/0301079v1 (2003)
4. Cormen, T., et al.: Introduction to Algorithms, 3rd edn (2009)
5. Vöcking, B., et al.: Algorithms Unplugged. Springer (2011)
6. Patel, A.: Quantum Algorithms: Database Search and Its Variations (2011), http://arxiv.org/abs/1102.2058v1
7. Charemza: An Introduction to Quantum Computing (2005)
8. Kaye, P., Laflamme, R., Mosca, M.: An Introduction to Quantum Computing. Oxford University Press (2007)
9. QCircuit: Quantum Circuit Designer and Simulato (July 27, 2012), http://www.cquic.org/Qcircuit/
10. Faro, S., Lecroq, T.: The Exact String Matching Problem: A Comprehensive Experimental Evaluation (2010), http://arxiv.org/abs/1012.2547v1
11. Faro, S., Lecroq, T.: Efficient Pattern Matching on Binary Strings (2008), http://arxiv.org/abs/0810.2390v2

12. Williams, C.P.: Explorations in Quantum Computing, 2nd edn. Springer, London (2011)
13. Nielsen, M.A., Chuang, I.L.: Quantum Computation and Quantum Information. Cambridge University Press (2000)
14. Shor, P.: Polynomial-Time Algorithms for Prime Factorization and Discrete Logarithms on a Quantum Computer. IEEE Computer Society Press
15. Brassard, G., Hoyer, P., Mosca, M., Trapp, A.: Quantum Amplitide Amplification and Estimation. AMS Contemporary Mathematics Series, vol. 305 (2002)
16. Hirvensalo, M.: Quantum Computing. Springer, Heidelberg (2001)

Design of a Scala-Based Monitoring Server for Web-Based Programming Environments

Koji Kagawa

ENIE, Faculty of Engineering, Kagawa University
2217-20 Takamatsu, Kagawa 761-0396, Japan
kagawa@eng.kagawa-u.ac.jp

Abstract. In learning programming, a Web-based programming environment could be helpful. We have constructed a platform called WappenLite to support Web-based programming environments for learning programming. In addition to avoiding installation time, Web-based environments offer potential advantages such as the real-time monitoring of learner activity. However, it is difficult to construct a Web server-side application that can support a variety of course designs and the grading criteria of various teachers. Therefore, in this paper, we aim to build a platform that can offer flexible customization capabilities to the server-side application of Web-based environments for learning programming.

1 Introduction

To date, a great many applications such as office suites and mailers have been offered as Web applications. Web-based programming environments could also be helpful for learning programming. There are already some Web-based programming environments such as codepad (http://codepad.org), wonderfl (http://wonderfl.net) and overtype (http://www.3site.eu/jstests/jhp). However, most are constructed as monolithic applications, making it difficult for teachers to customize them for their courses and to adapt to new programming language implementations.

To support the effective learning of programming, we have developed a platform for Web-based learning-support environments for programming [3,2], which can accommodate a variety of programming languages and custom libraries. We first constructed an EclipseRCP-based framework called Wappen [3]. However, it turned out to be difficult to program EclipseRCP plug-ins, making it challenging for Wappen to gain popularity. Moreover, the JavaScript platform for Web browsers has evolved and matured substantially since that time. Therefore, we are now switching over to a new AJAX-based client platform called WappenLite [2].

To take advantage of Web-based programming environments, the server should have several capabilities such as grading submitted programs automatically and sending messages. Grading criteria for programming exercises will vary according to the teacher and may not be determined simply by the execution outputs. Therefore, it is necessary to be able to customize grading criteria easily according to the teacher's policy and exercises. In addition, it is desirable to be able to send messages immediately to those learners whose submissions have met various conditions. Although we have already constructed

S. Nishizaki et al. (Eds.): WCTP 2012, PICT 7, pp. 143–150, 2013.
© Springer Japan 2013

a server-side application that uses JSP/Servlets as the platform and XML/JSON format for the configuration files, it has been difficult to support fine-grained customization of this kind.

So-called HTML5 technologies such as Canvas and WebSocket are beginning to be used extensively. Because WappenLite adopts ordinary Web browsers for the user interface, it should be possible to provide additional functionality such as real-time communication by taking advantage of such new technologies. It will then be possible for teachers and teaching assistants to give advice to learners effectively by receiving information about their progress during lab exercises. However, because it is difficult to make effective use of real-time messaging without scripts, such a feature has not yet been provided.

This study uses the Lift Web-application framework (`http://liftweb.net`), which is built on the Scala language (`http://www.scala-lang.org`), to construct a new server-side platform.

This platform should enable each teacher to define fine-grained scripts, including scripts for the customization of automatic grading and the visualization of learners' grades. It will also support real-time communications such as the collection of error messages issued by compilers and the automatic delivery of hints.

In this study, we use a system based on Scala, a language based on a JVM (Java Virtual Machine). The teacher can then use the same code as the programming environment (Wappen/WappenLite) to introduce scripts for automatic grading and other purposes. Scripts often use functions such as map/reduce/filter, which suit functional languages. An attempt was made to use JavaScript on the server side, which turned out to be impractical in terms of execution performance. By using Scala, it should be easier to compose fast and reliable scripts.

For this study, we aimed to construct a platform that was as loosely coupled as possible (providing a-la-carte features). Many platforms adopt plug-in schemes for extension, whereas in general, it is not intended that only a subset of the platform's functionality be used. Therefore, it is often reported that plug-ins cannot keep up with upgrades to the platform. In this study, we aim to avoid this problem as much as possible.

The remainder of this paper is structured as follows. First, Section 2 explains the motivations in more detail. Section 4 introduces WappenLite. Section 5 describes the design of the proposed system. Section 6 gives a summary and discusses future directions.

2 Motivation

Web-based programming environments have many potential advantages for education. Several features of Web-based development environments are desirable for learning programming. In addition to saving time and avoiding unnecessary effort by learners, we can collect much useful real-time information during lab exercises. For example, we can obtain:

- the source programs being edited by the learner
- details of key strokes

- error messages issued by compilers
- the timing of submission of files
- the input–output pair for running programs
- trace information about running programs.

By utilizing these data, teachers can not only judge correct/wrong answers but also identify those learners, for example:

- whose submission of exercise No.X is Y minutes later than the median,
- whose frequency of submission is too low or high,
- who continuously receive the same specific error messages from the compiler,
- whose program fails at a specific set of inputs,
- whose source code too much resembles others' in terms of Levenshtein distance,
- whose program has unnecessary code fragments
  (*e.g.* for (s=str; *s!='\0'; *s++)),
- whose source code has bad indentation,
- whose source code is too complex by some measure, or
- whose program requires too many steps to execute.

It would help to motivate learners if teachers could visualize these kinds of information and thereby recognize the learner's situation in real time, or if they could send real-time messages to those learners who have met a particular set of criteria. However, it would be difficult to provide all such functions in advance, because each teacher will have a different policy for courses in general, and the individual courses may deal with different programming languages. Therefore, it will be necessary for teachers to be able to write their own scripts for customizing the search conditions. For our purposes, the programming language for scripts will have to meet the following criteria.

- It must be efficient enough to enable searching among many submitted files, such as the files submitted by 100 learners over fifteen weeks.
- It must be a JVM-based language. WappenLite itself is implemented in Java and many of its supported language implementations are JVM based. Although it would be possible to launch language implementations as a separate process or to access other languages via network protocols, it is advantageous in terms of efficiency to invoke the methods in JVM-based implementations directly.
- It must be expressive enough to implement other programming languages. In particular, it must have lexer generators similar to Lex/Flex for the C language and parser generators similar to Yacc/Bison.
- It must include REPL (Read-Eval-Print-Loop). We would like to provide a Web-based REPL for teachers that would enable their scripts to be tested conveniently.

Our system adopts the Scala language [4], which is a JVM-based language that also meets the other criteria listed above. It also includes for-comprehension as an alternative notation for the map/reduce/filter functions often used in functional languages. This makes it easy to express enumerations of those items that meet certain conditions. Moreover, it has type inference, which makes it possible to describe scripts briefly and reliably.

It can not only use lexer/parser generators for Java such as JFlex, JavaCUP, Antler and Jacc, but is also equipped with expressive parser-combinator libraries. Although it is a compiler-based implementation, Scala does include REPL.

Lift (http://liftweb.net) is a Web-application framework built on the Scala language. It gives us straightforward access to cutting-edge Web technologies such as Comet and jQuery (http://jquery.com). It provides an environment within which we can edit and test scripts instantly by utilizing the REPL in Scala.

In summary, our system has the following features.

- It is based on Lift/Scala.
- It can grade submitted programs automatically.
- It can send real-time messages to those learners who satisfy specific conditions.
- It can enable visualization of the situation of learners via tables and graphs that teachers can comprehend easily.

Our system is a customizable server-side platform for observing the real-time state of a Web-based educational programming environment such as WappenLite.

3 Related Work

Some features of our system seem to overlap those of Web-based automated grading systems.

Web-CAT (http://web-cat.org) is an automatic Web-based grading system for programming exercises [1]. Web-CAT itself is Java-based but it can involve any language that can be installed on the server. The authors report that they mainly use Perl for scripting, which supports JUnit-like tests for submitted programs.

In contrast, our proposed system puts more focus on smaller (fine-grained) scripts. We are in a position whereby we cannot predict all possible kinds of wrong answers, and we often have to scan all submitted files. We also aim to detect inappropriate variable names, function names and indentation, which cannot be judged from the input and output of programs alone. Moreover, we would like to observe compilation errors in the middle of working on exercises. That implies an ability to analyze programs that cannot be compiled because of errors. For these purposes, we need expressive scripts that run not as external processes but inside the Web application itself.

There is also an overlap in terms of features such as messages to learners and the aggregation of grading with course management systems (CMSs) such as Moodle (http://moodle.org) and Sakai (http://www.sakaiproject.org). Moodle is a PHP-based CMS and Sakai is Java-based. It would be an option to develop our system as a plug-in for one of these CMSs. However, they are large systems with most of their features being irrelevant to programming exercises. We concluded that it would be unreasonable to write plug-ins to enable them to meet our requirements.

In future, it would be possible to delegate some overlapping functions to CMSs, but we currently consider CMSs as separate applications.

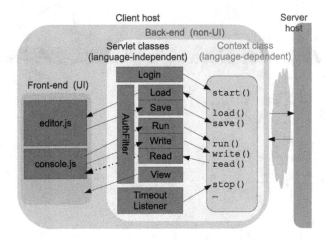

Fig. 1. WappenLite Architecture

4 WappenLite

WappenLite is a Web-application framework for educational programming environments, under development by the authors' research group. Like its precursor Wappen (**W**eb-based **a**pplications for **p**rogramming **p**aradigm **en**lightenment), it can be adapted to various programming languages. WappenLite applications can be launched from Web browsers without the installation of special software beyond a Java runtime environment.

A WappenLite application can use HTML + CSS + JavaScript at the front end and Java servlets at the back end. The front end comprises mainly the editor component and the console component. The front end and the back end communicate with each other via HTTP. The editor component can exploit JavaScript-based editor components. The console component can use the WebSocket protocol. Compared with Wappen, it is easier for teachers to produce and customize a WappenLite application according to the context in which it is to be used. To achieve load balancing, we also provide a Java applet to run servlets on the same computer as the Web browser.

Figure 2 shows a screenshot of WappenLite. In this screenshot, we use jQuery for the user-interface components and CodeMirror (`http://codemirror.net`) as the Web-based editor.

4.1 Applications of WappenLite

Providing Web-based programming environments for learning has several secondary advantages. Although they are not the focus of this paper, we would like to mention them briefly in this subsection.

First, we can make use of the various multimedia formats available in Web browsers, such as vector graphics, raster graphics, sounds and animation. We can provide custom

Fig. 2. WappenLite Screenshot

libraries to produce such formats even for minor programming languages for which no standard graphics library is available. Moreover, WebGL, a JavaScript API for 3D graphics, has been standardized recently, enabling 3D graphics and animations to be used conveniently in Web browsers. We could use the multimedia facilities of Web browsers to provide visual debuggers for novice programmers.

There are some Web-based structure editors for programs such as Waterbear (http://waterbearlang.com) and Google Blockly (http://code.google.com/blockly/). We can provide environments for beginners where they need not be too concerned about the syntax of programs.

5 Architecture

Figure 3 shows the architecture of our system. It receives and saves data such as submitted source code and error messages from the back end of a Web-based programming environment such as WappenLite. It applies various processes such as automatic grading, analysis of error messages and calculation of similarity according to scripts set in advance. The results of the process can be sent to learners as messages. Our system must therefore provide utility libraries for these features and should be able to register event handlers.

It communicates with the front ends (browsers) of learners and teachers directly (*i.e.* not via WappenLite). It receives scripts from the teachers' front ends and processes them in the REPL part. For messaging, scripts are used to identify the target learners and to compose the messages. Learners' browsers receive real-time messages via Comet or WebSocket. For aggregation and visualization, scripts are used to generate tables and graphs that are sent to teachers' browsers according to the stored data or to update visualizations according to submitted data. For these purposes, we need libraries for the various forms of visualization.

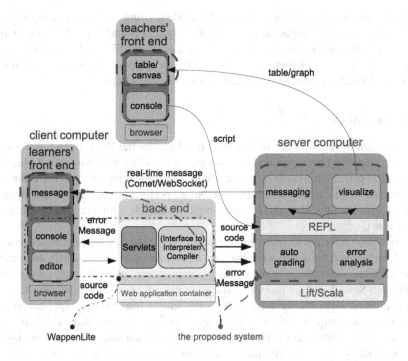

Fig. 3. The Architecture of the Proposed System

The system needs an authentication subsystem. We plan to use OpenID for the authentication of learners. In our university, students use Gmail, and we can therefore use Google as an OpenID provider. The advantage of using a single sign-on system is that it allows the system to coexist with other Web applications for learning without any additional burden on learners. We would aim to support other single sign-on systems if necessary.

The teachers' and learners' systems are loosely coupled so that even if the teachers' system becomes unstable, it does not affect the learners' system, particularly for critical components such as roll calls and file submission.

The common components of the teachers' and the learners' systems are shared as libraries. Otherwise, it would be difficult to develop the teachers' system while putting it into operation in the field.

6   Summary

This paper proposed a server-side customizable application for Web-based programming environments. It explained the reasons for selecting Lift/Scala as the development platform and described the current design of the system. By implementing it as an application in Scala/Lift, it becomes possible for teachers to develop scripts for the aggregation of learners' data that summarize the submitted exercises in real time.

The planned features of our system will include sending real-time messages generated by scripts to those target learners identified by scripts and the real-time visualization by teachers of aggregated learners' data such as source files and error messages.

We are currently implementing the visualization system for teachers. The system will be available as open-source software when it becomes stable.

6.1 Future Work

Scripts written by teachers can do anything permitted by the JVM. In fact, it would be possible to destroy files submitted by learners. In avoiding this, however, it would be unrealistic to prohibit file updates for the whole Web application. We will therefore need a mechanism to guarantee backup creation before performing operations on important and indispensable files. It would be necessary to cooperate with a revision-control system such as Git.

We will need different scripts for different programming languages, different courses and different course units. We will also need a mechanism to save, classify and arrange reusable scripts, so that the system can suggest appropriate scripts for relevant scenarios.

Although Scala/Lift may be considered an immature environment in some respects, when compared with older Java-based Web-application frameworks, experienced users would be able to compose scripts very rapidly. Scala is a language that has features both of object-oriented languages such as Java and of functional languages such as Haskell. It is syntactically somewhat complex. Therefore, inexperienced users can be confused even if they are teachers of programming in other languages. User interfaces that use a visual programming language such as Waterbear or Google Blockly, mentioned in § 4.1, would be necessary in future. We would like to be able to support other JVM-based language implementations such as JRuby and Jython as scripting languages to increase the pool of potential users.

Because WappenLite uses ordinary browsers (HTML + CSS + JavaScript) as the client, WappenLite applications can potentially be run on tablets and e-books. In thin clients, we often use a user interface that suggests using carousels and pull-down menus to compensate for the lack of a keyboard. To make learning programming using thin clients practical to some extent, we would like to provide assistance functions that are calculated by scripts on the server.

References

1. Edwards, S.H.: Improving student performance by evaluating how well students test their own programs. Journal of Educational Resources in Computing 3(3), 1–24 (2003)
2. Kagawa, K.: WappenLite: A Web application framework for lightweight programming environments. In: 9th International Conference on Information Technology Based Higher Education and Training (ITHET 2010), pp. 21–26 (April 2010)
3. Mimoto, Y., Kagawa, K.: A framework for Web-based applications for learning programming using Eclipse RCP. In: World Conference on Educational Multimedia, Hypermedia & Telecommunications (ED-Media 2008), pp. 2253–2264 (June 2008)
4. Odersky, M., Spoon, L., Venners, B.: Programming in Scala, 2nd edn. Artima Inc. (2010)

Recognizing Historical KANA Texts Using Constraints

Yuta Arai, Tetsuya Suzuki, and Akira Aiba

Graduate School of Shibaura Institute of Technology
Minuma, Saitama City, Saitama, 337-8570, Japan
{ma11011,tetsuya,aiba}@shibaura-it.ac.jp

Abstract. One of the first step for researching Japanese classical literature is reading Japanese historical manuscripts. But the reading process
is not easy and time-consuming since a set of characters used in those
manuscripts contain different from those currently used. There have been
several attempts to read Japanese historical manuscripts, but their targets are fixed-style documents. In this paper, we propose a new framework to assist human process for reading Japanese historical manuscript
in any style by formulating it as a constraint satisfaction problem. In any
Japanese historical manuscript, a sequence of hand-written characters is
constrained to form a valid word found in a dictionary for historical
Japanese. We implemented an experimental system for Japanese historical hiragana, one kind of Japanese characters, to verify the effectiveness
of our framework.

Keywords: character recognition, combinatorial optimization,
constraint hierarchy, constraint solving method.

1 Introduction

Before printing technologies became popular, all texts were hand-written, and
copied by a hand. Through a series of the copying process, texts were often
modified accidentally or intentionally. As a result, there were several versions
of the same text. For example, *Kokin waka shu* which was the first anthology
edited by an imperial order in 905 AD has more than 10 variants.

One of the first step of researching Japanese classical literature is comparing
those variants to determine the standard text. To do this, one has to read these
manuscripts, but this is time-consuming and requires training since they are
hand-written, and may contain characters different from what is currently used.
That is why we cannot use automatic character recognition system for current
texts.

In this paper, we propose a new framework for assisting the human process
for reading Japanese historical manuscript by employing constraint solving. A
sequence of characters is constrained to form a valid word which is found in a
dictionary for historical Japanese.

S. Nishizaki et al. (Eds.): WCTP 2012, PICT 7, pp. 151–164, 2013.

As the first step of our research, we experimentally implemented a system for reading Japanese historical manuscripts
using one kind of Japanese characters called hiragana.

In this paper, we use a word kana as well as hiragana for
the sake of convenience though hiragana is a part of kana in
general.

In section 2, we introduce hiragana, and we summarize existing research in section 3. We describe our motivation and
overall structure of our experimental system in section 4. In
section 5, we define the constraint satisfaction problem for
reading Japanese historical manuscripts. Experimental results
are summarized in section 6. Section 7 states concluding re-
marks.

Fig. 1. Historical
Japanese Text

2   Background of the Research

2.1   Japanese Classical Manuscripts and Hiragana

After introducing Chinese characters to Japan in the 4th, or 5th century, a
method to represent Japanese sentences using them were invented by the 8th
century. This is called *Man'yo-gana*. They were unmodified Chinese characters,
used to represent Japanese syllables according to their pronunciations. In the
8th century, it is said that there were 88 syllables in Japanese language different
from about 50 syllables at present, and over 900 Chinese characters were used to
represent them. During the 8th and 9th century, a new kind of characters called
hiragana was invented based on cursive form of hand-written Chinese characters.
They were gradually accepted by the end of the 9th century [3].

2.2   Reading Hiraganas in Japanese Historical Texts

Hiraganas used in historical texts (we call historical hiraganas hereafter) are
quite different from those currently used:

1. A set of historical hiraganas contain different characters from those currently
   used. There are several characters to represent one syllable.
2. A special symbol "" called odori-ji is used to represent a repetition of an
   immediately previous character.
3. Especially in hand-written texts, each occurrence of the identical hiragana
   may have different shape, persons by persons, texts by texts.

By these characteristics, reading Japanese historical text is difficult even for
Japanese. The Fig. 1 shows an fragment of Japanese historical text written in
hiraganas taken from *Tale of Ise*, originally written in 1234 AD, and copied in
1547 AD [1]. In this fragment, the 1st, 3rd, 5th and 6th characters are easy to
recognize for us since they are quite similar to those hiraganas currently used.

Actually, they are (mu), (shi), (to), and (ko), respectively. However, the 2nd, and 4th hiraganas are difficult since they are not currently used.

In the following, we will briefly explain how a human try to read them. When we focused in the 2nd character, we can suspect that this would be (tsu), or (ka) by its shape. We can determine that the 2nd character is (ka) since *mu-ka-shi* is a valid Japanese word having the meaning of the past. By applying the similar inference, we can determine that the 4th hiragana is (o) since *o-to-ko* is a valid word having the meaning of a male. By observing this small example, it is clear that the knowledge on shape of historical hiraganas is insufficient but the knowledge on historical Japanese words is necessary for recognizing Japanese historical manuscripts.

3   Existing Research

3.1   An Estimation Method of Unreadable Historical Character for Manuscripts in Fixed Forms Using n-gram and OCR

In this research, a system for reprinting characters in historical manuscripts by a combination of optical character recognition (OCR) and character n-gram is proposed [8].

Computer assisted reprinting is necessary because reprinting by reading historical manuscripts is a highly specialized work, and requires knowledge on cursive form of characters, formats of manuscripts, and history.

This system targeted to debt deeds in Edo-era, by the following two reasons:

1. Debt deeds usually take fixed format, and have many fixed form representations. Frequent appearance of fixed form suggests the usefulness of statistical information on manuscripts like n-gram.
2. Since debt deeds are official documents and written in a style of handwriting called "*O-ie-ryuu*", thus, there are regularities on cursive form of characters.

Fig.2 shows the process of the recognition. Unreadable characters are input to OCR, and it outputs the results. These results are associated with prepared n-gram information to output candidates of recognition.

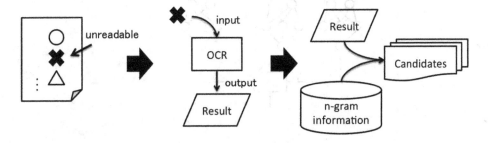

Fig. 2. Flow of n-gram and OCR

In the experiments, $231,161$ characters are taken from "Fushimi-ya Docu-ments ()", and $3,509$ characters are selected as recognition targets. As n-gram information, 2-gram, and 3-gram are used. As learning data, $24,244$ characters of $4,795$ kinds are used from a dictionary of Japanese cursive forms. As a result, candidates containing correct answer are obtained for 81.93% of $3,509$ test data. The average ranking is 4.69, and probability of containing correct answers within the ranking 20 is 79.77%.

3.2 A Study for Character Recognition of Ancient Documents Using Neural Network

In this research, characters are recognized by learning feature values of input images using Neural-Networks [7]. The structure of a Neural-Network is three-layer including an input layer as shown in the middle of Fig.3.

Feature values extracted from input images are classified in two stages. In the first stage, an average vector of learning data is compared with the vector of input image, and input images are roughly classified by euclidean distance to reduce the number of kinds of characters input to the second stage. By using the first stage, probability of miss-recognition is reduced. The second stage receives reduced kinds of characters as an input, and feature values of input are classified by Neural-Networks. In the second stage, one Neural-Network is assigned to one kind of characters, and it fires when an assigned kind of character is a candidate of recognition. Thus, the result of recognition is determined to a kind of character assigned to a modular network whose output is the highest.

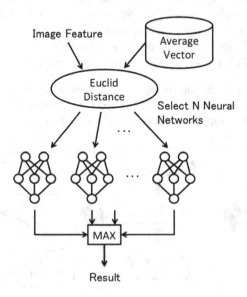

Fig. 3. Model of Neural-Networks

"Shu-mon-aratame-cho ()" is used for the experiments; 16 numerals in Chinese characters are used from the historical manuscripts images database. 200 samples of 15 kinds of characters are selected, 80 are used for learning, and resting 120 are used for recognition. As a result, rate for recognition is 97.05% in average.

3.3 A Constrained Approach to Multifont Chinese Character Recognition

In this research, a constraint-based approach to recognize multifont, multiple-size Chinese characters is proposed [4].

A class of images recognized as an identical character is described by a constraint graph model. A node and an edge of the graph correspond a sampling point on a character skeleton, which is an thinned image of the original character, and a constraint between nodes respectively. There are two kinds of constraints called connection constraints and position constraints. For two sampling points on a character skeleton, a connection constraint between them represents direct connectivity between the sampling points, and a position constraint is based on the distance between the sampling points. Satisfaction level of constraints is not boolean but numerical. These constraints for a class of images are obtained through a learning process.

Character recognition is formulated as an optimization problem. For a set of sampling points on a skeleton of an input image, the closest class of images is found by solving the optimization problem. A distance between a set of sampling points on a skeleton of an input image and a class of images is based on satisfaction level of constraints in the class. A heuristic solving method is given in the paper.

The authors implemented an OCR system with the recognition method. According to their assessments by many users for years, the recognition rate for multifont, multiple-size Chinese characters is between 95% and 99%.

4 Motivation and Configuration

4.1 Motivation

The existing researches described in the previous section recognize characters one-by-one. Even though n-gram is used, n is 2 or 3. Since characters form a word in a text, we think that using information on a sequence of characters is useful for getting generality of recognition. That is, recognition based on a sequence of characters means that information of other characters can be used with that of the character.

By taking into account a character sequence, we can handle an Odori-ji. Since an Odori-ji has the same reading as the previous one or two characters, recognition based on one character cannot handle it. Furthermore, by using this method, when one character can be determined, then characters having the same shape occurred at difference places can also be determined. Of course, these relationship among characters can be represented as *constraints*.

The aim of the system we propose is to assist a person on reprinting process. As described in [8], reprinting is a highly specialized task that requires knowledge on characters, language, grammar, history, etc. Even though general purpose full-automatic reprinting is quite difficult, we think that providing information to help reprinting process, such as candidates of reprinting is useful.

4.2   Configuration

Overall structure of our system is shown in Fig.4. In the figure, grey parts, *Constraint Solver* and *Word Dictionary* are newly added.

First, a feature value is extracted from an input image, then candidates are determined by using the *Image Feature Database* in the character recognition subsystem. Then, a *constraint satisfaction problem* is formed based on determined candidates. The *Constraint Solver* solves the constraint satisfaction problem by assigning possible reading using the *Word Dictionary*. Then the result of constraint solving is returned to the character recognition subsystem to revise the result of recognition. This feed-back is repeated if necessary.

Technical issues of the proposal is divided into two: one is those on image recognition, and the other is those on constraint solving. In the former issues, we have to think of extracting an image containing just one historical Hiragana from the original image of, for example, a page, and of the effective method to recognize an extracted historical Hiragana. In the latter issues, we have to think of the method of modeling of reprinting as a constraint satisfaction problem, and of efficient constraint solving. Our current main concern is on making constraint solving more efficient.

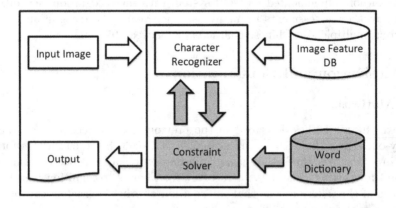

Fig. 4. System Configuration

5   Constraint Satisfaction Problem

5.1   Overview

A CSP for reading historical kana text, which we call a Reprint-CSP, consists of finite number of variables, their domains and constraints.

Each variable corresponds to a two-dimensional region on historical text which contains one character. Because of complex shapes of historical kana characters, the character recognizer may not be able to determine if a two-dimensional region corresponds to one character or two characters. In such a case, the character recognizer enumerates possible cases. As a result, the two-dimensional region of a variable may overlap the regions of different variables.

The domain of each variable is a finite set of possible characters for the two-dimensional region. In general, the character recognizer can not determine a unique character for a two-dimensional region of historical text.

A constraint is a relation over the domains. In Reprint-CSPs, some constraints are implicitly given by a pair of a dictionary of words and a directed acyclic graph (DAG) over nodes representing variables, and other constraints are explicitly given. The DAG represents reading order of variables. In general, the DAG consists of disjoint subgraphs. It is because reading order may stop at the end of sentences, paragraphs and so on. The dictionary constrains local character sequences and the DAG constrains combinations of the locally constrained character sequences. A path over the DAG determines a solution of a Reprint-CSP. Explicitly given constraints are, for example, used to declare constraints of Odori-ji. Satisfaction level of constrains is boolean.

Constraints are labeled with priority levels as not all constraints can be satisfied in general. Solutions are compared according to the priority levels. The solver finds maximally better solutions using a branch-and-bound method.

5.2   Definition of CSP

A constraint relevant to the variables x_1, ..., x_n, which is described as $C_{x_1,...,x_n}$, is defined as follows.

Definition 1. *For variables and their domains $x_1 \in D_1$ ..., $x_n \in D_n$, a constraint relevant to the variables x_1, ..., x_n is a subset of the product of their domains $D_1 \times ... \times D_n$.*

The following alphabet Σ is the universal set of variable domains.

Definition 2. *The alphabet Σ is a union of a set of all current hiraganas and a set of two special characters "?" and "!". The special character "?" denotes that a region on text recognized as one character does not correspond to any current hiragana. The special character "!" denotes that a recognition result of a region on text is incorrect.*

Each variable domain may include the two special charcters. For example, the character "?" is used for a region on text recognized as a Chinese character.

The character "!" will be assigned to variables whose domains do not include any appropriate current hiragana from view point of constraint satisfaction by a Reprint-CSP solver.

A pair of a dictionary and paths on a DAG over nodes representing variables determines a set of constraints. Words in the dictionary constrain local character sequences, and maximal paths on the DAG constrain combinations of the locally constrained character sequences.

Definition 3. *A* dictionary *is a finite subset of strings over the alphabet Σ, that is, a finite subset of $\bigcup_{i=1}^{\infty} \Sigma^i$.*

To deal with overconstraint situations, we use constraint hierarchies [2].

Definition 4. *A* strength *of a constraint is a priority level of the constraint. The priority levels are integers $0 \ldots m$. The smaller a priority level is, the priorer constraints with the level are. Constraints labeled with priority level 0 are required constraints which have to be satisfied. Constraints labeled with priority level $1 \ldots m$ are preferred constraints which only have to be satisfied as well as possible.*

Definition 5. *A set of constraints with strengths forms a* constraint hierarchy *$H = (H_0, H_1, \ldots, H_m)$ where H_i is a set of constraints labeled with priority level i.*

We define CSPs for reading historical kana text, which we call Reprint-CSPs, as follows.

Definition 6. *A* Reprint-CSP *consists of the following five components.*

- *a set of variables $V = \{x_1, \ldots, x_n\}$*
- *finite domains $D_1, \ldots, D_n$ where $x_1 \in D_1, \ldots, x_n \in D_n$*
- *a dictionary R*
- *a directed acyclic graph $G = (V, E)$ where $E \subset V \times V$*
- *explicitly-given constraints relevant to a subset of V*

For each maximal path over G, a constraint is implicitly given. Let p be a k-length maximal path $(x_{j_1}, \ldots, x_{j_k})$ over G. Then the implicitly-given constraint relevant to k-variables $x_{j_1}, \ldots, x_{j_k}$ is defined as follows.

$$C_{x_{j_1}, \ldots, x_{j_k}} = \{(s_1, \ldots, s_k) \in D_{j_1} \times \ldots \times D_{j_k} | \exists \alpha \in \Sigma^* \; \exists \beta \in \Sigma^* \; \alpha s_1 \ldots s_k \beta \in R^+\} \tag{1}$$

where $\Sigma^* = \bigcup_{i=0}^{\infty} \Sigma^i$ and $R^+ = \bigcup_{i=1}^{\infty} R^i$.

All of implicitly-given constraints are required constraints while explicitly-given constraints are either required or preferred.

We give some supplementary explanation about the definition of implicitly-given constraints. For each maximal path $p = (x_{j_1}, \ldots, x_{j_k})$ over G, we accept that the first variable x_{j_1} and the last variable x_{j_k} of p correspond to the middle characters of words. The strings α and β in the definition represent the head and the tail of the words respectively, which do not correspond to any variable sequence.

We use equality constraints and non-equality constraints as explicitly-given constraints. The equality constraints of the form $x_i = x_j$ are used to declare constraints of Odori-ji or constraints among characters with same shape. The non-equality constraints of the form $x_i \neq x_j$ are used to declare that each variable does not take the special character "!" as its value.

In the following, we define solutions of Reprint-CSPs. In general, solutions of CSPs are total functions, which maps variables to elements of their domains. Solutions of Reprint-CSPs are, however, partial functions. If there exist two variables whose regions on text overlap each other, we need not read both of them at the same time. In other words, if we have a solution which gives a value to one of such variables, we need not any value to another. For this reason, if a solution of a Reprint-CSP does not give any value to a variable, we regard constraints relevant to the variable as satisfied constraints.

Definition 7. *A solution of a Reprint-CSP with a set of variables V is a partial function that maps variables in V to elements of their domains. If a solution gives a value to a variable x, we say that the solution is relevant to the variable x.*

For a variable x of a Reprint-CSP and a solution θ relevant to x, $x\theta$ denotes the value which the solution θ gives to x.

For a solution θ and a constraint $C_{x_{j_1},\dots,x_{j_k}}$, $C_{x_{j_1},\dots,x_{j_k}}\,\theta$ denotes the boolean result of applying the solution to the constraint such that:

- $(x_{j_1}\theta, \dots, x_{j_k}\theta) \in C_{x_{j_1},\dots,x_{j_k}}$ *if the solution θ is relevant to all of the variables* $x_{j_1}, \dots, x_{j_k}$
- *true otherwise*

A solution θ satisfies a constraint $C_{x_{j_1},\dots,x_{j_k}}$ if $C_{x_{j_1},\dots,x_{j_k}}\,\theta$ is true.

Though an empty solution satisfies any constraint according to the definition above, it is not what we expect. We want to extract better solutions from admissible solutions defined below.

Definition 8. *Let G be the directed acyclic graph of a Reprint-CSP, which can be divided into disjoint subgraphs $G_1, \dots, G_m$ where a union of $G_1, \dots, G_m$ is equal to G and the undirected graph of G_i is connected for $1 \leq i \leq m$. A promising path of such Reprint-CSP is a path over G of the form $(p_1, \dots, p_m)$ where p_i is a maximal path over G_i for $1 \leq i \leq m$.*

Definition 9. *A promising solution of a Reprint-CSP is a solution of the CSP which gives values to variables represented by nodes in a promising path of the CSP.*

Definition 10. *A admissible solution of a Reprint-CSP is a promising solution of the CSP which satisfies all required constraints of the CSP.*

The better admissible solutions, which satisfy preferred constraints as well as possible, can be selected using solution comparators which determine partial orders over admissible solutions. Some solution comparators are introduced in [2]. One of the comparators is locally-predicate-better.

Definition 11. *For two admissible solutions θ, σ and a constraint hierarchy $H = (H_0, H_1, \ldots, H_m)$, a comparator locally-predicate-better(θ, σ, H) is defined as follows.*

$$locally\text{-}predicate\text{-}better(\theta, \sigma, H)$$
$$= \exists k > 0 \ such \ that \ \forall i \in \{1 \ldots k-1\} \ \forall p \in H_i \ e(p\theta) = e(p\sigma)$$
$$\wedge \ \exists q \in H_k \ e(q\theta) < e(q\sigma) \wedge \ \forall r \in H_k \ e(r\theta) \leq e(r\sigma)$$

The function e is an error function. For a constraint p and a solution θ, the function e is defined as follows.

- *$e(p\theta) = 0$ if the solution θ satisfies the constraint p.*
- *$e(p\theta) = 1$ otherwise.*

5.3 Solving CSP

We explain our solving method using an example. The solving method takes a Reprint-CSP as its input, and outputs a set of better admissible solutions.

We use a Reprint-CSP for another historical Japanese text of Fig.5, taken from *The tale of Ise* [1], as an example. It reads (*shi*) (*no*) (*hu*) (*su*) (*ri*) (*no*) (*ka*) (*ri*) (*ki*) (*nu*) (*wo*). For example, some characters can be recognized by our character recognizer as follows.

- The first character (*shi*) can be recognized as (*shi*), (*chi-tsu*), or (*chi-he*).
- The 4th character (*su*) can be recognized as (*su*), (*su-ro*), or (*su-tsu-tsu*).
- The last character (*wo*) can be recognized as (*wo*) or (*chi-to*).

Fig.6 shows a DAG representing the Reprint-CSP. Each node of the graph represent a pair of a variable and its domain. Directed edges of the graph represent reading order of the variables. Twenty-two variables $x_1, \ldots, x_{22}$ are used in the Reprint-CSP. For each variable x_i ($1 \leq i \leq 22$), a constraint $x_i \neq$ "!" is imposed. They, however, are omitted in Fig.6. The recognitions of the first, the 4th, and the last characters cause branches in the DAG.

Our solving method consists of the following three steps.

Step 1. We construct a reading-assignment graph using the dictionary R of a given Reprint-CSP. The graph is a DAG such that:
   - each node of the DAG is a pair of a sequence of variables and values to be assigned to the variables

Fig. 5. Another Historical Japanese Text

– solutions derived from maximal paths over the DAG satisfy implicitly-
    given constraints of the Reprint-CSP
  Fig.7 shows a reading-assignment graph constructed from the DAG of Fig.6.
Step 2. We simplify the reading-assignment graph constructed at the step 1
    using constraints to reduce both the search space and the depth of recursive
    calls at the step 3. Fig.8 shows the simplified DAG obtained from Fig.7.
Step 3. We search a set of better admissible solutions by branch-and-bound.
    We enumerates admissible paths over the simplified DAG visiting from the
    most upstream nodes to the most downstream nodes by depth first. If we find
    worse promising path than previously found admissible path, we abandon
    the promising path. Fig.9 shows DAGs representing two optimal solutions,
    which are two maximal path without "!" in Fig.8.

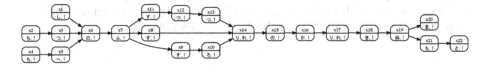

Fig. 6. A directed acyclic graph representing a Reprint-CSP

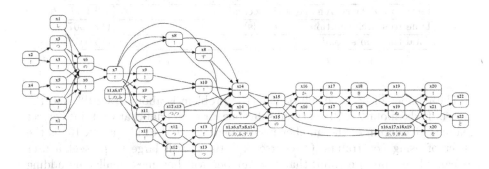

Fig. 7. An initial reading-assignment graph

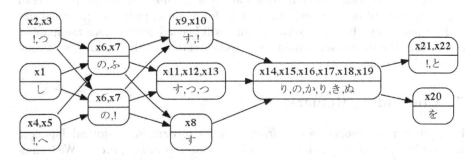

Fig. 8. A simplified reading-assignment

Fig. 9. Two optimal solutions

6   Experimental Results

A constraint satisfaction problem based on the result of character recognition produced manually is solved by a constraint solver we experimentally implemented. The problem has 108 constraints. A constraint graph of the problem consists of 100 nodes and 127 paths. For the experiments, we prepare a dictionary having 241 necessary words. Comparator is locally-predicate-better. Constraint solving of the problem is repeated 10 times, and the average is calculated by elimination a trial with maximum and minimum execution time on a computer having Intel Core i7 as a CPU of 2.7GHz with 4GB Memory.

Table 1. Time for executions

The number of assigned readings	2048
Time to load the dictionary and assign readings (in seconds)	0.5912
Time to reduce search space (in seconds)	2.1996
Time to search solution (in seconds)	112.8507
Total time (in seconds)	115.6419

The number of possible solutions of the problem is about 7×10^{10}, and the system finds about 2×10^3 optimal solutions for each trial that contains correct reading. Reduction of the number from about 7×10^{10} to about 2×10^3 is the effect of using constraints. Of course, 2×10^3 is still large to present a user as candidates, but we think that the number will becomes smaller by adding constraints of grammar, and conditions of pruning.

Table 1 shows that searching takes about 2 minutes, about 98% of overall execution time. When we apply the system for manuscript having practical size and a vocabulary, then the execution time will become longer since the number of words in the dictionary should be increased. Thus, further reduction of time for searching is necessary.

7   Concluding Remarks

In this paper, we propose a new framework for recognizing historical Japanese manuscript for realizing a system to assist human reprinting process. We formulate reading and reprinting historical Japanese manuscripts as a combination of a single character recognition and constraint solving. The single character recognizer constructs a constraint satisfaction problem (CSP) according to recognition results. The constraint solver then solves the CSP and send the solution back to the recognizer. After that, the recognizer revises the CSP. The system repeats this cylce until it finds a satisfactory solution. We define constraints on sequences of characters as a word dictionary for historical Japanese, since all historical Japanese texts should consist of valid historical Japanese words.

Through preliminary experiments, the number of candidates for character recognition is reduced by introducing constraints: for a manuscript consist of about 100 characters, 7×10^{10} candidates obtained in the single character recognition are reduced to about 2×10^3.

Our future works about constraint solving are as follows. Reduction of the number of optimal solutions is our future work because the number of optimal solutions obtained in our experiment is not a small number. Use of another solution comparator and addition of more constraints will reduce the number of optimal solutions, but they should preserve correct solutions. The reduction will also contribute to speed-up of constraint solving. Instead of implementing several ideas to our solver, it may resolve both of the reduction and the speed-up to solve a SAT problem or a MAX-CSP translated from a given Reprint-CSP by a state-of-the-art SAT or MAX-CSP solver [5,6]. If we adopt the strategy, we also have to consider computational cost of the translation.

At this moment, the character recognition subsystem is not yet combined with the constraint solving subsystem, but the result of the preliminary experiment shows the effectiveness of our approach for realizing a system to assist human reprinting process.

When we combine the character recognition subsystem and the constraint solving subsystem, we have to consider feedback from the latter to the former, which suggests how to correct the result of character recognition, and computational cost of the entire system. These are also our future works.

Thus most important feature of our approach is that the overall quality of reprinting is compared to the system using just character recognition. If we use the best character recognition in our framework, then the quality of result will be better.

To conclude, our framework will be quite useful for assisting human process on reprinting in the near future, and will be a driving force for researching on Japanese classical literature.

References

1. Tales of Ise (photocopy). Kasama Shoin (1994-4-30)
2. Borning, A., Feldman-Benson, B., Wilson, M.: Constraint hierarchies. In: Lisp and Symbolic Computation, pp. 48–60 (1992)
3. Frellesving, B.: A History of the Japanese Language. Cambridge University Press (2010)
4. Huang, X., Gu, J., Wu, Y.: A constrained approach to multifont chinese character recognition. IEEE Transactions on Pattern Analysis and Machine Intelligence 15, 838–843 (1993)

5. Simon, L.: Glucose's home page (2012),
   http://www.lri.fr/~simon/?page=glucose
6. Tamura, N.: Sugar: A sat-based constraint solver (2012),
   http://bach.istc.kobe-u.ac.jp/sugar/
7. Waizumi, Y., Kato, N., Nemoto, Y., Yamada, S., Shibayama, M., Kawaguchi, H.:
   A study for character recognitoin of ancient documents using neural network. IPSJ
   Reports. IPSJ SIG Computers and the Humanities Reports 2000(8), 9–15 (January
   21, 2000), http://ci.nii.ac.jp/naid/110002930070/
8. Yamada, S., Shibayama, M.: An estimation method of unreadable historical
   character for manuscripts in fixed forms using n-gram and OCR. IPSJ Reports. IPSJ SIG Computers and the Humanities Reports 2003(59), 17–24 (2003),
   http://ci.nii.ac.jp/naid/110002911078/

PedInfoSys: An OpenMRS Based Pediatric Information System

Ryan Steven D. Caro, Dinah Marie A. Maghuyop, Ingrid R. Silapan,
and Rommel P. Feria

Web Science Group, Department of Computer Science,
University of the Philippines, Diliman
{ryansteven.caro,dinahmaghuyop,ingridsilapan}@gmail.com,
rpferia@dcs.upd.edu.ph

Abstract. Since the emergence of electronic medical records (EMRs) several implementations have been faced with problems such as technical issues, high implementation costs and usability issues. In the case of specialized medicine, healthcare institutions are finding out that the one-size-fits-all approach to EMR deployment does not work. In the field of pediatrics, studies show that general EMRs have limited usefulness in pediatrics because they are generally designed for adult care.

In previous years, several EMRs, Ped-Files, PEDiA and PedSync included, have been developed with the aim of providing an EMR designed to address the specialized needs of pediatrics, which conforms to the standards set by the Philippine Pediatric Society. However these EMRs have encountered problems with database management and synchronization issues.

With the aim of addressing the issues encountered, PedInfoSys, an OpenMRS-based Pediatric Information System, was developed. It is a web-based medical records system built on OpenMRS, which is an open source EMR application that provides the 'building block' to the creation of customized EMRs. The use of OpenMRS has provided advantages over previous systems like PEDiA and PedSync. These include the elimination of complications presented by having different databases, the ease of database communication through simple calls, the restriction of invalid data to maintain data integrity and the use of standards like HL7 to promote reusability and interoperability. More importantly, it has a highly scalable model and uses an approach that allows ease of use and flexibility with regards to the users' needs.

This paper aims to present the system design modifications to create PedInfoSys, and its capabilities in addressing the shortcomings of previous systems. It also analyzes the issues encountered with the implementation of the application and presents recommendations regarding future improvements.

Keywords: PedInfoSys, Medical Records, Information Systems, EMR, Philippine Pediatric Society.

1 Introduction

Medical records are important documents which include patients' personal information, demographics, medical history, prescriptions, and other important

S. Nishizaki et al. (Eds.): WCTP 2012, PICT 7, pp. 165–177, 2013.

information primarily meant for planning patient care and providing for continuity in information about a patient's medical treatment [1]. Information found in medical records is considered crucial and must therefore be complete and accurate to prevent medical errors, ensuring patient safety [2]. Studies show that paper-based records are susceptible to data inaccuracy, loss and fragmentation. Since these records are paper-based, medical records tend to be bulky, leading to lack of overview and poor data mobility. These identified weaknesses of paper records could impede continuity and quality of healthcare [3]. As a solution to this problem, a system that allows computerized storage, retrieval, and modification of medical information was implemented. This system is more commonly referred to as the electronic medical record (EMR) system and has been rapidly growing in Europe and the US in recent years. This has been driven by the belief that EMRs promise significant advances in the quality of patient care by enhancing readability, availability and data quality. Providing potential for automating, structuring and streamlining clinical workflow, EMRs are becoming important tools in reducing medical errors, improving patients' health data management and reducing paper usages [4].

1.1   OpenMRS

In 2004, the Regenstrief Institute, Inc., a world-renowned leader in medical informatics research together with Partners in Health (PIH) recognized the need for basic clinical data management as a response to pandemics such as HIV/AIDS and TB in developing countries [5]. Their attempt to create a foundation for a collaborative EMR development within these countries paved the way to the creation of a general purpose EMR system: the OpenMRS. OpenMRS is an open source EMR application which provides the 'building block' to the creation of customized EMRs to meet specific needs. More than just software or data model, OpenMRS is a community. With a global network of software designers and developers, the OpenMRS community has ensured that their forms reflect clinical best practices, international standards, and research. It is now in use in clinics in Argentina, Botswana, Cambodia, Congo, Ethiopia, Gabon, Ghana, Haiti, Honduras, India, Indonesia, Kenya, Lesotho, Malawi, Malaysia, Mali, Mozambique, Nepal, Nicaragua, Nigeria, Pakistan, Peru, Philippines, Rwanda, Senegal, South Africa, Sri Lanka, Tanzania, The Gambia, Uganda, United States, Zanzibar, Zimbabwe, and many other places [6]. However, despite the success of OpenMRS implementations in several countries, computerized information systems have not achieved the same degree of market penetration in healthcare as that seen in other sectors such as finance, transport, and the manufacturing and retail industries [8]. Widespread implementation of EMRs has been hampered by perceived barriers such as technical issues, high cost, culture change, and usability issues which lead to loss of productivity and steep learning curves. [9]. Based on a study on the use of EMRs in urban primary care in Boston, however, appropriately designed systems, attributed to its presentation of structured elements to enhance completeness, will be accepted and can improve quality in healthcare [10].

2   Related Work

There are hundreds of EMR systems available today. There are customized systems for outpatient care, inpatient care, solo practices catering to a wide range of specialties, each with its own set of unique features as well as its own limitations.

2.1   Top EMRs of 2012

According to the 2012 Black Book Rankings Survey on Top Ambulatory Electronic Medical Records Vendors by Brown-Wilson Group [11] conducted on 12,000 EMR users, the Top 2 EMR Vendors for Family Practice, General Practice, Pediatrics and Geriatrics are Practice Fusion and Greenway Medical.

Practice Fusion. Practice Fusion is a free web-based EMR based in San Francisco, California. Its features include prescriptions, billing, charting, scheduling, lab integrations and patient records. Based on the Black Book Rankings survey on client satisfaction on EMRs, Practice Fusion ranked 1st overall with a mean score of 9.42, and also ranked 1st across all primary care specialties [12]. It is the largest and fastest growing EMR community in the US with over 130,000 doctors to track records for 25 million patients [13].

Practice Fusion offers a hosted EMR service, which requires users to have an internet connection to be able to access the EMR. Since the EMRs are hosted in Practice Fusions data centers, it does not provide the users with the option to set-up the EMR locally. This may pose a problem for practices who wish to set-up their EMRs in a clinic's local area network or in a single machine. Also, Practice Fusion has yet to release its mobile EMR.

PrimeSUITE: Greenway EMR. PrimeSUITE is a .NET-based EMR system by Greenway Medical Technologies Inc., based in Carrolton, Georgia. Its features include patient records, prescriptions, scheduling, messaging, speech understanding and many more. PrimeSUITE also has its mobile counterpart, PrimeMOBILE. It currently provides for more than 33,000 healthcare providers in the US, in 30 specialties and subspecialties [14].

However, PrimeSUITE is not cross-platform as it only runs on Windows and requires the use of Internet Explorer 8 or earlier versions. Having a cross-platform application poses several advantages, such as flexibility, as it allows the use of different kinds of resources and a choice as to which platform the user prefers. The ability to run on Linux or other open source platforms can decrease costs.

2.2   CHITS

In the Philippines, the University of the Philippines Manila National Telehealth Center (UPM-NThC) developed the Community Health Information Tracking System (CHITS) as a low cost computerization initiative for local health centers [15]. It has been reported that with the use of CHITS, patient care has improved and patient visits are more efficient, having reduced the time needed to search paper records. As of June 2011, CHITS has been deployed in 50 rural health units. However, like most EMR implementations, several challenges were also faced in its implementation. Training health workers in using CHITS was one of the biggest challenges faced. Most participants were not computer-literate, therefore requiring a basic computer course training before the actual training for CHITS. Lack of funding and resources also became a problem [16].

According to a study on CHITS' database design, there are still several issues and flaws affecting the overall performance of the application. Although designed as an open source project, the absence of proper documentation and a flawed database design make CHITS difficult to maintain and extend [17], which makes it unlikely that CHITS can be easily extended to EMRs for use in other healthcare units such as clinics and for specialized fields such as pediatrics.

2.3   EMR Implementations Using OpenMRS

There are currently 9 reported OpenMRS implementations in locations such as Kenya, Rwanda, Lesotho, Tanzania, Malawi and Cape Town. However, there are currently more than a hundred clinical and research sites for OpenMRS all over the world, including the Philippines.

OpenMRS AMPATH Project (AMRS). Founded in 2000 as a response to the HIV epidemic in western Kenya, AMPATH is one of the largest HIV treatment programs in sub-Saharan Africa. With a catchment area that has over 2 million people and provides care to more than 130,000 HIV-positive patients across 55 urban and rural clinics, AMPATH uses OpenMRS and Open Data Kit (ODK) to provide care at this scale. Customized OpenMRS modules have been developed to meet the needs of the AMPATH project. AMPATH has an extensive home-based and counseling program where community health workers go house to house to identify and enroll persons in need of care (i.e., pregnant women not in antenatal care, orphaned children, and persons at high risk for tuberculosis infection). In their need for mobile data collection, AMPATH used devices such as Palm TX, eTrex, and in 2009, Android devices such as the HTC Dream [18].

Millennium Villages Project. The Millennium Villages Project (MVP) is a project of the Earth Institute at Columbia University, the United Nations Development Programme, and Millennium Promise which is an approach to meeting the Millennium Development Goals which address eight globally endorsed targets which include healthcare. This project focuses on community health workers delivering basic healthcare services to strengthen health systems. With around 90% of community health workers staying in these rural areas, this had led to a dramatic reduction in conditions such as malaria prevalence, malnutrition, and improvements in infant vaccination programs. The Millennium Global Village-Net (MGV-Net), the EMR developed for MVP runs on OpenMRS. The MGV-Net infrastructure was based on the AMRS infrastructure and is currently in use in two Millennium Villages Project sites; Sauri in Kenya and Mayange Rwanda [19].

OpenMRS in Capiz, Philippines. OpenMRS is currently in use in 2 hospitals in Capiz, Philippines; the Mambusao General Hospital and Bailan District Hospital. The implementation was assisted by the University of the Philippines Manila National Telehealth Center [7]. In an OpenMRS Spotlight interview with Jonathan Galingan, who was the project head for the OpenMRS implementations in the Capiz Hospitals, he mentioned that several custom modules have been developed and modified for the project, including a module which will enable the system to connect with Philhealth, the national health insurance program of the Philippines. The systems in these hospitals are said to be in use in emergency rooms, laboratories and out-patient departments for patient registry, lab results, etc.

These implementations of OpenMRS prove that it is indeed scalable as it is in use for more than 100,000 patient records in the AMPATH and Millennium Villages projects. The variation in the setting on which OpenMRS was used, HIV treatment for AMPATH, community health units for Millennium Villages, and hospitals in Capiz, Philippines, show the flexibility of OpenMRS to be implemented for use in different healthcare environments.

2.4   Mobile EMRs

The increasing popularity of the use of mobile technologies in health services and information has paved the way to the emergence of Mobile Health (mHealth). In a survey conducted in 2011 by CompTIA, an information technology association, 56% of the surveyed physicians are using smartphones and 25% are using tablets for work [20]. With this, several mobile EMR apps for tablet devices and phones have been developed and are currently available in the market. Based on the 2012 Survey on Healthcare Mobility Trends by Aruba Networks, EMR is the most commonly supported mobile application [21].

Epic Canto. Canto provides authorized clinical users of Epic's Electronic Health Record with secure access to clinic schedules, hospital patient lists, health summaries, test results and notes. Canto mobile app is free and works on iPads running iOS 4.2 or greater [22].

This system relies access to Epic's Electronic Health Record and the type of users that could access it is limited. This poses some advantages and drawbacks. This approach allows maintenance of a high level of security but would hinder interoperability. Although the iPad app is free, a license from Canto is needed and one must be on Epic's Summer 2009 version which will determine the feature set to be provided corresponding to some amount of charges.

drchrono. drchrono is certified as a complete EHR (Electronic Health Record) in accordance with ONC-ATCB stage 1 meaningful use criteria. Its basic features include the creation of records pertaining to patient's history, drugs taken, physical exams, allergies and condition notes. It also has customizable clinical notes, clinical forms, medical billing, and drug prescriptions. drchrono is also compatible with the iPad running an OS of version 4.2 or later [23].

Use of the SOAP standard provides a way to communicated between applications of different operating systems, with different technologies and programming languages. Use of additional multimedia features such as speech to text, photo and video taking, as well as chat and messaging provides convenience in documentation and communication. Its many features allow it to accomplish many tasks at once. The app is offered for free, but this version limits storage size, number of patient records and staff logins, with only one iPad that can have EHR access. Its additional features such as prescription, speech to text, uploading of demographics, billing and many more are available only in the paid version, with features depending on the pricing scheme chosen.

OfficeEMR. OfficeEMR Mobile features include Rounds and Remote Scheduling, Patient Medical Records: medications, known allergies, active problems, past progress notes, recent vitals, immunizations, family and social history; e-prescribing with contraindications alerts; charging and billing; tasks lists; office communications, patient demographics: patient contact information, insurance files, and responsible party information; mobile imaging for patient's record; practice management [24].

OfficeEMR not only allows the management of patients but the management of its staff as well. It is also available for different iOS devices ranging from the iPhone, iPod touch and the iPad, ensuring portability. However, like Epic's Canto, it relies on a web-based electronic medical record system. Although OfficeEMR Mobile is free of charge, the acquisition of the needed web EMR called iSALUS OfficeEMR comes at a cost.

Application downloads for Epic Canto, drchrono and OfficeEMR are free but they all rely on a web-based electronic medical record that may come with a certain price. However, it is undeniable that its mobility and the provision of many essential functions is an edge to providing usability and assistance.

3   PedInfoSys

There are a lot of EMR systems available in the market. However, there are certain functions not available in those systems that are critical for the care of infants and children. Such functions include immunization management, growth tracking, medication dosing, etc.

Recently, two projects namely PEDiA, and its mobile extension for mobile devices named PedSync, were developed to serve the specialized needs pediatricians while conforming to the standards set by the Philippine Pediatric Society [27]. The clinical workflow and drug database used in PEDiA and PedSync became the basis for the Pediatric Information System, or PedInfoSys.

3.1   System Architecture

PedInfoSys is a web-based application implemented using the OpenMRS standalone version 1.8.1 The OpenMRS standalone version comes with both an embedded database and webserver that allows a local copy of OpenMRS to be running within minutes. OpenMRS supports Linux, Mac OS X and Windows with Java 6+ installed.

In the Application Layer, users are able to perform tasks using two interfaces, one for desktops and one for mobile tablet devices. The desktop interface was developed using CodeIgniter, an open source framework for use in building web applications with PHP. Meanwhile, the mobile interface was developed using Sencha Touch (version 1.1), a mobile application framework based on HTML5, CSS3 and Javascript.

The Application layer communicates with the OpenMRS API in the Service Layer using Representational State Transfer (REST) web service calls. REST is an architectural style where web services are viewed as resources and can be identified through URLs. Clients specify methods, such as GET and POST, to describe what action is to be done on the resource [28]. This communication mechanism underlying each user task was developed following the format specified by the OpenMRS Rest Webservices Technical Documentation.

The Open MRS API then uses Hibernate to retrieve or post information to the back end database, which is in MySQL, then sends back the necessary data in XML form to the Application Layer. A process was developed to decode the XML and present the data properly to the user.

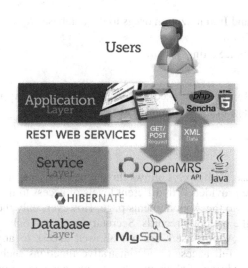

Fig. 1. PedInfoSys System Architecture

3.2   Main Features

The main feature of PedInfoSys is Patient Records, which holds all the patient's medical records such as:

- Recent Medical Cases which can be associated with Clinical Examinations, Prescriptions and Lab Exams
- Clinical Examinations which include Vital Signs, Physical Examination, and Anthropometric Measurements

- Laboratory Examinations
- Past Medical History such as Allergies, Past Illnesses, Family Illnesses

PedInfoSys also features reports generation for:

- Prescriptions
- Medical Certificates
- Doctor Referrals
- Lab Requests

To address the specific needs of Pediatricians, PedInfoSys has additional features such as:

- Immunization Records
- Developmental Milestones Tracking
- Growth Charts for weight, length and BMI, based on information supplied by the Philippine Pediatric Society and the World Health Organization

Using the administrative features of PedInfoSys, users are also able to:

- Add Prescription and Immunization drugs to the database
- Manage Clinical Profile
- Change User Account Settings
- Manage User Access

Designed especially for clinic use, PedInfoSys also features the Appointments module where users can manage clinical appointments.

3.3   Users

Designed for clinic use, PedInfoSys allows three types of users, the Doctor, Secretary and Admin which are provided with different levels of access to the system. All three users are given access to the Appointments page. However, only the Doctor users are given full access to the patient's records. Secretary users can only access reports and personal information. Admin users are not allowed access to any of the patient's records, but are given full access to administrative functions such as the adding of drugs to the database, editing of clinic profile, and management of user accounts.

3.4   Workflow

Designed for pediatric clinics, the workflow of PedInfoSys is based on a standard clinical visit (see Fig. 2).

3.5   PedInfoSys Mobile

PedInfoSys features a mobile component, which is a web-based application developed using Sencha Touch. It supports almost all the features of its desktop counterpart,

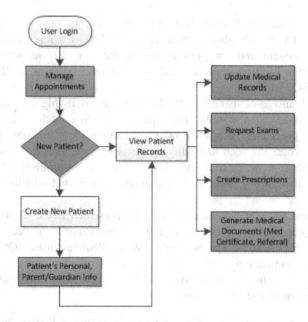

Fig. 2. PedInfoSys Clinical Workflow

with exception to the graphs and reports generation features. It is also accessible by only one type of user, the Doctor user. Its design is optimized for use in tablet devices running on Android or the iPad.

As of the moment, access to the mobile application requires the mobile device to be constantly connected to the system via internet or LAN for local set-ups. The offline feature, which will allow the doctor to add or update records without being connected to the system, is yet to be implemented in PedInfoSys.

4   Discussion

The workflow and basic functions of PedInfoSys were patterned after its predecessors, PEDiA and PedSync. But additional features were developed to improve the previous systems. Among these include the monthly, weekly, daily and agenda views for Appointments, the ability to print reports like Medical Certificates, Prescriptions, Lab Requests and Doctor's Referral and the ease of managing Developmental Milestones Records through dragging and dropping. Moreover, what sets PedInfoSys apart from its predecessors is its use of the open-source EMR platform called OpenMRS.

OpenMRS features an API, which enables the developer to communicate with the database layer through simple calls without having to understand its complex data model. It is tightly constrained to guard against invalid data to maintain data integrity. OpenMRS conforms to the Health Level Seven International (HL7) representations of observations, encounters, etc. Its use of HL7 as the primary mode of transmitting data between the external application and the repository, promotes re-usability and interoperability. OpenMRS also relies heavily on XML for representation of data outside of the system and for presentation of data. It also supports and stores mappings between local concepts and existing standards, such as LOINC, ICD-10, SNOMED, and CPT [25].

The main advantage of using OpenMRS, however, is its highly scalable design. OpenMRS is based on a concept dictionary-based data model which stores all diagnosis, tests, procedures, drugs and other general questions and potential answers. This approach avoids the need to modify the database structure to add new diseases, making the system more flexible and easy to use for customized EMRs such as PedInfoSys. It also allows easier analysis, summarization and addition of new data items.

Another advantage that the use of OpenMRS presented is the elimination of the complications presented by the difference in the databases of its predecessors, PEDiA and PedSync. PEDiA uses Apache Derby, while PedSync uses SQLite 3 which recognizes different data types, which may consequently lead to problems.

In the implementation of OpenMRS in the development of PedInfoSys, the developers encountered several issues:

1. In order for a user to be able to access a patient's records, a Patient Query based on a search string must be performed. A list of patients matching the search criteria would then be returned. This, however, meant that OpenMRS does not allow the user to retrieve a full list of patients.
2. Similar to the issue on the list of patients, OpenMRS also does not allow the retrieval of the list of concepts. This particular limitation became an issue in the Prescription and Immunizations features of PedInfoSys. Since the list of prescriptions and immunizations are stored as concepts, the user cannot gain access to the full list of drugs in the system. The user is limited to doing a query on the list of concepts, returning the list of concepts matching the search string.
3. In accessing the list of encounters, OpenMRS does not allow a query. This limitation greatly affected the speed in the processing of encounters, as the system would have to constantly retrieve the full list of a patient's encounters and filter the list for the needed encounter types. The longer loading times for the patient's records pages is attributed to this limitation.
4. OpenMRS initially provides a list of basic Person Attribute Types. However, additional attributes were needed in PedInfoSys, which could be added using the built-in OpenMRS web application. However, the UUID (Universally Unique Identifier) of the newly added attribute types aren't shown. Since PedInfoSys uses REST Web Services, which is greatly reliant on UUIDs, this posed a big problem in adding patient's information to the system.

Given these limitations of OpenMRS, several workarounds had to be done. For the issues on the retrieval of the full list of patients and concepts, the system had to make do with performing queries. For the issue on additional attribute types, the developers had to manually search through a patient's XML data in order to retrieve the needed UUID.

On the use of the OpenMRS REST Web Services, the developers also encountered a problem on how the data, that was to be sent through REST Web Service calls, should be formatted. No complete documentation regarding this was available, and the developers had to figure out the correct format through trial and error.

The developers encountered problems in the implementation of the offline feature using the HTML5 offline database. This is because the data stored in the HTML5 offline database is decoded by the PHP code in both desktop and mobile applications. The PHP code will then send the decoded data to be loaded onto the HTML files. Without an active connection, the data stored in the offline database cannot be decoded by the PHP code, preventing the HTML files from being loaded onto the application.

5   Conclusion and Recommendation

The use of OpenMRS has provided advantages over previous systems like PEDiA and PedSync. These include the elimination of complications presented by having different databases, the ease of database communication through simple calls, and the restriction of invalid data to maintain data integrity among others. More importantly, OpenMRS has a highly scalable model and uses an approach that allows ease of use and flexibility making it easier for developers to use the data model for customized EMRs.

However, a few drawbacks were encountered like the limitation of showing complete a complete list of patients and concepts which are attributed to the limited web service calls supported by the REST Web Services API. Also, in the processing of encounters which affect process speed, the developers recommend the search for a workaround which will store all patients' encounters once, and perform the call for the full list of encounters only when new encounters have been added. This workaround might increase processing speed significantly.

For the problem encountered with the UUIDs of newly added attributes, the developers reported the issue to the OpenMRS Developers Community. An OpenMRS developer suggested the creation of a ticket addressing the need for this feature. With this, the developers recommend the reporting of all encountered bugs and needed features to the online OpenMRS Community to help improve OpenMRS as well as future developers who might need the same features.

The developers also recommend participation in the development of the modules needed. Pertaining to the improvements with the overall functionality of PedInfoSys, future developers must continuously consult with pediatricians to determine what functions are needed in their practice.

References

1. Jethani, J.: Medical Records - Its Importance and the Relevant Law. AECS Illumination 4(1), 10–12 (2004)
2. Stimpfel, N.: Quality Medical Charts: The Importance of Proper Medical Record Documentation. TransforMed (2007)

3. Roukema, J., et al.: Paper Versus Computer: Feasibility of an Electronic Medical Record in General Pediatrics. Pediatrics 117(1), 15–21 (2006)
4. Hsieh, S., Hou, I., Cheng, et al.: Design and Implementation of Web-Based Mobile Electronic Medication Administration Record. Journal of Medical Systems 34, 947–958 (2009)
5. Fraser, H., Choi, S., Galipot, M., et al.: Successful transfer of a Web-based TB medical record from Peru to the Philippines. In: AMIA Annual Symposium Proceedings (2006)
6. Floss Manuals: OpenMRS, OpenMRS Guide, http://en.flossmanuals.net/openmrs-guide/ (accessed April 4, 2012)
7. Firaza, P.: Capiz Province Embraces Hospital Informatics, Rolls out Electronic Medical Records. University of the Philippines - Manila, National Telehealth Center (2011), http://bit.ly/Hrb912 (accessed April 6, 2012)
8. Open Clinical. OpenClinical: Knowledge Management Technologies for Healthcare, http://www.openclinical.org/publicreportsEMR.html (accessed April 3, 2012)
9. Smelcer, J.B., Miller-Jacobs, H., Kantrovich, L.: Usability of Electronic Medical Records. Journal of Usability Studies 4(2), 70–84 (2009)
10. Adams, W., Mann, A., Bauchner, H.: Use of an Electronic Medical Record Improves the Quality of Urban Pediatric Primary Care. Pediatrics 111, 626 (2003)
11. Black Book Rankings, Survey Findings of Top EHR Vendors. Primary Care Physician Practices, http://bit.ly/IcQYqw (accessed April 5, 2012)
12. Practice Fusion, Top EMR for Physician Satisfaction, http://bit.ly/IaypAn (accessed April 4, 2012)
13. Constine, J.: Practice Fusion, # 1 in EMR With 25M Electronic Medical Records, Debuts iPad App. (2011), http://tcrn.ch/HiqEbx (accessed April 4, 2012)
14. Yahoo! Finance, Healthcare Systems Nationwide Select Greenway's PrimeSUITE EHR, http://yhoo.it/I0PCgK (accessed April 1, 2012)
15. Health Market Innovation, Community Health Information Tracking System (CHITS) (2012), http://bit.ly/HrbfJl
16. Fernandez-Mark, P., Sy, M., Bañez, N.: Case Study: Community Health Information Tracking System (CHITS). Center for Health Market Innovation (CHMI) (2010)
17. Domingo, C., Ramirez, N., Feria, R.: Evaluationof CHITS' Database. In: Proceedings of IADIS International Conference e-Health 2011. IADIS Press, Rome (2011)
18. Open Data Kit, AMPATH Improving Care At Scale With ODK and OpenMRS, http://bit.ly/HrrHJ6 (accessed April 2, 2012)
19. World Health Organization, Community Health Workers at the Millennium Villages Project increase access to the health workforce, http://bit.ly/HrrIN7 (accessed April 2, 2012)
20. Dolan, P.: Use of mobile devices by physicians influences EMR purchases. American Medical News, http://bit.ly/I5lJzq (accessed April 3, 2012)
21. Hernandez, P.: Survey: EMR most commonly supported mobile app in hospitals. Aruba Networks 2012 Healthcare Mobility Trends Survey, http://bit.ly/HirIw4 (accessed April 3, 2012)
22. Epic Canto, Epic Canto for iPad in iTunes (2011), http://bit.ly/Ikv8i0 (accessed April 8, 2012)
23. DrChrono, DrChrono for iPad in iTunes, http://bit.ly/Ifh3aq (accessed April 8, 2012)

24. OfficeEMR, OfficeEMR for iPad in iTunes (2011), http://bit.ly/I6bgQW (accessed April 8, 2012)
25. Chavis, S.: Pediatric EMRs: Big Considerations for Small Patients. For the Record 20(2), 14 (2008)
26. Spooner, S.A.: Special Requirements of Electronic Health Record Systems in Pediatrics. Pediatrics 119(3), 631–637 (2007)
27. Mendoza, C., Pajarito, K., Valentino, M.: PEDiA: Streamlining Pediatric Healthcare Delivery Using an End-to-End Mobile and Desktop Information System. University of the Philippines - Diliman (unpublished)
28. Tyagi, S.: RESTful Web Services (2006), http://bit.ly/Rz5bDB (accessed September 1, 2012)

Correlation of Stress Inducers and Physiological Signals with Continuous Stress Annotations

Gemilene Uy[1], Charlene Frances Ngo[1], Rhia Trogo[1], Roberto Legaspi[2], and Merlin Suarez[1]

[1] De La Salle University, Manila, Philippines
[2] Osaka University, Osaka, Japan

Abstract. This study proposes a methodology in building a multimodal stress-level model using different non-invasive physiological signals: Galvanic Skin Response (GSR), Blood Volume Pulse (BVP), and Respiratory Variability (Resp). Paced Stroop, Mental Math, and Game were used to induce stress to 4 subjects. A fixed window size of 5 seconds and 1 second sliding size were used to segment the self-annotated data. Six statistical features were extracted from each physiological signals, giving a total of 18 features. Different classification and regression algorithms were used to get the correlation of the features and the stress labels. The results show very high correlation for all three stress inducers with Paced Stroop Test as the highest which in term means that all 3 activities are effective in inducing stress. Results also show a good correlation for the 3 physiological signals with Respiration as the highest among the 2 others, which indicate that Respiration provides better correlation even by just using low-level features.

1 Introduction

Stress is commonly known or defined as a change in a person's state from calm to excited. It can also inhibit personal happiness and productivity [1]. Stress is not just one dimensional rather it encompasses different levels and has distinctions. Studies have shown two distinct stress characteristic namely eustress and distress. Eustress is known as good stress where the person perceive stress mixed with feelings of joy or any emotional state that benefits the person in a positive way. Distress is the opposite of eustress wherein the type of stress that the person is feeling is harmful to him/her and can even impair performance [2].

There are a few stress detection systems that have already been created. One such research made use of the Paced-Stroop Test otherwise known as the color matching test to induce stress. This test requires the test subject to match color words to its font color and not what the word actually says. The physiological signals that were gathered were galvanic skin response (GSR), blood volume pulse (BVP), pupil diameter (PD) and skin temperature (ST). The classification labels used are stressed and relaxed and the paper's conclusion is that the four aforementioned physiological signals are relevant in determining a person's level of stress, however pupil diameter gave the highest accuracy[3].

S. Nishizaki et al. (Eds.): WCTP 2012, PICT 7, pp. 178–183, 2013.

Another research in stress detection involved a different technique in stress inducement. To invoke stress, the researchers made use of Hyperventilation and Talk Preparation. The hyperventilation stress inducer is an activity wherein the researchers require the subject to inhale and exhale deeply and fast for 2-3 seconds each. This represents an obvious behaviour of the physiological signals when a person in under stress; it results to changes in his/her corporal sensations. The talk preparation stress inducer on the other hand is a method where the researchers ask the subject to deliver an extemporaneous speech on a certain topic in front of a recording camera, given one or two minutes of preparation. The subject most likely experiences social stress during the stress inducing activity. In this research, only the galvanic skin response and heart rate were monitored and collected. The research concluded with 95% accuracy using the model built based on data gathered using the Talk Preparation inducer [4].

In another research the test subjects were asked to participate in real life driving scenarios to induce stress. The physiological sensors that were used were electrocardiogram, electromyogram, skin conductance and respiration. The activity scenarios were divided into three levels of stress: no stress, medium stress and high stress. The model resulted in a 97% accuracy, concluding that the physiological signals collected were adequate and the stress inducement scenarios were successful in eliciting signs of stress [2].

For this research the focus lies mainly on mental stress elicited while doing indoor non-physical activity such as computer or academic work. With this premise the inducers such as real-world driving, hyperventilation and talk preparation cannot be used since they provide a different form of stress which may have a mix of social as well as physical stress.

2 Data Collection

For this research, three activities were chosen to serve as stress inducers. They are Paced Stroop Test, Mental Math Test and Game. These 3 activities were chosen among all other stress inducers because among all other stressors mentioned before, Paced Stroop Test, Mental Math and Game are stress inducers that best mimics the academic and/ or computer stimulated mental stress, which is the premise for the extension of this research.

The data was collected using the Infiniti Biograph physiology device. The set-up included three physiological sensors: galvanic skin response (GSR), otherwise known as skin conductance, blood volume pulse (BVP) and respiration sensors. Screen captures were also taken while they were doing the activities. A baseline was collected before the activity started. The test subject was asked to rest for 1 minute to obtain the baseline (e.g. close their eyes and keep calm). After the rest period the test subject proceeded to do the three stress inducing activities. The test subjects were asked to do three activities in order to elicit stress. These three activities were all done on a computer and were automated with instructions for the test subjects to read so as to have minimal human intervention during the data collection. The first activity is the Paced Stroop Test which requires the test subject to correctly match different names of colors with the color of the text and not what the words themselves say. The second activity comprises of a basic arithmetic test where the test subjects either added, subtracted or multiplied randomly generated one digit to two digit numbers. Both activities have increasing levels of difficulties and are also timed per question in order to add more pressure. There will also be a buzzer noise when the timer runs out or if a wrong answer is given. Both activities will also be last two to three minutes depending on how fast the test subject answers each question. The third activity is a flash game on the internet called Save the Cursor wherein the test subjects played for 3 minutes. The goal of the game is to basically keep the cursor from being taken away or destroyed by a monkey. In between the three activities there will also be another minute of baseline where the test subjects were asked to close their eyes and relax before the next activity starts so as to have a fresh start with the stress readings.

After the data collection, the test subjects were asked to annotate their levels of stress by viewing the video of the screen capture. The annotation labels are from 1 to 5; 1 being the lowest stress and 5 to the highest stress. These annotations were used as labels in the corpus building phase.

3   Building the Corpus

A total of 4 test subjects participated in this experiment to build a model. The data collected were then segmented into 5 second windows with an overlap of 1 second per window. 5-second window size was used because it resulted to high correlation of the physiological signals and the stress level, compared to the 7-second and 8-second windowing. A paper by [5] used 1-second window size for their experiment using Galvanic Skin Response and Heart Rate Variability. However, this was not used because the minimal time interval would not be able to provide adequate or significant changes in the signals. Features were then extracted from these data. The stress level annotations were then averaged per segment to represent a continuous label for that instances of physiological values.

The corpus was divided by means of the modality or physiological signals in order to figure out which of the three represents or correlates with the stress levels best. The original corpus was also divided in terms of the inducers, in order to determine which of the inducers actually influence stress.

3.1   Feature Extraction

Six statistical features were extracted from each modality which makes it a total of 18 features. These six features are means of the raw signal, standard deviation of the raw signals, means of the absolute values of the first and second differences of the raw signals and the means of the absolute values of the first and second difference if the normalized signals. The means and standard deviation are common statistical computations while the first and second differences approximates a gradient. These statistical features were suggested by [6] as a general low-level form of physiological feature extraction.

3.2 Modelling the Corpus

The corpus was trained on different types of machine learning algorithms. Since the labels of the corpus are continuous values that represent stress levels, regression algorithms were used. The algorithms that showed high results were Multilayer perceptron and SMO regression. Please refer to the tables shown below for the results. There are a total of 331 instances, 72 of which represent Paced Stroop Test, 115 for Mental Math Test and 144 for Game. The stress level annotations that were averaged per segment to represent a continuous value were used as the model labels. The corpus was unbalanced in terms of stress inducers wherein the number of instances for the Game is twice as much as the Paced Stroop Test. This is because the test subjects had spent more time in the Game activity than the Paced Stroop Test.

4 Results and Analysis

The tables below show the results of each experiment for the model along with the sub models created. The first model that was experimented was the combined model wherein data from all 4 test subjects were combined. The first two tables show the results for the combined model using MLP and SMO regression algorithms respectively. The results show very high correlation coefficient, which means that the physiological signals used as are greatly correlated to the stress level annotations. Please refer to Table 1 for the results of the combined model which makes use of all 3 inducers as well as all 3 physiological signals.

Table 1. Combined Results

	MLP	SMO Reg
Correlation Coefficient	0.9661	0.9847
Mean Absolute Error	0.2126	0.0961
Root Mean Squared Error	0.3922	0.264
Total Number of Instances	331	331

Tables 2 and 3 show the results for the sub models which were divided based on the stress inducers. Both algorithms show that each inducer has a high correlation coefficient with the Paced Stroop Test as the highest for both MLP and SMO regression algorithm. Even though the number of instances of the game is twice as much as the Paced Stroop Test inducer, which provided much more training for the game inducer. The results also indicate that the stress annotations provided by the test subjects are in fact correlated with their own physiological responses. Please refer to Tables 3 and 4 for the MLP and SMO reg results.

For the models that were divided in terms of modality or physiological signals. In the MLP results, Respiration has the highest correlation coefficient while BVP the lowest. In the SMO algorithm, Respiration also has the highest correlation

Table 2. Stress Inducers MLP Results

	PacedStroop	Mental Math	Game
Correlation Coefficient	**0.9957**	0.9902	0.9948
Mean Absolute Error	**0.0265**	0.0922	0.1145
Root Mean Squared Error	**0.0394**	0.1355	0.1808
Total Number of Instances	**72**	115	144

Table 3. Stress Inducers SMO Reg Results

	PacedStroop	Mental Math	Game
Correlation Coefficient	**0.9783**	0.9576	0.9766
Mean Absolute Error	**0.0396**	0.0941	0.1229
Root Mean Squared Error	**0.0904**	0.2841	0.3792
Total Number of Instances	**72**	115	144

while GSR has the lowest, which implies that BVP and GSR poses a lower correlation based on stress levels using statistical features than Respiration. Please refer to Tables 4 and 5 for the physiological model results of MLP and SMO respectively.

Table 4. Physiological Modalities MLP Results

	BVP	GSR	Resp
Correlation Coefficient	0.7845	0.8048	**0.8956**
Mean Absolute Error	0.6639	0.6997	**0.5452**
Root Mean Squared Error	0.9589	0.9044	**0.7201**
Total Number of Instances	331	331	**331**

Table 5. Physiological Modalities SMO Reg Results

	BVP	GSR	Resp
Correlation Coefficient	0.8344	0.7802	**0.9433**
Mean Absolute Error	0.4907	0.5738	**0.2509**
Root Mean Squared Error	0.8403	0.9511	**0.5027**
Total Number of Instances	331	331	**331**

5 Conclusion

This paper experimented on 3 types of stress inducers along with 3 physiological modalities and determined which among them correlates with stress. Based on the experiment results, a high correlation was found among all the stress inducers, best of which came from the Paced Stroop Test. These results indicate that all three stressors: Paced Stroop Test, Mental Math Test and Game are effective in inducing stress. A high correlation was also found with all 3 physiological signals used in terms of the statistical features extracted with respiration rate as the highest. Possible future works for this research is to obtain higher level features to extract for the physiological signals as well as gather more data in order to build the general model. By doing so, another possible future work is to be able to characterize positive stress and negative stress from the stress levels of the model.

References

1. Boonnithi, S., Phongsuphap, S.: Comparison of Heart Rate Variability Measures for Mental Stress Detection Stress: The International Journal on the Biology of Stress (2011)
2. Healey, J.A., Picard, R.W.: Intelligent Transportation Systems. IEEE Transactions on Detecting Stress During Real-World Driving Tasks using Physiological Sensors, 156–166 (2005)
3. Zhai, J., Barreto, A.: Stress Detection in Computer Users Based on Digital Signal Processing of Noninvasive Physiological Variables Engineering in Medicine and Biology Society. In: 28th Annual International Conference of the IEEE 2006, EMBS 2006, pp. 1355–1358 (2006)
4. de Santos Sierra, A., Sanchez Avila, C., Casanova, J.G., del Pozo, G.B.: A Stress-Detection System Based on Physiological Signals and Fuzzy Logic. IEEE Transactions on Industrial Electronics, 4857–4865 (2011)
5. de Santos Sierra, A., et al.: Two Stress Detection Schemes Based on Physiological Signals for Real-Time Applications. Intelligent Information Hiding and Multimedia Signal Processing, 364–367 (2010)
6. Picard, R.W., Vyzas, E., Healey, J.: Toward Machine Emotional Intelligence: Analysis of Affective Physiological State. IEEE Trans. Pattern Anal. Mach. Intell., 1175–1191 (2001)

Towards the Design and Development of Anticipation-Based Event Selection Modeling for Survival Horror Games

Vanus Vachiratamporn, Roberto Legaspi, and Masayuki Numao

The Institute of Scientific and Industrial Research, Osaka University
{vanus,roberto,numao}@ai.sanken.osaka-u.ac.jp

Abstract. With the preference toward more action games from audiences, survival horror games need more of the fearful quality that makes it distinct in order to once again become attractive to the market. In this paper, we investigate the impact of game events, which consist of several visual and audio elements, on player affect as opposed to looking at the effects of individual game elements. In our preliminary experiment, we collected player emotion intensities using a brain-computer interface while playing a survival horror game. Our results provide insights as to how emotions transit during events. Based on these insights, we introduce the concept of an anticipation-based event selection modeling that aims to choose an event with maximum fear-inducing capability using anticipative emotion.

Keywords: Emotion, fear, anticipation, video games, survival horror.

1 Introduction

Survival horror is a sub-genre of action-adventure game where the player leads an avatar character through uncanny and hostile environments, reminiscent of horror fiction or movie. The character is usually being in a helpless situation and has to utilize limited resources thoughtfully in order to survive any upcoming threat. Survival horror games is said to impact most a gamer's emotion more than any other genre since the player has expectations of actually getting frightened or scared [1]. Nevertheless, for over a decade now, developers and audiences seem to continue losing their interest in survival horror genre due to the shift of players' preferences toward more action style games. For example, the most famous survival horror franchise Resident Evil (Capcom, 1996) is turning away from its original genre by revising its game to have more action gameplay which consequently lessens the horror. Recently, survival horror seems to have its place left only on small-budget games from independent video game developers. One example is *Amnesia: The Dark Descent* (Frictional Games, 2010), a successful survival horror game from an indie developer that lit up some hope to the genre and showed that there are still a lot to explore in this genre.

S. Nishizaki et al. (Eds.): WCTP 2012, PICT 7, pp. 184–194, 2013.

Designing survival horror games is all about how to elicit emotional responses from players, particularly fear. Game designers have to consider and make use of each game element (i.e. visual, sound, gameplay or story) wisely in order to induce the most fearful experience to players. Adjusting game elements will obviously change play experience, so it should be possible to consider this adjustment as an optimization problem which tries to manipulate some game elements in order to optimize the player's fear level. The manipulation also favors a problem of different fear experience profile and variation between individuals [2] which could be solved by real-time game element adaptation.

By looking at a higher level, games consist of an event, the specific sequence of visual and audio which aim to lead any player to the same experience that the developer desired. While sound, for example, is considered to be an element that carries more emotional content than any other part of a game [3], sound is normally used with some visual or gameplay-related element for inducing fear, which makes it difficult to distinguish the influence of sound alone. Analyzing the effect of how the player's emotion is translated by specific events rather than individual game elements, which might not fully affect the player's emotion, should give game designers clearer insights on how to optimize player's emotion to desired states.

In this paper we conducted a preliminary experiment to visualize player emotion transitions while playing a survival horror game. Using the insights we gathered, we then proposed a methodology for building an anticipation-based event selection model that makes use of a player's anticipative emotion to predict and select the best event that could sustain or stimulate further the player's emotion intensities. Player's anticipation is related heavily to the outcome of an event, for example, the player might show a huge startle reaction to a monster when he has some anticipation on the monster presence but might not be surprised at all if he does not anticipate. This model could aid game developers in examining the effect of an event and choosing the event that suit their demand in controlling the players' emotion. It should be noted that, here, we refer to an *event* as any sudden change of visual and/or audio that is not caused by the player's action directly, but rather by the level designer who wants the player to experience something at some point of the game.

In the succeeding sections, we first discussed survival horror as a genre and why it is interesting to study it. We then reviewed literatures related to the affective state fear in survival horror games. Afterwards, we discussed the methodology of our preliminary experiment. Finally, we discussed our initial modeling results and we concluded with our future work.

2   Review of Related Literature

There is some confusion in classifying the survival horror game from other genre. Although this normally falls under the action-adventure category due to the fact that it tends to have action-adventure elements, it is not unusual for a survival horror game to draw upon elements from another genre, e.g., the first-person shooter. Fahs [4] pointed out a good survival horror concept, stating that "survival horror is one of the only genres defined not by gameplay mechanics, but by theme, atmosphere, subject matter, and design philosophy." When players typically considered a survival horror game as not scary, hence, a bad game, designers opt for those that are commonly used in many survival horror games such as, main character arms with only inefficient or limited weapons, presence of horrific monster in an environment or forewarning of danger which aims to create feeling of anxiety and uncertainty. These techniques help build signs of threat and consequently stimulate fear.

With the preference toward more action games from audiences, traditional developers slowly lost their attention in survival horror leading to a big transition of the genre. Resident Evil 4 (Capcom, 2005) has made a huge impact on its established genre by broadening the gameplay toward more action and thrilling combat which obviously lessen the scary atmosphere. Although many questions on the direction of survival horror were raised, it turned out that Resident Evil 4 was a major success in both sale figures and reviews. After the success of Resident Evil 4, relying gameplay heavily on action rather than horror seems to be a wiser marketing strategy. An upcoming title of the series, Resident Evil 6 (Capcom, 2012), is using the term "Dramatic Horror" as its genre instead of survival horror which indicated that the game will feature more on dramatic action and lessen a helpless-survival situation. We then question if bringing about only fearful experience is not anymore enough for the market.

Nevertheless, many games are also finding their way in this market while keeping true to survival horror genre although it might not be very commercially successful compared to action-horror games. One notable game that brought back some attention to this is *Amnesia: The Dark Descent* from the independent video game company, Frictional Games. The game was praised broadly from both critics and gamers bringing the scariest experience to them; some critic went as far as to say that Amnesia is the most successfully frightening game to have been made [5]. Without using any cheap-jump scares, Amnesia creates tension through its dark atmosphere, monster awareness or random sound cues which keeping players on their toes all the time. This game has showed that many players still hunger for more of this genre and it is not yet exposed to its maximum capability.

There is evidence to suggest that there is value to leveraging event-level in horror game design. Parker and Heerema [3] stated that sound carries more emotional content than any other part of the game, in that, sounds could trigger feelings and memories directly. Still, there has been surprisingly little work on the relation between emotion and sound, other than music and speech [6], even though sound was used as an important element in many survival horror games, especially, as forewarning cues for an upcoming danger which puts the player in the state of uncertainty about an outcome that he is not yet faced [7][8]. Grimshaws's [6] study on an uncanny sound, pointed out the problem that a player will be familiar to uncanny sounds that do not change across multiple plays and suggested that real-time sound synthesis or use of biofeedback might solve this issue. Garner et al. [9] tried to investigate properties of sound, particularly pitch, loudness and 3D positioning, that affect the player's intensity most, using players' in-game action and real-time vocal response as data. However, this study could not find any conclusive evidence that those properties actually cause any important effect on players. They argued that the different levels of player experience and confidence might have an effect on results. However, we think that the data used for evaluating player's intensity might not be good enough as well. Kang et.al. [1] also tried to solve the problem in identifying user fear levels for future use in a horror game context. Brain-Computer Interface (BCI) was used to collect brain signals while the participant watched a clip of horror movie. Still, fear has not been totally isolated from any other emotion. Garner and Grimshaw [2] have proposed a framework that classifies the potential of audio properties within a fearful scenario in a computer video game context by dividing fear into caution, terror and horror states, based on Fanselow's [10] defensive behavior system of pre-encounter, post-encounter and circa-strike defenses, respectively. However, this framework has yet to be proven usable on other scenarios.

3   Preliminary Experiment

We aimed to visualize the player's emotion data while playing a survival horror game so that we can obtain insights on affect behavior and build a model based on it. We first gave the details of the game environment and then proceeded in describing the experiment procedure and how the data was collected. Finally, the result of the experiment is shown.

3.1   Game Environment

Choosing a commercial survival horror game as a test environment seems to be our best choice instead of developing a game from scratch. Furthermore, this game has the quality that we cannot duplicate in a short amount of time. We are cognizant that commercial game design is normally abstracted, i.e, we cannot really look into the mechanisms on how the game works, retrieve information on how the game is played, or change a game element that we want to adapt. Still, we deem that building a survival horror game from scratch will pose more problems in our investigation. While using a good game engine might allow us to create some playable survival horror game, questions can be raised as to its quality and viability, especially in terms of it induces varied emotions. We need some fine line between quality and control of the game to investigate our theories. Hence, we decided to employ an existing game that allowed us to extract information while making pertinent changes.

Amnesia: The Dark Descent, as mentioned before, is the most successfully frightening survival horror games [5] that got positive reviews on its fear-inducing quality from across many critics. Developers also released various tools that allow the creation of custom level and components to edit the levels that were used in real game as well. With the access to game scripts and resources, it is possible to change how the events in the game will occur. Although the engine allows us to make only high-level changes, and not on over its underlying features such as its physics or AI, it is enough for us to use this game as our experimental environment.

Fig. 1. Gameplay screenshot of Amnesia: The Dark Descent

Fig. 1 showed an event where the player is encountering a monster in *Amnesia*. Player's avatar is partially absent and only the left hand can be seen when the player is holding a lantern. Distortion in the screen indicates that player's sanity is dropping down because he is looking to the human-like monster on the top of stairs directly. If sanity continues to drop, the player's avatar may go crazy and cannot move for a short amount of time. In this situation, the player should turn back, run, and find some dark place to hide. Although, player can run straight through the monster as well, he will have higher chance to get attacked.

To give the same experience as the original, which already does really well, we decided to edit the original game levels instead of creating a new one. Our custom levels have an introduction similar to the original where the player can spend some time to get used to the game control and atmosphere. However, our levels skipped some part of the original to reach the intense levels earlier; otherwise it will take too long in the experiment with the original pace. We also adjusted the presence of important items and secret ways to make it easier for them to be seen. Our early tests revealed that inexperience players were likely to get stuck for a long time and get bored, which, consequently, made them feel unscared. Although this change obviously made our custom levels easier than the original, we argue that players who participate in our experiment are different from the players who normally play games with eagerness, which may not necessarily be true with our subjects, who may have less will to play and are easier to get bored if they encounter problems coping with the game. It should be noted that all of the changes we applied are not related to diminishing the impact of scary events, but rather meant to make the players more immersed easily in our game and have lower chances of getting bored.

In Amnesia, events are mostly consist of visual distortion (e.g. blurry screen or reddish screen), scary sound (e.g. person scream), super natural force (e.g. door opened itself) or monster presence. Some events are story related which have no potential to expose any explicit fear reaction but it can create some anxiety over an upcoming threat. The subject player will go through at least 26 events in the custom levels. There are some additional events that the player may or may not find, but they exist in order to create more anticipation as the player wanders around in the game. It should take around 30 to 45 minutes to finish the custom levels.

3.2   Experimentation

The data that we collected during our preliminary experiment came from two sources – the game and the player. For *Amnesia*, information on when each event occurred was retrieved through an in-game debugging log.

The Emotiv EPOC Neuroheadset (Emotiv Systems, 2009) was placed on the player's head. The headset is a BCI device that uses electroencephalography (EEG) data collected from 14 sensors touching the scalp. Aside from raw EEG data, emotions could also be measured as a value ranging from 0 to 1 which included *long-term excitement, short-term excitement, meditation, frustration* and *boredom*. Additionally, player's face and voice were also recorded using a webcam and a microphone.

Before playing the game, brief details on the game and experiment procedure were given to the player while the Emotiv EPOC, webcam and microphone was set up. Then all data collection tools were started with the game. The player was then left alone in a considerably dark room wearing an earphone and the Emotiv EPOC. The player played the game until the game credits were shown. After the player finished the game, he had to rate the overall scariness of the game that he just played, indicated the scariest event and game element based on his opinion.

Only one subject was investigated in the preliminary experiment. The subject has known Amnesia: The Dark Descent before but never played it because he does not like to play a scary game. The subject has experience in first person shooter game control so he got familiar with the Amnesia in no time. He was asked to play the same session twice so that we can get some comparative data and to see how the subject's emotions change between both experiments.

3.3 Results

Table 1 shows statistical information on player emotion data retrieved from the Emotiv EPOC with the same subject playing the game twice on separate occasions. Long-term excitement, short-term excitement and frustration have considerably higher variance compared to meditation and boredom in both runs. This indicates that these three emotions were more sensitive to the game and made it more interesting for further use in finding the effects of events. However, those three emotions were also decreased significantly in the second experiment which can be indicative of the player were getting more familiar to the game and remembered where some events are going to occur. This obviously lessens his excitement and frustration. Nevertheless, the three emotions still have high variance in the second experiment.

Table 1. Mean and standard deviation of emotions data

Emotion Intensity	1st Experiment			2nd Experiment		
	Average	SD	[Min, Max]	Average	SD	[Min, Max]
Long-Term Excitement	0.66	0.15	[0.28, 0.97]	0.51	0.13	[0.22, 0.85]
Short-Term Excitement	0.66	0.23	[0.00, 1.00]	0.51	0.23	[0.01, 1.00]
Frustration	0.66	0.17	[0.25, 1.00]	0.54	0.17	[0.20, 1.00]
Meditation	0.37	0.06	[0.23, 0.67]	0.36	0.03	[0.30, 0.53]
Boredom	0.65	0.09	[0.49, 0.99]	0.64	0.11	[0.49, 1.00]

To get more understanding on the behavior of emotions and the effects of events, Fig. 2 visualizes some parts of the long-term excitement (LTE) value with the line and bars indicating the time when the ith event E_i occurred. Fig. 2 shows some of the interesting changes in the LTE during the events.

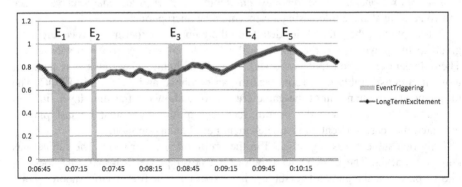

Fig. 2. Long-term excitement and event triggering from 1st experiment

From the start point to the end of E_1, the LTE trend was going downwards until something that the player saw or heard during the event caused the LTE to rise. Looking at E_4 and E_5, the LTE trend was going upward from before E_4 until it reached E_5 with maximum intensity. However, the LTE trend was not sustained by E_5 and eventually dropped. While there are some unknown causes of interesting changes of LTE between E_3 and E_4, these are too difficult and too inconsistent to define because it is at these points that the player acted and experienced the game world by himself without any specific control of game elements/events from designers. Different players may experience different things *between* E_3 and E_4 but they will always get the same or, at least, similar experience *in* the E_3 and E_4 events.

Fig. 3 shows the same portion of the events from the second experiment. Although the player entered E_1 with the similar anticipation trend as the first experiment, the event did not cause the trend to rise anymore which indicated that familiarity has some effects to the player's LTE. While familiarity might also affect the downward trend in the latter events, we argued that different levels of anticipation when the player was entering the events might be the factors as well, i.e. playing through E_4 in low-excitement state might not be very well exciting as when playing in high-excitement state. Nevertheless, getting data from more subjects would reveal more insights in the behavior of emotions that related to game events.

By examining the transition of emotion intensities at an event, we can label the effect of a certain event as to whether it can leverage or stimulate the emotion intensities or not. The player's emotion that related to an event E_i could be divided into three parts, namely, *anticipative emotion*, i.e. before E_i occurred, *on-the-event emotion*, and *lingering emotion*, i.e. after E_i. By looking at certain windows in time, the emotion trend could be defined as upward, downward or stable. So, it should be possible to define the effects of events to emotions as results of emotion trends during and after the event, using the anticipative emotion as input.

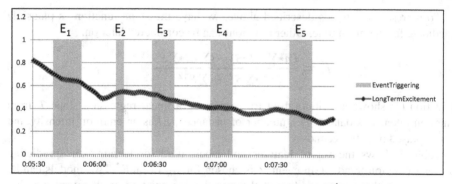

Fig. 3. Long-term excitement and event triggering from 2nd experiment

4   Anticipation-Based Event Selection Model

From our preliminary experiment, the in-game event shows some potential and usability in controlling the player's emotion. In order to optimize the use of player's emotion, particularly fear in this context, in game design, we propose a methodology for creating an anticipation-based event selection model that should be capable of predicting how an event will affect the player's emotion trend and choose the event that fit best to the demand of the design.

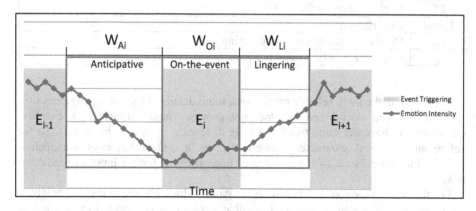

Fig. 4. Emotion-Event graph

Fig. 3 shows a simplified version of an emotion-event graph similar to Fig. 2 where we visualized the data from our preliminary experiment. From the graph, based on the ith event E_i, W_{Ai} is the time interval of the anticipative emotion between E_{i-1} and E_i, W_{Oi} is the on-the-event time interval, and W_{Li} is the time interval of the lingering emotion between E_i and E_{i+1}. Using the emotion state intensities collected from a player while playing the game, emotion trends T_{Ai}, T_{Oi} and T_{Li} can be calculated from W_{Ai}, W_{Oi} and W_{Li}, respectively. So, based on the graph, T_{Ai} should be some negative value to indicate a downward trend, while the opposite should hold for T_{Li}.

To represent the trends described above, we have the *slope* function as plausible candidate for the mean time. Other functions can be considered later on.

$$T = \frac{(n*\sum(EI*S)) - (\sum EI * \sum S)}{(n*\sum S^2) - (\sum S)^2} \tag{1}$$

Equation (1) shows a formula used for computing a linear regression slope T from emotion intensities data. n is a number of instances, EI is an emotion intensity and S is a played time by seconds.

Table 2 shows the list of features and labels used for learning and predicting player's emotion trends if an event k is chosen for i_{th} event E_i. Two models can be created using these features: first is the model for predicting on-the-event emotion trend T_{Oi} and the latter is for predicting the lingering emotion trend T_{Li}. It is possible that in some situations, the system may want to automatically lower the player's emotion intensity during the event in order to raise it up further after the event. That is why on-the-event trend and lingering trends should be approximated separately.

Table 2. List of features and labels

Feature	Description
T_{Ai}	Anticipative emotion trend between E_{i-1} and E_i
I_{Ai}	Representative of emotion intensities between E_{i-1} and E_i
E_i^k	Event k used in the ith event E_i.
Label	**Description**
T_{Oi}	On-the-event emotion trend during E_i
T_{Li}	Lingering emotion trend between E_i and E_{i+1}

In order to make more reliable predictions, more details of the anticipative emotion have to be given since trend T_{Ai} does not provide enough information. Emotion intensities can have the same trend, but huge difference in term of intensities. So we define an additional parameter, namely, I_{Ai}, as a representative of anticipative emotion intensities, such as the average of intensities, as further input to prediction task.

For the development of a survival horror game, which obviously aims to heighten up the player's fear state as much as possible, each event is expected to produce an upward trend or sustain the trend if fear intensity is already high enough. Heightened lingering emotion should be targeted by the model regardless of on-the-event trend. So, we got

$$AESM(T_{Ai}, I_{Ai}) = (E_i^k \mid ET_L(T_{Ai}, I_{Ai}, E_i^k) = \max_{x \in N}(ET_L(T_{Ai}, I_{Ai}, E_i^x))) \tag{2}$$

Equation (2) describes the anticipation-based event selection model (*AESM*) which chooses the event E_i^k, from N possible events, that expects to maximize lingering emotion trend ET_L at the ith event, given anticipative emotion trend T_{Ai} and anticipative emotion intensity I_{Ai}.

This model will open up the possibility of building an adaptable survival horror game that can automatically switch in-game event to match the player's profile and current emotion trend, in order to maximize the intensity of the player's fear state. Nevertheless, the model has not been evaluated so we still not have an evidence to prove it efficiency yet. An empirical experiment needs to be conducted to get data for validating this model.

5   Conclusion

In this paper we conducted the preliminary experiment to have some insights on player's emotions while playing the commercial survival horror game called *Amnesia: The Dark Descent*. Player emotion intensities were extracted using the Emotiv EPOC and some part of the data were visualized together with game events data. The visualization showed many interesting changes in emotion intensities that were caused by events, whether stimulated further or cannot be sustained, which leads us to the conceptual design of the anticipation-based event selection model that makes use of a player's anticipative emotion to predict the event that maximizes emotion intensity. The model should benefit survival horror games most because they normally play with players' emotion and to maximize fear-inducing capability is the target of all games in this genre. Though, the model has not been evaluated yet in this work.

In our future work, we plan to make some more adjustments to the game's events so that it can give us more meaningful and easier to examine emotion trends. Experiments will be conducted to gather data for building the predictive models which will be further used to create an adaptive survival horror game using our anticipation-based event selection model. There are some more features we intend to use later on to make robust the prediction task, such as physiological feedbacks which can give a comparative results to the Emotiv EPOC or extracting emotion from the game video and audio which can be used together in constructing the models.

References

1. Kang, T.R., Perez, I., Matias, G.L.E.: Design and Development of an Affect-Sensitive Horror Game. In: 12th Philippine Computing Science Congress, Laguna, Philippines, March 1-3 (2012)
2. Garner, T., Grimshaw, M.: A Climate of Fear: Considerations for Designing a Virtual. Acoustic Ecology of Fear. In: Proceedings of the 6th Audio Mostly Conference: A Conference on Interaction with Sound, pp. 31–38 (2011)
3. Parker, J.R., Heerema, J.: Audio Interaction in Computer Mediated Games. International Journal of Computer Games Technology 2008, Article ID 178923 (2008)
4. Fahs, T.: IGN Presents the History of Survival Horror (2009), http://retro.ign.com/articles/104/1040759p1.html (accessed July 10, 2012)
5. Walker, J.: Wot I Think: Amnesia – The Dark Descent (2010), http://www.rockpapershotgun.com/2010/09/07/wot-i-think-amnesia-the-dark-descent (accessed July 10, 2012)

6. Grimshaw, M.: The audio Uncanny Valley: Sound, fear and the horror game. In: Proceedings of the 4th Audio Mostly Conference: A Conference on Interaction with Sound, pp. 24–26 (2009)
7. Krzywinska, T.: Hand-On Horror. Spectator 22:2, 12–23 (2002)
8. Perror, B.: Sign of a Threat: The Effects of Warning Systems in Survival Horror Games. In: Proceedings of 4th Conference on Computational Semiotics for Games and New Media, Croatia, September 14–16 (2004)
9. Garner, T., Grimshaw, M., Abdel Nabi, D.: A Preliminary Experiment to Asses the Fear Value of Preselected Sound Parameters in a Survival Horror Game. In: Proceedings of the 6th Audio Mostly Conference: A Conference on Interaction with Sound, Article No. 10 (2010)
10. Fanselow, M.S.: Neural Organization of the Defensive Behaviour System Responsible for Fear. Psychonomic Bull. Rev. 1, 429–438 (1994)

Sidekick Retrospect: A Self-regulation Tool for Unsupervised Learning Environments

Paul Salvador Inventado[1,2], Roberto Legaspi[1], Rafael Cabredo[1,2], and Masayuki Numao[1]

[1] The Institute of Scientific and Industrial Research Osaka University
Ibaraki, Osaka 567-0047, Japan
[2] Center for Empathic Human-Computer Interactions, College of Computer Studies,
De La Salle University
2401 Taft Avenue, Manila, 1004, Philippines
{inventado,roberto,cabredo}@ai.sanken.osaka-u.ac.jp,
numao@sanken.osaka-u.ac.jp

Abstract. Self-regulation is an important skill for students to possess. It allows them to learn more effectively and it has been shown to cause better learning gains. Self-regulation is not an easy task especially for poor learners. This is the motivation behind researches that use computer-based learning environments to promote self-regulation through embedded tools that help students keep track of their self-regulation processes. Although these researches have shown promising results, they focus on self-regulation processes inside controlled learning environments. Not much research has been done on learning in unsupervised learning environments where students learn on their own, introducing additional challenges. In this research, we developed software to help students perform self-regulation in this setting. Results showed that the software was able to help students set goals, monitor their activities and evaluate their learning behavior. Students who used the software reported that it made them more aware of the activities they did when they were learning and it also helped them identify what to do in order to improve their learning behavior in succeeding learning sessions.

Keywords: self-regulation, self-monitoring, unsupervised learning environment.

1 Introduction

Self-regulation is "the process wherein students' cognition, behavior and affect are systematically oriented toward attaining their goals [8]". It involves the use of component skills including setting one's goals, adopting strategies to attain one's goals, monitoring one's progress, adjusting the physical and social context to make it compatible with one's goals, managing one's time efficiently, evaluating one's self, attributing causation to results and adapting future strategies [12]. Among the different strategies for self-regulation, self-monitoring is probably the most important [6,13]. Self-monitoring allows students to identify which goals

S. Nishizaki et al. (Eds.): WCTP 2012, PICT 7, pp. 195–205, 2013.

were met and what activities were actually performed, enabling them to evaluate their behavior and adopt better future strategies.

Although students benefit from self-regulation, it is no easy task. According to Zimmerman [11], self-regulation involves temporally delimited strategies and students' efforts to initiate and regulate these strategies proactively, which require preparation time, vigilance and effort. Unless the results of the efforts required for self-regulation are attractive enough to students, they will not be motivated to self-regulate [6]. It is even more difficult for poor self-regulators because they can't change easily even if they know what to do [2]. However, self-regulation is teachable and can lead to an increase in motivation and achievement [9,10].

In this research, we developed a system that helps students self-regulate in an *unsupervised learning environment*, which we define as an environment where students learn on their own without guidance from a human or automated teacher. An example of such an environment is research, wherein students are required to seek information on their own. In each learning session, students need to manage their own goals, manage their time, overcome internal and external distractions and manage their emotions in order to complete their task. Learning to self-regulate is not only important in academic settings but also beyond formal education, empowering students to become lifelong learners [7].

2   Related Work

Self-regulation can be viewed in terms of three cyclical phases: *forethought phase*, *performance phase* and *self-reflection phase* [12]. In the forethought phase, students set their goals and strategies for the learning session. In the performance phase, students then apply the learning strategies to accomplish their goals. Students also monitor their activities and take note of significant observations such as the type of activities they did, the amount of time they spent doing an activity and the problems they encountered during the activity. In the self-reflection phase students compare their learning performance with their expectations, their previous performance or another person's performance. This usually leads them to attribute the causes of the problems they encountered during the session such as distractions affecting performance, the complexity of the task requiring more time and effort and the inability to complete the task.

The importance of self-regulation in learning has spawned research in providing automated support for self-regulation while learning through computer-based learning environments. MetaTutor for example, is a hypermedia learning environment for Biology which not only presents content but also helps students self-regulate [1]. The system helps students become aware of their own self-regulation strategies by providing them with a list of self-regulation processes that they can select to indicate and track which process they used. It also has a pedagogical agent to help students with questions regarding the content. Process Coordinator is another system developed to help students self-regulate as they learned about Physics [5]. The tool provided students with the content they needed to learn and also helped them perform different self-regulation processes by presenting to them a set of goals for the current topic, strategies for accomplishing these goals, guide questions to help them check their understanding of the topic and prompts that would remind them to reflect on the guide questions.

3   Sidekick Retrospect

The Sidekick Retrospect software was developed to help students self-regulate while learning in an unsupervised learning environment. Learning is not constrained with the use of particular software so students can use any tool or activity they prefer for learning. Students can also do activities outside of the computer such as writing notes or reading a printed paper as long as they are done in front of the computer. The software's design was based on Zimmerman's three self-regulation phases discussed earlier and consists of three modules namely, the goal setting module, the recording module and the evaluation module. Each one is discussed in detail in the following subsections.

3.1   Goal Setting Module

Students begin the process by first setting their goals for the current learning session using the interface shown in Figure 1. These goals are expected to be related to the final goal which students want to accomplish. For example, if a student's final goal is the creation of a research report on an assigned research topic, possible goals for the session would include searching for related literature about the topic or designing a methodology for the research. Students have the freedom in defining their goals and are allowed to identify as many goals as they wish. However they are encouraged to list only goals that they think can be completed within the session.

Fig. 1. Screenshot of the goal setting interface

3.2   Recording Module

The recording module is responsible for recording the activities done by the students while learning. Currently, the software records a screenshot of the desktop and the webcam feed every second. The webcam is focused on the student so that activities outside the computer and their facial expressions and gestures can be recorded for later evaluation. The different software applications used throughout the learning session are also logged together with corresponding timestamps. In this initial version of the software, there is still no feedback given to the student while learning. Throughout the duration of the session, a window containing the students' goals are shown to help them manage their activities. Students can also add more goals in case they feel the need to. The list of goals helps students keep track of their progress, prioritize their activities and keep non-learning related activities under control so they can complete their learning goals. Figure 2 shows a screenshot of the recording module.

Fig. 2. Screenshot of the recording module's interface

3.3   Evaluation Module

After completing the learning session, students are asked to review and annotate the session. As seen in the annotation interface screenshot (see Figure 3), a timeline is used to represent the student's entire learning session. When the mouse hovers over the timeline, the screenshot of the desktop and the students' webcam feed is displayed on the screen to help students recall what happened at that point in time. Screenshots can also be displayed continuously just like a video to make it easier to review. The play, stop, step left, step right and playback speed settings give more playback control to make annotation easier.

Students then select time spans by clicking and dragging on the timeline or using the playback controls. The selected time span is then annotated by the student based on three aspects namely, the *intention, activity* and *emotion*. Intention refers to students' current intention which can be *goal related* or *non-goal related*. Goal related intentions refer to performing activities done to achieve

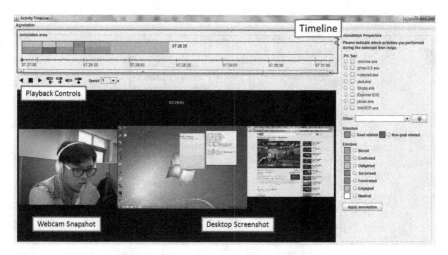

Fig. 3. Screenshot of the annotation interface in the evaluation module

any of the goals defined for the learning session. Performing any other activity is considered a non-goal related intention. In our methodology, goal related and non-goal related activities are synonymous to learning and non-learning related activities because the goals set by the student are for learning. Activities refer to what was performed by the student and is composed of a *primary activity* and *secondary activities*. Primary activities refer to activities that the student did during the selected time span (e.g., winword.exe, reading a technical paper). Secondary activities refer to other activities done together with the primary activity (e.g., listening to music, drinking coffee). Emotion refers to the emotion felt by the student while performing an activity. Although a single emotion set can be used for annotating students' emotions when engaging in goal-related and non-goal related activities, important contextual information may be lost. Craig et al. [3] reported that students commonly experienced delight, engagement/flow, confusion, frustration, surprise and fear when learning, which are affective states that do not only denote basic emotions but also indicate significant points in a learning session. Knowledge of these affective states enables the provision of appropriate feedback for future systems and motivated us to use such an annotation set for learning related activities. However, these affective states are not as meaningful and may not be observed in non-goal related contexts so Ekman's six basic emotions were used instead. Ekman's six basic emotions include anger, disgust, sadness, delight, fear and sadness [4]. In both annotation sets, the neutral state was added to denote a state wherein no particular emotion was experienced. Figure 4 shows the annotation labels used for a time span.

The annotation process lets students view and analyze their activities during the learning session to identify their intention, activity and emotion thus, perform self-monitoring. This may help students become more aware of instances when they were distracted or spent less time learning and their possible causes.

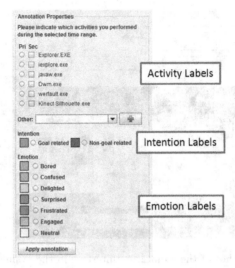

Fig. 4. Labels used for annotating time ranges in the evaluation module

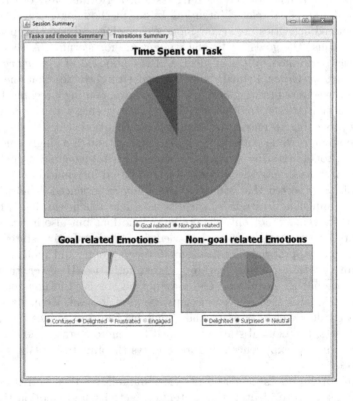

Fig. 5. Statistics on intentions and emotions during the learning session in the evaluation module

After annotating, students are also presented some statistics about their intentions and emotions during the learning session which gives them a more general view of their learning behavior (see Figure 5).

Lastly, students were asked to identify how much they completed their goals, which activities they thought were helpful in completing their goals and the productivity rating of their learning session (see Figure 6). These questions encouraged the students to self-evaluate. The rating may be based on the difference between their current learning session and a previous learning session, other people's learning behavior or their self-expectations. Evaluation helps students become aware of what they need to change in their learning behavior. Hopefully, in the next learning session students will bring with them their revised learning strategies so they can have better learning outcomes in their succeeding learning sessions.

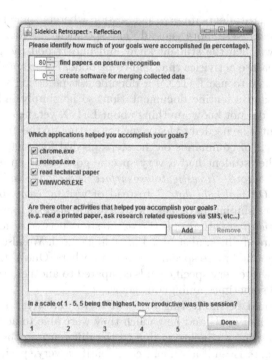

Fig. 6. Interface for self-evaluation in the evaluation module

4 Preliminary Data Gathered from Sidekick Retrospect

We collected data from one male undergraduate student, one male masteral student and two female doctoral students. These students were performing research to accomplish the requirements of their degrees. We considered this an unsupervised learning environment because these students were doing research individually and were only given feedback by their advisers at least once a week. The rest of the time, these students set their own goals and accomplished them using the tools and information sources of their choice, at the location and time they were most comfortable working in. This kind of environment introduced many challenges for them such as time management, managing motivation, managing emotional states and avoiding distractions. Students used the software for two hours each day for five days in the span of one week. This resulted in a total of 10 hours of data per student. The concept of self-regulation was not discussed with the students at any point during the experiment. The researchers asked each student to answer a survey after every learning session and conducted a personal interview after the whole experiment was completed to investigate the effects of the software on their self-regulation process.

4.1 Goal Setting

The goals set by the students differed per session wherein some goals were very specific, while others were more general. The specificity of goals seemed to be dependent on how much the student knew and understood the goal as well as the number of separate activities that had to be done. For example, one student wanted to learn how to use LaTeX[1], a document typesetting system used for writing technical and scientific documentation, so he simply indicated *"Learn LaTeX."* Since he did not know anything about LaTeX yet, he could not specify what exactly about it he needed to learn such as setting up the program, learning about specific kinds of commands or the steps needed for compilation. On the other hand, another student had a very specific goal about an experiment she was doing so she wrote *"Investigate isosurface value level indicator specified in vesta through O2 molecule data."* Instead of writing each specific activity, students sometimes summarized these with a single goal such as *"Create input files for next calculation."* Here, all the other steps necessary for creating the input files have already been abstracted by a single goal. We also observed that some students identified more specific goals than others. One of the students for example, usually wrote very specific goals compared to another student who had more general and sometimes ambiguous goals.

Although it would be interesting to see if there were relationships between how students set their goals and how much they were able to accomplish them, it was difficult to draw a conclusion due to the difference in the complexity of the tasks being done. Even if a student can identify a very specific goal, it does not mean it can be completed since it might actually be a complex task. On the other hand, even though a general goal is identified, it can be completed if it involves simple individual tasks.

4.2 Self-observation and self-reflection

When the students performed annotation, they inherently performed self-monitoring and self-reflection as they were identifying the appropriate labels for

---

[1] http://www.latex-project.org/

intention, activity and emotion. This was verified in the survey which showed that students were able to discover positive and negative aspects of their learning behavior. One student said *"I actually do work once I'm absorbed into it. I get distracted less than I thought,"* which showed that he felt contented about his current learning behavior. On the other hand, another student said *"I need to be more focused in doing my learning activity. When I did multitasking, none of my goals were completed."* Here the student was able to identify that her learning strategy was ineffective and attributed it to working on multiple goals simultaneously. Most of the students identified that one major factor hindering them from accomplishing their goal was spending too much time in non-learning related activities such as Facebook or YouTube. They also indicated that feeling boredom and lacking motivation made them work slower than was needed to accomplish their goals.

The annotation and reflection interfaces of the software were able to help the students easily review their learning behavior, identify how much of their goals were accomplished and identify what hindered them from accomplishing their goals.

4.3 Self-evaluation

In the evaluation module, students were asked to evaluate how productive their learning session was. Based on the survey and additional interviews conducted with the students, the basis for their evaluations also differed. Some students based their rating on their previous performance, the accomplishment of their goals and their beliefs about their own potential. The students' self-reflection and self-evaluation helped them identify what to change and how to change them. One student said *"When reading scientific paper, I should finish reading one paper first then continue to another paper. I should not read two or more papers at the same time because I would fail to understand those papers comprehensively."*

5 Conclusion and Future Work

In this research we were able to develop a tool that helped students self-regulate in an unsupervised learning environment. The different modules of the system helped the students perform self-regulated strategies including goal planning, self-monitoring, self-reflection and self-evaluation. Even though the software was not able to provide content related information, the system helped the students become more aware of their learning behavior and helped them identify what helped them accomplish their goals or hindered their goals. It helped them evaluate their learning behavior and enabled them to reflect and find ways to improve on it. From the survey, we were able to identify that the inability to fend off distractions and lack of motivation were some causes that hindered students from accomplishing their goals in unsupervised learning environments.

The subjects in the study were mostly graduate students who were mature, likely to be self-regulated and had more self-control. It would be interesting

204 P.S. Inventado et al.

to observe how different less mature students would behave and what effect feedback in this kind of environment would result to.

More research needs to be conducted especially regarding goal setting in an unsupervised learning environment to see what kind of support will be effective for students. We also feel that providing support through reminders and prompts while learning will be helpful to students aside from just helping them reflect after the learning session. It will also be important to keep track of the solutions that students identified to overcome the problems they discovered from their learning session. These solutions can be used to help them set their goals, select learning strategies and avoid what previously caused the hindrances to accomplishing their goals in succeeding learning sessions. Lastly, it would also be helpful to track the changes in the students' learning behavior over time to help them further evaluate and refine their learning strategies.

Acknowledgment. This research is supported in part by the Management Expenses Grants for National Universities Corporations through the Ministry of Education, Culture, Sports, Science and Technology (MEXT) of Japan, by the Global COE (Centers of Excellence) Program of MEXT, and by KAKENHI 23300059. We would also like to thank the students who participated in our data gathering session.

References

1. Azevedo, R., Johnson, A., Chauncey, A., Burkett, C.: Self-regulated learning with MetaTutor: Advancing the science of learning with MetaCognitive tools. In: Khine, M.S., Saleh, I.M. (eds.) New Science of Learning, ch. 11, pp. 225–247. Springer, New York (2010)
2. Brooks, D.W., Nolan, D.E., Gallagher, S.M.: Web-Teaching. Innovations in Science Education and Technology. Kluwer Academic Publishers (2001)
3. Craig, S.D., Graesser, A.C., Sullins, J., Gholson, B.: Affect and learning: An exploratory look into the role of affect in learning with AutoTutor. Journal of Educational Media 29(3), 241–250 (2004)
4. Ekman, P.: Are there basic emotions? Psychological Review 99(3), 550–553 (1992)
5. Manlove, S., Lazonder, A.W., Jong, T.: Software scaffolds to promote regulation during scientific inquiry learning. Metacognition and Learning 2(2), 141–155 (2007)
6. McMahon, M., Oliver, R.: Promoting self-regulated learning in an online environment. In: Montgomerie, C., Viteli, J. (eds.) Proceedings of World Conference on Educational Multimedia, Hypermedia and Telecommunications 2001, pp. 1299–1305. AACE, Chesapeake (2001)
7. Schraw, G., Crippen, K.J., Hartley, K.: Promoting Self-Regulation in science education: Metacognition as part of a broader perspective on learning. Research in Science Education 36(1), 111–139 (2006)
8. Schunk, D.H., Zimmerman, B.J. (eds.): Self-regulation of learning and performance: Issues and educational implications. Erlbaum, Hillsdale (1994)

9. Schunk, D.H., Zimmerman, B.J.: Self-Regulated Learning: From Teaching to Self-Reflective Practice, 1st edn. Guilford Press (1998)
10. Symons, S., Snyder, B.L., Cariglia-Bull, T., Pressley, M.: Why be optimistic about cognitive strategy instruction. In: McCormick, C.B., Miller, G.E., Pressley, M. (eds.) Cognitive Strategy Research, pp. 1–32. Springer, Berlin (1989)
11. Zimmerman, B.J.: Self-regulated learning and academic achievement: An overview. Educational Psychologist 25(1) (1990)
12. Zimmerman, B.J.: Becoming a Self-Regulated learner: An overview. Theory Into Practice 41(2), 64–70 (2002)
13. Zimmerman, B.J., Paulsen, A.S.: Self-monitoring during collegiate studying: An invaluable tool for academic self-regulation. New Directions for Teaching and Learning 1995(63), 13–27 (1995)

Extensible Network Appliance Platform (e-NAP)

Russell Cua, Wrenz Go, Kaiser Wee, Mary Joy Yu, and Karlo Shane Campos

De La Salle University, Manila, Philippines

Abstract. Computer-based devices such as network appliances have become quite a norm lately. These appliances are equipped with a specific special ability. Some examples includes being able to store multimedia files into the appliance as a storage device or being able to share files through the network or internet as a shared resource and many more. As technology advances, people's expectations continue to grow. Existing network appliances either do not have the ability to merge with other appliances or are not able to work fully functional when used together with other appliances. This is because different network appliances have different frameworks. With this kind of problem, software developers will have a difficult time coding new applications or merging all the different appliances 1.The Extensible Network Appliance Platform (e-NAP) is a kind of network appliance that allows software developers to use it as a multi-purpose network appliance and add new functions to the device for future purposes. The e-NAP, like computers, is able to support different peripherals connected to the appliance, including foreign exchange office (FXO) components. Also, an extensible application programming interface (API) is given for other software developers to be able to make new codes for adding new functions to the e-NAP. With the e-NAP, both software developers and enthusiasts are able to interface different programs and appliances to the e-NAP without making codes of different framework.Based from the findings, each module in the e-NAP is working using the APIs provided. For the Camera class of the Media module, the client can capture images using the web cam connected to the e-NAP Server or their own computer. For the Audio class of the Media module, the client can play audio files through speakers connected to the e-NAP Server or their own computer. In addition, the client can record sounds using microphones connected to the e-NAP Server or their own computer. For the Network API module, clients are able to connect sequentially or simultaneously to the e-NAP Server without any problems. The e-NAP FXO module allows the clients to make and receive outside calls.

Keywords: e-NAP client, Telephony, Service Provider.

1   Introduction

Network appliances are computer-based devices that serve a specific purpose. Some examples of network appliances are routers, switches, NAS, and many more. Most network appliances are connected to a network or the Internet. Current network

S. Nishizaki et al. (Eds.): WCTP 2012, PICT 7, pp. 206–220, 2013.

appliances have limited features and can only do a specific task 1. Some of the specific features of network appliances are to provide storage, additional shared resource, and telephony solutions.

Some of the existing network appliances out in the market nowadays are Xtreamer Side Winder, Western Digital ShareSpace, Edimax Broadband Router with Print Server, FreeSWITCH, and Asterisk. Xtreamer Side Winder is a storage solution network appliance that enables users to upload and download multimedia files into the appliance and play it for viewing on the TV 10. Western Digital ShareSpace is a network attached storage solution that expands the storage needs of users and allows different users to share the storage space through the network 7. Edimax Broadband Router with Print Server is a kind network appliance that makes the printer a shared resource in the network by allowing the printing of documents to a printer connected to the router 9. Lastly, FreeSWITCH11 and Asterisk 8 are IP-based telephony solution that offers voice-gateway services and digital facsimile services for a local area network (LAN). It can be installed into a server or a network appliance.

These different kinds of network appliances are made from different frameworks. If there are certain scenarios where different network appliance purposes are needed, connecting these different kinds of network appliance together may lead to interoperability complications and require software developers to write tons of codes to make applications work on these appliances 4.

This paper is about e-NAP, a network appliance platform that supports different peripherals and provides an API to allow creation of new or additional codes. Linux operating system is used for the platform since Linux can access all drivers connected in the platform through the command line. Java is universally used to support both the server and the client 3.

2   Extensible Network Appliance Platform (e-NAP)

The e-NAP is an undergraduate thesis project of the proponents. Its main objective is to produce a Java API wherein users can access different peripherals connected to the server through the use said API. The e-NAP is not a commercial-ready product, it is only a trial concept.

e-NAP has two elements. First is the e-NAP Server which serves as the server side program. Second is the e-NAP Client API which is provided for the clients to connect to the e-NAP.

The e-NAP allows different components to be connected to it. It can also connect with other computers through the network. However, the highlight of the e-NAP is that it has an extensible API (e-NAP Client API) which allows software developers to extend or to add functions to the e-NAP.

The e-NAP is modular and the interaction of different component depends on the client. When the client only asks to capture images from the camera, it does not need to interact with other components such as the FXO module or the Audio module. Since it is not asking any data from the other components thus it does not need to access hardware devices such as the speaker or microphones.

The e-NAP supports peripherals such as audio jack microphone input and speaker output, FXO ports, Local Area Network (LAN) ports, and Universal Serial Bus (USB) powered devices. USB powered devices only include keyboards, mouse, web cams, microphones, and speakers.

2.1   e-NAP Server

The e-NAP server runtime is the software running inside the e-NAP server. It is made from the server API. The server API is composed of the network, camera, speaker, and microphone API. Each of these has corresponding managers that handle their process. The managers are responsible for initializing the devices. The data of each manager are stored in the e-NAP collection so that different users will be able to access the data at anytime. The e-NAP Client Handler is responsible for handling the processes and relaying to the device drivers what to do or process.

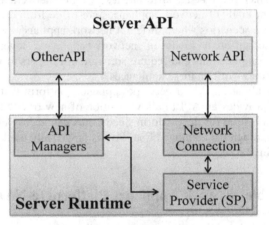

Fig. 1. e-NAP Server Overview

Figure 1 shows an overview diagram of the e-NAP Server. The Server API is the API of the e-NAP Server.

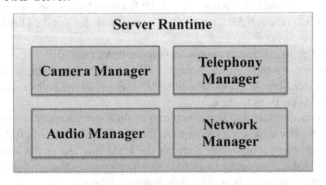

Fig. 2. e-NAP Server API Managers

APIs found inside the e-NAP Server API has a corresponding API Manager that handles the work in each of the API. The list of different API Managers can be seen in Figure 2.

Once the e-NAP Server API managers have gathered all the necessary data and processes, it will then send them to the e-NAP collection. The e-NAP client handler handles all the incoming connections to the e-NAP server. If needed, the e-NAP client handler will send the required data from the e-NAP collection to the e-NAP client through the network with the use of the e-NAP Network API.

The e-NAP Server Runtime is the engine of the e-NAP Server where the e-NAP Server API will be running on. It is composed of threads that handle both the e-NAP Server API managers and e-NAP client handler. The threads are responsible for continuously running the e-NAP service provider, e-NAP API managers, and the e-NAP Server API.

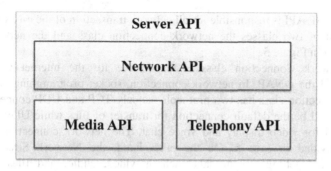

Fig. 3. e-NAP Server API

As seen in Figure 3, the e-NAP Server API is composed of the media API, network API, and telephony API. Each e-NAP Server API class consists of different methods and each has its own methods in order to access the corresponding device.

The Media API is responsible for all input and output media. It is composed of three classes namely, camera class, and   audio class.

The camera class of the Media API utilizes the Video 4 Linux (V4L) programming interface to access web cams connected to the system and to capture images. The video display class of the Media API also uses the V4L programming interface in order to capture video using a web cam. The video display class is only responsible for capturing videos and displaying it to the users.

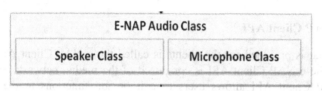

Fig. 4. Audio Class Diagram

The Audio class of the Media API is composed of two components, the microphone class and the speaker class, as shown in Figure 4. The microphone class utilizes the Java sound in order to access microphones connected to the e-NAP Server and to receive audio streams. On the other hand, the speaker class can use either Java Sound or Java Media Framework (JMF) to access the speakers connected to the e-NAP. Then with the use of the JMF, audio can be produced.

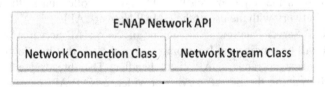

Fig. 5. Network API Diagram

The Network API is responsible for all network transaction of the e-NAP system. It is composed of two classes the network connection class and the network stream class, as seen in Figure 5.

The Network Connection class is responsible for the internet and network connection of the e-NAP. In network connection, socket programming is used. The network connection class has two protocols, namely TCP and UDP connection. TCP connection will be the default connection for transfer of files while UDP connection will be used for video display and voice chat. The network connection class uses Sockets provided in Java 5. On the other hand, the Network Stream class is responsible for getting stream data such as video, audio, and files. It is also responsible for sending these data stream to the other network or host.

The Telephony API utilizes Asterisk-Java to access the FXO device and in order to make FXO transactions work. The telephony API requires that a telephone and a telephone line must be connected to the FXO ports.

Lastly, the Shared Object API is responsible for the sharing and accessing of files stored in the e-NAP to the e-NAP clients. For example, images captured by the e-NAP can be shared to the e-NAP clients through the shared object API.

The e-NAP Server API will also be provided to the users however, changes cannot be made upon it. The design is made as such in order to avoid accidental changes made by other software developers. If software developers want to add new codes or make minor changes to the codes, the e-NAP Client API is provided for that purpose. The e-NAP Client API serves as an extension of the e-NAP Server.

2.2   e-NAP Client API

The API that is provided for the clients is called the e-NAP Client API. As seen in Figure 6, the e-NAP Client API is composed of the media, network, and telephony. The e-NAP Client API allows users to edit or add functions to the e-NAP. The functionality of the e-NAP Client API is almost the same as the e-NAP Server API except that changes made to the e-NAP client API will not affect the original codes of the e-NAP since the e-NAP Client API is only an extension.

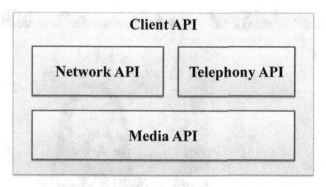

Fig. 6. e-NAP Client API Overview

2.3   Web Management Interface

The Web Management Interface or WMI is a user interface provided in the e-NAP Server. The WMI is an avenue where users which has access to the server can stop devices, modify camera and audio settings, see list of audio and camera devices connected to the server, and change the default settings of the devices when devices are instantiated without specific values.

3   Experiments and Results

Currently, the e-NAP Modules are fully functional. The group has made sample programs that use the different modules individually and collectively in order to verify that each module is individually working and can be integrated with each other. The Camera, Speaker, Microphone APIs can be used by the client to access their own devices. While to access devices found on the server, the Network API should be used. The initial configuration of the e-NAP Server in the experiments it runs on a Linux operating system and uses Sun Java JDK 6.

The group verifies that the Camera API of the e-NAP is working by running a sample program. This sample program also uses the Network API of the e-NAP. The sample program can be run on both the e-NAP Server and the client. The sample program allows clients connected to the server and running the same program chat with each other and also view the web cameras of each other. The server is able to save frames sent by the clients and send it to other clients when requested.

Figure 7 shows the GUI of the sample program. The window contains 5 parts: The message board, the camera frame, the chat box, the dropdown box, and the send button. The message board displays all messages sent by all the clients running the program. The camera frame displays the chosen client of the user. The drop down box is where the user can choose which camera the user wants to view. The chat box is where the user will type the message the user wants to send. The send button sends the message to the server for it to be displayed in the message board.

Fig. 7. Camera class Sample Program

With the use of the sample program, the group is able to determine that the Camera API of the e-NAP is working. Client can access web cams connected to the e-NAP Server through the network by choosing the IP address of a specific client in the drop down box. The group is also able to determine that the Network API of the e-NAP works because messages inputted in the chat box were sent, also the camera frames that each client was sending was sent to the e-NAP server and was viewed by other clients.

By using a code to detect how many frames the e-NAP server is sending and how many frames the client is receiving, the group is able to determine that the video quality is good. When tested, the e-NAP server sends a maximum of six frames wherein the client receives approximately four to six frames.

The group also verifies that the Telephony API of the e-NAP is working by running a sample program. This sample program also uses the Speaker and Microphone APIs of the e-NAP since the Telephony API uses the Speaker and Microphone APIs for the audio. The sample program allows clients connected to the server to make and receive calls through the use of the Telephony API and the FXO port of the e-NAP server which is connected to the telephone line.

Figure 8 shows the sample program for the FXO API of the e-NAP. The window visible upon starting the program shows eight parts, the dial pad, the text boxes, the call button, answer call button, hang up button, the busy here button and the status panel. The first text box is where user will input the telephone number of the one the user is going to call. The text box after the "@" symbol is where the IP address of the server should be inputted. The Call button allows the user to call the dialed number/ the number in the text box. The answer call button allows the user to answer incoming calls. The hang up button allows the user to stop the current call. The busy here button allows the user to send a busy tone on the incoming call. The dial pad allows the user to input extension numbers to, in case of calling phone lines which make use of extension numbers.

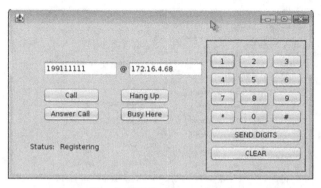

Fig. 8. FXO Sample Program

With the use of the FXO sample program, the group was able to determine that the FXO API is fully functional. Users were able to make calls and receive calls using the sample program. The users were also able to converse with the person on the other line. The group also determined that since the Telephony API use the Microphone and Speaker APIs, the said APIs work because the user and the person on the other line were able to converse.

To ensure that the Speaker and Microphone APIs of the e-NAP Media Module is working, the group tested the API using two sample programs, the first program is the FXO sample program which is explained above, the second sample program lets the user choose whether the user wants to record or play audio locally or through the network. The group has tested that it can play audio files using speakers connected to the client and the speakers connected to the e-NAP Server using the sample program. The sample programs were run using the client. The sample program was able to access speakers and microphones connected to the client's machine and also the e-NAP server's speaker and microphone.

Figure 9 shows the GUI of the sample program of the Audio Sample Program. To test if the client is able to access the speakers and microphone device connected to it. The sample program also tests whether the client can connect to the e-NAP Server remotely and use the audio devices connected to the e-NAP Server. The sample program enables clients to record and play back sounds using either their own audio devices or the e-NAP Server's audio device. The Audio Sample Program has four tabs. When the program is first run, only two tabs are available for use. The Set Audio Device tab and the Play Audio File tab. The Set Audio Device tab allows the client to initialize the audio settings of the microphone and speaker device.

The Play Audio File tab allows the user to play audio files found in their repository. As shown in Figure 10, click on the "Open file to play" button to select an audio file to play. Then a new menu will appear asking for an audio file to open. Once an audio file is selected, click on the "Click to Select Audio File to Play" button and file chooser menu will appear. The client can then choose an audio file and click "Open". Once an audio file is chosen, the file name will be displayed on the sample program, as shown in Figure 10.

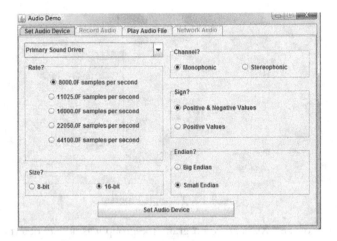

Fig. 9. Audio Sample Program-Set Audio Device Tab

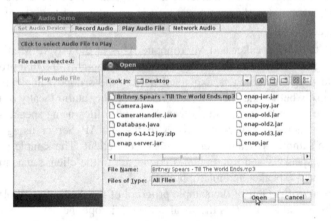

Fig. 10. Play Audio File Tab

The Record Audio tab and Network Audio tab will only be enabled once the device settings have been set in the Set Audio Device tab. The Record Audio tab is shown in Figure 11. The Record Audio tab allows the user to record sounds using the microphone device connected to their computer system. In this tab, the client can choose the type file he wants to record, the file path, file name, and duration of recording. Once the client has decided on the settings, simply click on the "Record Audio" button to record sounds. To stop recording, simply click on the "Stop Recording" button.

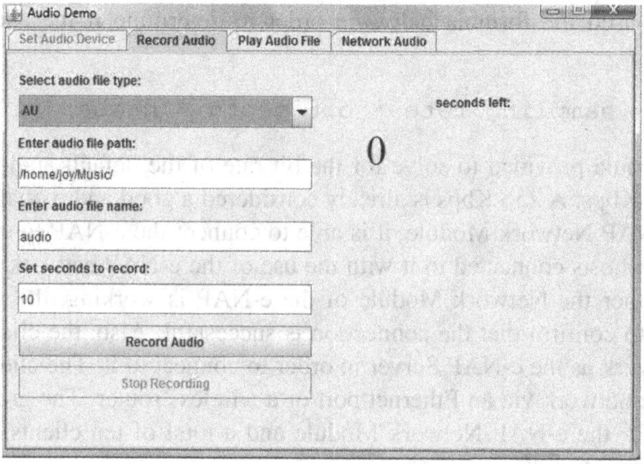

Fig. 11. Record Audio Tab

The Network Audio tab allows the client to connect to the e-NAP server and use the e-NAP server's microphone and speaker. As shown in Figure 12, the client must enter the IP address of the e-NAP Server and Select the type of protocol he wants to use before connecting to the e-NAP Server. Once the client is connected to the e-NAP Server, the buttons "Use Server Audio Device" and "Stop Server Audio Device" will be enabled. To record and play back audio using the audio device connected to the e-NAP Server, simply click on the "Use Server Audio Device" button. To stop the recording and playing sounds, simply click on the "Stop Server Audio Device" button.

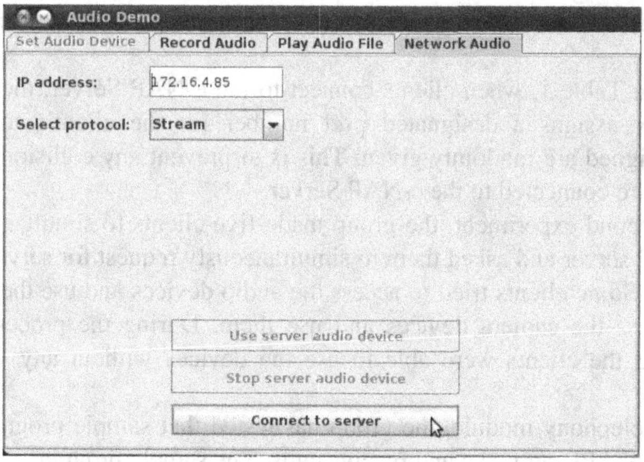

Fig. 12. Network Audio Tab

The group used the formula below in order to determine if the audio quality is good.

```
Bit rate = sampling rate * bit depth * number of channels
```

Using the formula provided to solve for the bit rate of the default audio settings, the bit rate is 256 Kbps. A 256 Kbps is already considered a good audio quality.

For the e-NAP Network Module, it is able to connect the e-NAP to the clients and can handle ten hosts connected to it with the use of the e-NAP network API. In order to check whether the Network Module of the e-NAP is working, the group tries to send a string to confirm that the connection is successful. Also, the client must be in the same network as the e-NAP Server in order to connect to it. The clients can either connect to the network via an Ethernet port or a wireless router. The group made two experiments for the e-NAP Network Module and a total of ten clients are tested for the two experiments.

Laptop-PT
e-NAP

Linksys-WRT300N
myWifi

Laptop-PT
CLIENT

Fig. 13. Network Topology for One Client Testing

For the first experiment, the clients try to access the e-NAP through the network sequentially. He group used the network topology shown in Figure 13. Both the e-NAP server and the client are connected to a wireless network connection called "myWifi".

As seen in Table 1, when clients connect to the e-NAP Server, the e-NAP Server automatically assigns a designated port number for the client's usage. The port numbers assigned are randomly given. This is to prevent any collision between other clients who are connected to the e-NAP Server.

For the second experiment, the group made five clients to simultaneously connect to the e-NAP server and asked them to simultaneously request for services from the e-NAP server. Some clients tried to access the audio devices and use them while others tried to access the camera devices and use them. During the process, no problem occurred and the clients were able to use the devices without any network failure occurring.

For the Telephony module, the group has tested that sample programs made with the Telephony API can make and receive calls. The sample program can be run with a client which is connected to the e-NAP server. The sample program can make, hang up, and receive calls.

Table 1. Network Connection for One Client Test Results

Client #	Result	Port Number
1	Successfully Connected	14083
2	Successfully Connected	17624
3	Successfully Connected	11840
4	Successfully Connected	19338
5	Successfully Connected	24842
6	Successfully Connected	9890
7	Successfully Connected	9436
8	Successfully Connected	25870
9	Successfully Connected	2445
10	Successfully Connected	2450

Table 2. FXO test results

FXO Test		
Trial	Receive/Make	Pass or Fail
1	Receive	Pass
2	Receive	Pass
3	Receive	Pass
4	Receive	Pass
5	Receive	Pass
6	Make	Pass
7	Make	Pass
8	Make	Pass
9	Make	Pass
10	Make	Pass

As shown in table 2, two modules of the FXO API were tested, the receiving and calling modules. Five trials each were made and the sample program made using the FXO API was able to receive and make calls.

For the Web Management Interface (WMI), the group has tested that the functions that the WMI can do does work. Several tests with multiple trials were made to ensure that the modules of the WMI really work. The modules that were tested were the following: Active Devices, Kill thread, Configuration. To test the Active Devices and Kill thread modules, sample programs that access the camera and the audio devices of the e-NAP server are needed. Once the client has connected to the e-NAP server, the settings of which the client has specified will be visible in the WMI. To test the Active Devices module, the tester edits the settings provided in the webpage. After which, a client must connect in order for the new settings to be recognized. The new settings were implemented in the sample program after recognizing the new settings. For the Kill device module, the tester just chooses a specific device to stop. Once the tester has chosen and submitted what device to stop, a new client connection is needed for it to be recognized by the e-NAP server. The device stopped working after the new client connection. To test the Configuration module, the tester just changes the settings of a selected device. Once the new setting is submitted, the configuration text file found in the desktop will now be changed.

Table 3. WMI Active Devices test results

WMI Test (Active Devices)		
Trial	Device	Pass or Fail
1	Camera	Pass
2	Camera	Pass
3	Camera	Pass
4	Camera	Pass
5	Camera	Pass
6	Mic	Pass
7	Mic	Pass
8	Mic	Pass
9	Mic	Pass
10	Mic	Pass
11	Speaker	Pass
12	Speaker	Pass
13	Speaker	Pass
14	Speaker	Pass
15	Speaker	Pass

As shown in table 3, multiple device types were used in the testing of the WMI Active Devices test, and these devices had 5 trials each. The table also shows that the tests made all resulted into successes, meaning the settings of the devices were changed.

Table 4. WMI Configuration Test Results

WMI Test (Config)	
Trial	Pass or Fail
1	Pass
2	Pass
3	Pass
4	Pass
5	Pass

Table 4 shows that the WMI can change the default settings of the Configuration file.

Table 5. WMI Kill Device Test Results

WMI Test (Kill Device)		
Trial	Pass or Fail	Device
1	Pass	Camera
2	Pass	Camera
3	Pass	Camera
4	Pass	Camera
5	Pass	Camera
6	Pass	Mic
7	Pass	Mic
8	Pass	Mic
9	Pass	Mic
10	Pass	Mic
11	Pass	Speaker
12	Pass	Speaker
13	Pass	Speaker
14	Pass	Speaker
15	Pass	Speaker

As shown in table 5, multiple device types were used in the testing of the WMI Kill Device test with 5 trials per device. The results show that the WMI Kill device is able to stop the currently running devices.

4   Future Work

Since the e-NAP API is extensible, future users may add new modules to the e-NAP. The future users may also choose to improve and/or add new features to the existing modules of the e-NAP Server and Client APIs.

Acknowledgements. We would like to acknowledge the help of Mr. Alexis Pantola, Ms. Jocelynn Cu, Mr. Gregory Cu, Ms. ArlynOng, and Mr. Martin Villarica. Mr. Alexis Pantola has helped us clarify and understand some issues regarding the FXO/ FXS port. We would like to thank Ms. Jocelynn Cu, Ms. ArlynOng, and Mr. Gregory Cu in evaluating and interpreting some of the system's test results. We would also like to thank Mr. Martin Villarica for giving us some advice on how to setup the FXO port.

References

1. Network appliance (n.d.), Webopedia: http://www.webopedia.com/TERM/N/network_appliance.html (retrieved March 14, 2011)
2. The Java Native Interface (JNI) (August 31, 2009), think-techie.com: http://www.think-techie.com/2009/08/java-native-interface-jni.html (retrieved July 14, 2011)
3. JAVA, The Java Programming Language (December 16, 1997), engine: http://groups.engin.umd.umich.edu/CIS/course.des/cis400/java/java.html (retrieved November 27, 2011)
4. Why Network Appliances Suck and What to Do About It (March 14, 2007), Jon's Network: http://jonsnetwork.com/2007/03/why-network-appliances-suck-and-what-to-do-about-it/ (retrieved March 12, 2011)
5. Socket Programming (n.d.), buyya: http://www.buyya.com/java/Chapter13.pdf (retrieved August 6, 2011)
6. What is Videoconferencing and Telepresence? (n.d.), dimensiondata: http://www.dimensiondata.com/RGN/UK/SOLUTIONS/VISUALCOMMUNICATIONS/Pages/Home.aspx (retrieved November 27, 2011)
7. Rasnake, D.: Western Digital ShareSpace Network Storage System Review (March 10, 2008), Notebook Review: http://www.notebookreview.com/default.asp?newsID=4625 (retrieved December 5, 2011)
8. Solomon: Asterisk IP PBX review (April 23, 2006), Solomon's VoIP World: http://solokay.blogspot.com/2006/04/asterisk-ip-pbx-review.html (retrieved December 5, 2011)
9. Wireless 3G Broadband Router. (n.d.), Edimax: http://www.edimax.com/en/produce_detail.php?pl1_id=1&pl2_id=2&pl3_id=170&pd_id=312 (retrieved December 5, 2011)
10. Xtreamer Side Winder (n.d.), Xtreamer: http://shop.xtreamer.net/products/Xtreamer-SideWinder.html (retrieved December 5, 2011)
11. FreeSWITCH. (n.d.). Welcome to FreeSWITCH, FreeSWITCH: http://www.freeswitch.org/ (retrieved December 5, 2011)

Marker-Less Gesture and Facial Expression Based Affect Modeling

Sherlo Yvan Cantos, Jeriah Kjell Miranda, Melisa Renee Tiu,
and Mary Czarinelle Yeung

De La Salle University-Manila 2401 Taft Avenue, Manila 1004

Abstract. Many affective Intelligent Tutoring Systems (ITSs) today use multi-modal approaches in recognizing student affect. These researches have had achieved promising results but they have their own limitations. Most are difficult to deploy because it requires special equipment/s which are disruptive to student activities. This work is an effort towards developing affective ITSs that are easy to deploy, scalable, accurate and inexpensive. This study uses a webcam and Microsoft Kinect to detect the facial expressions and body gestures of the students respectively. A corpus for 8 students were built and SVM PolyKernel, SVM PUK, SVM RBF, LogitBoost, and Multilayer Perceptron machine learning algorithms were applied to discover patterns over the facial expression and C4.5 was used for body gesture features. The body gestures and facial point distances data sets were used to build user-specific models. The range of f-Measure produced for fusion of gesture and face is 0.017 to 0.342.

1 Introduction

Humans are effective tutors because they see what specific emotions the student feels while studying. This is the reason why they are able to give the proper response to them. This is important because proper response contributes to the improvement of the student's learning [10]. Particularly, in the field of ITS, in Boyer et al.s research, they asked human tutors to behave like ITS to prove how effective it can be [2]. The things that the tutor will say to the students, both the lessons and the side comments, are a read-only basis. How much the students have learned is measured and the Kappa statistics reached 0.86. Given this, Boyer et al. was able to show that ITS also has the capability to teach effectively [2]. The existing systems that target recognition of affect through observable behavior uses limited features. For Li et al., and Butko et al., only the facial expressions of the students are observed. For Bustos et al, Gupta et al., only the body gestures of the students were observed [3] [4] [13] [9]. This is a problem because genuine emotions are expressed not only through facial expressions but also through the movements of our body. Non-verbal actions reveal how a person feels even without saying it [8]. Emotions are manifested through the facial expression, speech, and body movements of a person. When the computer sees the gestures of a person, it will be able to tell what emotions that the user is feeling [7]. There are many researches that focused on studying effective ways

S. Nishizaki et al. (Eds.): WCTP 2012, PICT 7, pp. 221–241, 2013.
© Springer Japan 2013

on how to make the computer understand affect through gestures. It is not enough to focus solely on gestures or solely on facial expressions in detecting emotion. As explained in Ekman et al., for some, accurate emotions are detected through body movements. As for others, it is seen in the facial expressions [7]. The observed limitations of the other researchers will be addressed by using a webcam to record the face and Microsoft Kinect to record the body gestures of the students. The fusion of these two modalities will also be done. As explained in [3] there are certain emotions that cannot be seen solely in the body gestures. Through the observation of their data, they were able to see that the emotion frustration is not seen in the body gestures, but in the face.

2   Significance of the Research

Emotions are expressed in a multimodal fashion that is why it is important to take into consideration both facial expressions and body gestures combined. Also, Bustos et al. recommended that facial expressions be explored because some of the emotions, such as frustration had very expressive faces [3]. By detecting both the facial expressions and body gestures of a student, the system will be able to recognize the student's emotion more accurately. Since only a webcam and Kinect will be needed, this research also offers easy deployment and duplication. Furthermore, because the means will be non-obtrusive, it will offer a comfortable environment for the studying student. Aside from the improvement of affect-aware intelligent tutoring systems, the final output of this research could also be applied in other systems that require affect detection.

3   Corpus Building

8 subjects are asked to study in front of a computer while they are recorded by the Microsoft Kinect and a webcam. The students are asked to study for a minimum of 45 minutes. Two students had the chance to be recorded for three sessions, three students had the chance to be recorded for two sessions, and three students had the chance to be recorded for a single session. So far, for all the recordings that were made, 45 minutes to an hour is just enough to finish a particular lesson. The materials that they are studying are their academic requirements so that the five academic emotions (flow or engagement, boredom, confusion, frustration, neutral) is felt and shown naturally during the recording [11]. Right after the recording session, the students are asked to view the complete recorded video from the webcam and to list down the starting and the end time of the emotions that they felt from the 4 academic emotions, boredom, confusion, flow, and frustration (including neutral), during a specific period. For example, for time 00:00:00 to 00:01:00 the student felt bored. The student will take note of that and type it in a text pad. They are also asked to list the level of difficulty of what they are studying if it is easy, medium, and during a specific period of time. The level of difficulty was determined by consulting the screen capture video which simultaneously recorded all their activities on the computer while they were studying. It was explained to

the students that there are four emotions that they could choose from (boredom, confusion, flow, frustration). The annotation period takes about 15-30 minutes because the video was either played normally or fast forwarded. Usually, the student asks the researchers to fast forward the video, then ask them again to pause or play it normally when there is a significant change in gesture, or when he feels that a change in emotion occurred. This method is based on what D'mello et al. and Bustos et al. experimented on [3] [6]. The segmentation of gestures was done by the researchers manually. Segments will be according to what the researchers consider to be a gesture (e.g. scratching of head, raising of brow, leaning forward). Throughout the data collection, there were 2-3 sessions for each student.

Fig. 1. Data collection set-up

Microsoft Kinect is positioned 1.5 meters, and at an of angle 45 degrees, to capture the student's frontal view. Video acquisition using the webcam and Kinect is done simultaneously. Figure 1 shows an example of the studying space of the student. The student is required to study in front of a laptop, with Microsoft Kinect positioned inside the cabinet, and the computer below it is the one used to manipulate Microsoft Kinect.

3.1   Feature Extraction

Body Gesture Extraction by Microsoft Kinect. Microsoft Kinect is used to extract the upper body joints namely the head ,shoulders (center, left and right), elbows (left and right), wrist (left and right), and hands (left and right). Each joint has its X, Y, and Z coordinates where X is the horizontal coordinate, Y is the vertical coordinate, and Z is the depth. All the coordinates of the joints are written in a CSV (comma separated values) file format. These are the joints that are collected because these joints produces more accurate location compared to the lower half of the body because the student is sitting down. These joints are then processed to produce extracted gestures such as right hand on face, left hand on face, right hand on back of head, left hand on back of head, lean forward, lean backward, body facing right, and body facing left. The upper body joints were extracted with the use of Kinect Natural User Interface (NUI) Skeleton

API. This API contains methods that automatically generates a Skeleton Data for every person it detects. The Skeleton Data consists of a collection of joints, these joints are the head, shoulders, elbows, wrist, hands, hips, knees, ankles and feet. Each joint has its X, Y and Z coordinates which Kinect automatically generates. The values of x can range from approximately -2.2 to 2.2, 0 being the center and -2.2 to 0 being the range of values on the left side from the view of Kinect and 0 to 2.2 on the right side.Values of y can range from approximately -1.6 to 1.6, 0 being the center and -1.6 to 0 being the range of values on the lower half from the view of Kinect and 0 to 1.6 on the upper half. Values of z can range from 0.0 to 4.0, 0 being the nearest point from kinect and 4.0 as the farthest. Since the gestures needed by the researchers are the upper body gestures only the coordinates of the Head, shoulders, elbows, wrists and hands are recorded.

Extracted Gesture Feature Extraction. Kleinsmith et al. discussed in their study that before recognizing emotions using body expressions, the gestures should first be identified [12]. Using the coordinates of skeletal points gathered by Microsoft Kinect, the extracted gestures are computed. The extracted gestures using the X, Y, Z points produced by Kinect are: RIGHTHANDONFACE, LEFTHANDONFACE, RIGHTHANDONBACKOFHEAD, LEFTHANDONBACKOFHEAD, LEANFORWARD, LEANBACKWARD, BODYFACINGLEFT, and BODYFACINGRIGHT. These attributes were classified using the same emotions (boredom, confusion, flow, frustration) which were used in the XYZ data model building. Each attribute will contain only yes and no as values.

RIGHTHANDONFACE, LEFTHANDONFACE, RIGHTHANDONBACK OFHEAD, LEFTHANDONBACKOFHEAD.Each X and Y coordinate of both hands are first checked if it is within the threshold of "hand near head". Based on the test data collection, as well as observations of values from the actual data from the 8 students, the hands of the subject are usually within the threshold of 0.2 units away from the coordinate of the head. The Z coordinate is then used to evaluate whether one or both hands are in front or at the back of the head. If the Z or the hand is less than the Z of the head, this means that the hand is nearer to the Kinect camera compared to the head. Therefore, the hand is labeled RIGHTHANDONFACE or LEFTHANDONFACE. On the other hand, if the hands Z coordinate is more than that of the head, it is labeled RIGHTHANDONBACKOFHEAD or LEFTHANDONBACKOFHEAD.

LEANFORWARD and LEANBACKWARD. The leaning forward and backward is based only on the depth of the center-shoulder. The center-shoulder for Kinect is basically the lower part of the neck, or literally the center of the shoulder points. Before computing the leaning position of the students, the neutral position of the center shoulder is first recorded. This neutral position is taken only from the very first instance of the dataset. For determining whether the students are leaning forward or backward, the depth or Z coordinate of the center shoulder is compared to the recorded neutral position. If the center shoulder is less than the neutral position (or nearer to the camera), it is LEANFORWARD; else, it is LEANBACKWARD.

BODYFACINGRIGHT and BODYFACINGLEFT. The researchers used the Z coordinate of the left and right shoulders to determine if the body is facing to the left or the right. When the Z coordinate of the left shoulder is less than that of the right shoulder, then the body is possibly facing to the right. Otherwise, if the right shoulder's Z coordinate is nearer to the Kinect camera, then the body facing to the left.

Facial Feature Extraction by SAM-D2. SAM-D2 uses the points and distances of the face; it can also recognize emotion [5]. For this particular research, we only used the facial point distance extraction feature of SAM-D2. The emotion recognition of SAM-D2 recognized the six (6) basic emotions, as opposed to the four (4) academic emotions used in this research. Figure 2 shows these 170 distances drawn from the seven points that were used; right eyebrow center, left eyebrow center, right eye center, left eye center, midpoint of the eyes, nose tip, and mouth center, Figure 3 shows the image representation of these points.

Fig. 2. Facial distance used by SAM-D2 (Cu, 2011)

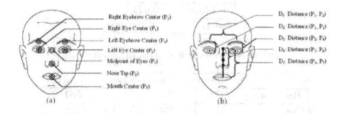

Fig. 3. Important facial points and distances

4   Results and Analysis

Model Building Results and Analysis of extracted Gesture. A user-dependent model was built using the gestures extracted from the upper body joints produced by Microsoft Kinect. There were eight (8) data sets gathered during the data collection phase. Each set of data will be used to build a model user-specific model. The classification was done by Weka. Since the data set is binary, C4.5 was used to build the models. The data consisted of yes or no values that answer whether or not the student is doing a particular gesture. For example, right hand on face: yes, if the student is doing that, and no if not. To test the accuracy of the model, the researchers used 10-fold cross-validation. The data set of all the students were balanced depending on the least number of emotion label on the data set.

4.1   Student 1

The accuracy of the model for Student 1 is 45.94%, the results are shown in Table 1. The results seem to indicate that the model classified all of the emotions equally well. All four (4) emotions were exhibited during the data collection phase. The emotion flow is the most misclassified gesture, some of the gestures for flow were classified as boredom and confusion because the gestures for flow are similar to the gestures that the student has for boredom and confusion. This can be seen in 3. The highest gained f-measure is 0.516 for the emotion confusion as seen in Table 2. The reason for this is because confusion has an action that is not present in the other emotions (tilting of head) which made it is less likely for confusion to be misclassified.

Table 1. Result of C4.5 for Student 1

Accuracy	Kappa Coefficient
45.94%	0.28

Table 2. f-Measure of each emotion of Student 1 using C4.5

Emotion	f-Measure
Boredom	0.389
Confusion	0.516
Flow	0.471
Frustration	0.421

Table 3. Confusion matrix of Student 1

	Boredom	Confusion	Flow	Frustration
Boredom	4084	3343	5414	3581
Confusion	11	8838	7341	232
Flow	321	3573	11624	904
Frustration	170	2080	8542	5630

Fig. 4. The gesture of Student 1 for boredom, confusion, flow (from left to right: hand on face, hand on chin)

4.2 Student 2

All four (4) of the acadamic emotions were exhibited during the data collection phase. As shown in Table 4, the accuracy of the model for student 2 was 55.77%. The highest gained f-Measure was 0.596 for the emotion confusion as seen in Table 5. However, the rest values are not far from 0.555 and 0.534 for boredom and flow respectively. The reason why the f-Measure of the emotions are not far from each other is because they share few similar gestures. The confusion matrix can be seen in Table 6. For example, for boredom and flow, the gestures they have are both left hand on chin and hand on face as seen in 5 and for confusion and flow, both have right hand on chin as seen in 6.

Table 4. Result of C4.5 for Student 2

Accuracy	Kappa Coefficient
55.77%	0.34

Table 5. f-Measure of each emotion of Student 2 using C4.5

Emotion	f-Measure
Boredom	0.555
Confusion	0.586
Flow	0.534
Frustration	0

Table 6. Confusion matrix of Student 2

	Boredom	Confusion	Flow
Boredom	2181	879	1674
Confusion	436	2896	1402
Flow	511	1380	2843

Fig. 5. The gesture of Student 2 for boredom or flow (from left to right: left hand on chin, hand on face)

228 S.Y. Cantos et al.

Fig. 6. The gesture of Student 2 for confusion or flow (right hand on chin)

4.3 Student 3

Student 3 was able to achieve a moderately high accuracy of 75.31%, as illustrated in Table 7. Table 8 shows the f-Measure per emotion of student 3. As observed, the highest attained f-Measure is 0.962 for frustration and the lowest is 0.237 for flow. The reason for this is because frustration had a gesture which none of the other emotions have, which was looking to the side. The reason why flow had the lowest f-Measure is because the gesture that the student does for flow was lean forward and this was also what the student does when he is confused. This gesture can be seen in 7. The confusion of the classifier can examined in Table 9.

Table 7. Result of C4.5 for Student 3

Accuracy	Kappa Coefficient
69.7113%	0.6

Table 8. f-Measure of each emotion of Student 3 using C4.5

Emotion	f-Measure
Boredom	0.577
Confusion	0.605
Flow	0.237
Frustration	0.962

Table 9. Confusion matrix of Student 3

	Boredom	Confusion	Flow	Frustration
Boredom	1163	243	482	17
Confusion	0	1341	518	46
Flow	0	133	1744	28
Frustration	0	4	837	1064

Fig. 7. The posture that student does for flow or confusion (lean forward)

4.4   Student 4

The model for Student 4 was able to attain an accuracy of 63.92% as shown in Table 10. The emotion bordeom had the highest f-Measure because the student had a lot of unique gestures for boredom which were right hand on chin, looking down, playing with hand, and playing with hair. The lowest f-Measure is 0.636 for flow as seen in Table 11. This is because all the gestures that the student did for flow were also present for boredom. By looking at the confusion matrix at Table 12, it is noticeable that for flow, it was mostly classified as confusion. The reason was the one prominent gesture of the student was neutral position and this is true for both flow and confusion. This can be seen in 8.

Table 10. Result of C4.5 for Student 4

Correctly Classified	Kappa Coefficient
63.92%	0.52

Table 11. f-Measure of each emotion of Student 4 using C4.5

Emotion	f-Measure
Boredom	0.758
Confusion	0.74
Flow	0.636
Frustration	0.695

Table 12. Confusion matrix of Student 4

	Boredom	Confusion	Flow	Frustration
Boredom	443	473	0	0
Confusion	51	865	0	0
Flow	122	605	123	66
Frustration	4	0	1	911

Fig. 8. This is the gesture of the student when he is engaged or confused (neutral position)

4.5 Student 5

The model of Student 5 was able to attain an accuracy of 66.92% as shown in Table 13. The highest f-Measure is 0.839 for frustration as seen in Table 14. For all of the emotions except frustration, the neutral position was exhibited. The lowest f-Measure is 0.354 for flow, because all the gestures of the student for flow was also present in boredom, frustration and confusion. These gestures were right hand on face, neutral position, and left hand on face. This can be seen in 9 and 10. The confusion of the classifier can be examined in Table 15.

Table 13. Result of C4.5 for Student 5

Accuracy	Kappa Coefficient
66.92 %	0.56

Table 14. f-Measure of each emotion of Student 5 using C4.5

Emotion	f-Measure
Boredom	0.714
Confusion	0.655
Flow	0.354
Frustration	0.839

Table 15. Confusion matrix of Student 5

	Boredom	Confusion	Flow	Frustration
Boredom	1056	147	88	596
Confusion	14	1698	127	48
Flow	0	1414	452	21
Frustration	0	42	0	1845

Fig. 9. The gesture that Student 5 does for boredom or flow (from left to right: neutral position, right hand on chin)

Fig. 10. The gesture that Student 5 does for frustration or flow (left hand on chin)

4.6 Student 6

Student 6 was able to achieve a 75.31% accuracy, as illustrated in Table 16. The highest f-Measure was 0.917 for confusion and the lowest was 0.625 for flow. This is seen in Table 17. Confusion had the highest f-Measure because neutral position is present for all emotions except for confusion. Flow was able to achieve a lower f-Measure because the gestures done by the student for flow was also present in boredom. These gestures were neutral position, right hand on face, and lean forward. The gestures are displayed in 11. The confusion of the classifier can be seen in Table 18.

Table 16. Result of C4.5 for Student 6

Accuracy	Kappa Coefficient
75.37 %	0.63

Table 17. f-Measure of each emotion of Student 6 using C4.5

Emotion	f-Measure
Boredom	0.693
Confusion	0.917
Flow	0.625

Table 18. Confusion matrix of Student 6

	Boredom	Confusion	Flow
Boredom	2102	213	622
Confusion	7	2916	14
Flow	1021	293	1623

Fig. 11. The gesture that the student does when he is bored or engaged (from left to right: neutral position, right hand on face, lean forward)

4.7   Student 7

Only three (3) academic emotions were exhibited during the data collection phase. As shown in Table 19, the model of Student 7 was able to reach 50.76% accuracy. The highest f-Measure attained was 0.641 for flow and the lowest f-Measure attained was 0 for frustration, as seen in Table 20. The f-Measures for the emotions were quite low because all the emotions shared similar gestures. The neutral position was present in all three emotions; confusion and flow shared the lean backward gesture. These gestures can be seen in 12 and 13. The confusion matrix of the classifier can be examined in Table 21.

Table 19. Result of C4.5 for Student 7

Accuracy	Kappa Coefficient
50.76 %	0.26

Table 20. f-Measure of each emotion of Student 7 using C4.5

Emotion	f-Measure
Boredom	0.271
Confusion	0.539
Flow	0.641

Table 21. Confusion matrix of Student 7

	Boredom	Confusion	Flow
Boredom	1066	4121	466
Confusion	558	4267	528
Flow	581	1796	3276

Fig. 12. This is the gesture that the student does when he is confused or engaged (lean back)

Fig. 13. This is the gesture that the student does for all the emotions (neutral position)

4.8   Student 8

The accuracy for Student 8's model was of 44.11% as seen in Table 22. The highest f-Measure attained was 0.581 for boredom and the lowest was 0 for frustration as seen in Table 23. Frustration was not classified correctly because all of the gestures for frustration was also present in the other emotions. These gestures were hand on chin, neutral position, and lean backward. These gestures is displayed in 14 and 15 and 16.

Table 22. Result of C4.5 for Student 8

Accuracy	Kappa Coefficient
44.11 %	0.25

Table 23. f-Measure of each emotion of Student 8 using C4.5

Emotion	f-Measure
Boredom	0.581
Confusion	0.13
Flow	0.571
Frustration	0

Table 24. Confusion matrix of Student 8

	Boredom	Confusion	Flow	Frustration
Boredom	7448	163	1281	0
Confusion	4154	652	4086	0
Flow	1202	102	7588	0
Frustration	3955	193	4744	0

Fig. 14. This is the gesture that the student does when he is confused or frustrated (hand on chin)

Fig. 15. This is the gesture that the student does when he is bored, confused, or frustrated (lean back)

Fig. 16. This is the gesture that the student does when he is engaged or frustrated (neutral position)

4.9   Analysis of Results of Extracted Gestures

Gesture Classification. The models for all the students were able to achieve accuracies ranging from 44.11% to 75.37%. For most students, flow was the most incorrectly classified emotion because the gestures that the students have for flow is also present in the other emotions. Most of the time, students were in a neutral position when expressing the emotion flow. However, the neutral position was also present when the students were expressing other emotions. On the other hand, it was observed that some students had gestures that were unique only for a certain emotion, which gave the model less confusion.

5   Cross Model Comparison

5.1   Comparing a Cross Model Using Face Features

For the Facial Feature Model Cross Comparison, the model for student 4 was used as the basis to test the data of student 2, the model for student 4 was used as a basis to test the data of student 8, and so on. (See Table 25). The diagonal boxes that are in bold represents the results attained when the data set of the model of the student was tested against the data set of the same student. Table 25 shows the result of the cross model comparison for MLP. One could observe that doing a cross comparison model does not yield a high result. This shows that the different students differ in the way they manifest a particular emotion. This means that if Student A frowns when he or she is bored, Student B, or Student C, does not necessarily do the same.

Table 25. Cross model comparison using MultiLayer Perceptron. The table contains the Kappa coefficient of the cross model comparison.

	student 4	student 2	student 8	student 1	student 5	student 3
student 4	**0.9113**	0	0.0186	0.0171	-0.2529	-0.2529
student 2	0	**0.9939**	0	-0.006	0	-0.0549
student 8	0	0	**0.7146**	-0.0034	0	0.0005
student 1	-0.333	-0.2716	0.0434	**0.7553**	0	0.0355
student 5	0	0.1852	0.0207	0.0034	1	-0.0097
student 3	-0.0349	0.0617	-0.01	0.0051	-0.1264	**0.6704**

5.2   Comparing a Cross Model Using Extracted Gestures

The raw X, Y, and Z coordinates that were gathered from Microsoft Kinect were very fine-grained. That is why the raw data was used to compute for the extracted gestures data set (left hand on face, left hand at the back, left hand on chin, right hand on face, right hand at the back, right hand on chin, leaning forward, leaning backward, orientation to right, orientation to left, tilting to the left, tilting to the right). As explained earlier, different students have different gestures that is why when the model of Student A was used as a basis for testing the data of Student B, the result will not be good.

Table 26. Cross Model Comparison of Extracted Gestures using C4.5. The values inside the table are the kappa coefficient of the comparisons of the models.

	student 4	student 2	student 6	student 8	student 7	student 1	student 5	student 3
student 4	**0.5189**	0.0292	-0.1103	-0.1235	0.2159	-0.0043	-0.2701	0.1227
student 2	0.1154	**0.3365**	0.142	-0.2199	0.2392	-0.0776	-0.2206	0.0392
student 6	0.0972	0.087	**0.6306**	-0.2262	0.2469	-0.0951	-0.1994	0.0572
student 8	-0.167	0.0732	0.1874	**0.2548**	-0.2471	-0.04	0.1339	-0.0142
student 7	0.0335	0.1274	-0.1633	-0.1236	**0.2615**	0.0912	-0.1922	0.1108
student 1	-0.2325	0.328	-0.1003	0.079	-0.0977	**0.2792**	0.1318	0.0784
student 5	0.0244	-0.0464	0.3333	0	-0.1395	-0.0195	**0.5589**	0.1563
student 3	0.0957	0.0911	0.3333	0.375	0.2181	0.047	-0.1168	**0.5962**

18 and reffig:student1C4.5 shows the C4.5 pruned tree created by C4.5 classifier of weka. Table 26 tells that the cross model comparison of student 1 and student 2 gave a kappa coefficient of 0.328. The kappa is higher compared to the other cross model comparison. It achieve this Kappa coefficient because of the similarities of some decision nodes in order to have a corresponding emotion.

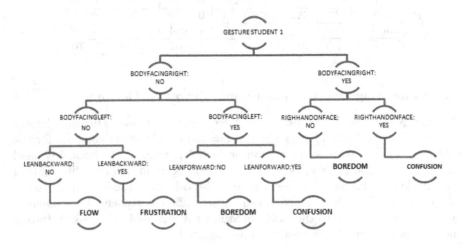

Fig. 17. Shows the C4.5 Pruned Tree of student 1

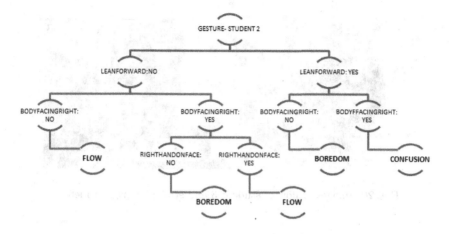

Fig. 18. Shows the C4.5 Pruned Tree of student 2

For example, in order for student 2 to have the emotion FLOW, the nodes it must traverse are LEANFORWARD: NO, and BODYFACINGRIGHT: NO. On the other hand, for student 1 to have the emotion FLOW the nodes it must traverse are the following: BODYFACINGRIGHT: NO, BODYFACINGLEFT: NO, and LEANFORWARD: NO. Both of them involves LEANFORWARD: NO, and BODYFACINGRIGHT: NO to have the emotion FLOW, making the model of student 1 and student 2 fairly similar. Both of them should not be leaning forward and not facing to the right to have the emotion flow as shown in ref-fig:s1s2.

Fig. 19. Images of flow of student 1 and student 2 (right to left)

Table 26 shows that the cross model comparison of student 3 and student 8s cross model comparison gained a kappa coefficient of 0.375. The reason why it gained a fairly high kappa coefficient is similar with the case of student 1 and student 2. The C4.5 pruned tree of student 3 and student 8 is shown in 21 and 22 respectively.

Fig. 20. Images of flow of student 3 and student 8 (right to left)

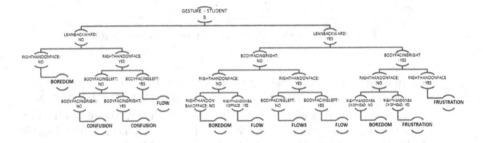

Fig. 21. Shows the C4.5 Pruned Tree of student 3

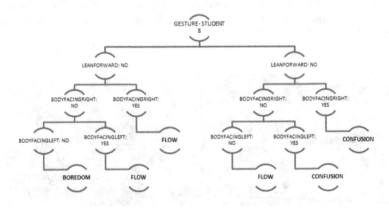

Fig. 22. Shows the C4.5 Pruned Tree of student 8

For example, in order for student 8 to have the emotion FLOW, the nodes it must traverse are LEANFORWARD: NO, BODYFACINGRIGHT: NO and BODYFACINLEFT: NO. While for student 3 to have the emotion FLOW the nodes it must traverse are the following:LEANBACKWARD: YES, BODYFAC-INGRIGHT: NO, RIGHTHAND ONFACE: YES and BODYFACINGLEFT:

NO. Both of them involves BODYFACINGRIGHT: NO and BODYFACIN-LEFT: NO to have the emotion FLOW making the model of student 3 and student 8 somewhat similar. Both of them should not be facing to the right and facing to the left to have the emotion flow as shown in 20.

6   Fusion Model Building and Emotion Recognition Results and Analysis

After extracting the features for body gestures and facial expressions separately, they were fused using decision-level fusion. Decision-level fusion combines the decision of two modalities and evaluates which emotion is more likely fit for a specific instance. The model will be fused using decision-level fusion using sum rule as shown in formula (1). The decision of the gesture model is multiplied to a specific weight which is added to the decision of the facial expression model multiplied to a specific weight. The weights are determined based the error rate or the modality. The higher the error rate, the less weight will be given to it. The weights are computed using formula (2).

$$fusedprob(x) = (gestureprob(x) * w) + (facialprob(x) * w) \qquad (1)$$

$$w = \frac{\dfrac{1}{ErrorRate_m}}{\sum_{m=1}^{N} \dfrac{1}{ErrorRate_m}} \qquad (2)$$

Table 27 contains the f-Measure of the emotion recognition produced using the extracted gestures data, the emotion recognition f-measure of facial expression and the fusion of the 2 modalities. It is observed that fusion does not necessarily yield a higher result when the data is not clean. The only thing that could be concluded given the data and models that the researchers were able to build is that in order to have a good result for fusion, one has to take into consideration both modalities. Kinect does capture body gestures accurately but the other modality which is the capturing of facial expressions is more problematic. Upon reviewing the video recorded, the researchers observed that there are a lot of times when only half of the face of the student is visible from the camera. Also, there were a lot of times when the hands of the student is obstructing the face that is why there are a lot of missing values in the data. The data collected using the webcam was fewer because it had produced a lot of missing values which the researchers removed because it was deemed useless. By monitoring and making sure that the face of the student is seen all throughout the video recording, this will most probably improve the performance of fusion.

Table 27. f-measure of the fusion of Extracted Gestures and Face

Name	Gesture	Face	Fused (Gesture+Face)
Student 1	0.34	0.188	0.333
Student 2	0.283	0.25	0.315
Student 3	0.423	0.25	0.342
Student 4	0.25	0.25	0.25
Student 5	0.007	0.25	0.017
Student 6	0.207	0.069	0.276
Student 7	0.27	0.327	0.313
Student 8	0.153	0.209	0.266

7 Conclusion

The corpus for facial expressions and gestures was built through testing and experimenting with various corpus of spontaneous gestures and facial expression with corresponding emotion labels, an acceptable results for emotion recognition were produced. The feature set was identified by observing the gestures that the eight (8) students were doing during the data collection phase. The features for gestures and facial expressions were extracted using a webcam and Microsft Kinect respectively. Eight (8) participants were asked to study or do any academic requirement in front of Microsoft Kinect and a webcam. At least one (1) data gathering session for each of the eight (8) participants were conducted. Each session is composed of 45 minute data collection and 15-30 minute annotation. The data sets produced by SAM-D2 and Kinect were then fed to Weka and various classification algorithms were used to build models. A separate recording session was then conducted to collect data from these students which as used to test the model.

The model for face and gesture were then fused using the sum rule formula and the weights for each model was computed using formula number 10. The algorithm that is most fit for emotion recognition using body gestures and facial expressions is C4.5. Binary values that describe extracted gestures are preferred because the data set which consisted of points extracted from Microsoft Kinect will produce overfitted models.

The researchers observed that there were some students that have similar gestures while studying. However, this is still not enough to create a general model. The researchers conclude that each user should have a unique model. Fusing will not produce good results if one of the modality is noisy. By monitoring and making sure that the face of the student is always visible in the webcam, as to prevent noisy data, this will most probably improve the performance of facial feature extraction which will also improve the performance of fusion.

References

1. Alepis, E., Stathopoulou, I.-O., Virvou, M., Tsihrintzis, G., Kabassi, K.: Audio-lingual and visual-facial emotion recognition: Towards a bi-modal interaction system. In: 2010 22nd IEEE International Conference on Tools with Artificial Intelligence (ICTAI), vol. 2, pp. 274–281 (October 2010)
2. Boyer, K.E., Phillips, R., Ingram, A., Ha, E.Y., Wallis, M., Vouk, M., Lester, J.: Characterizing the effectiveness of tutorial dialogue with hidden markov models. In: Aleven, V., Kay, J., Mostow, J. (eds.) ITS 2010, Part I. LNCS, vol. 6094, pp. 55–64. Springer, Heidelberg (2010)
3. Bustos, D.M., Chua, G.L., Cruz, R.T., Santos, J.M., Suarez, M.T.: Gesture-based affect modeling for intelligent tutoring systems. In: Biswas, G., Bull, S., Kay, J., Mitrovic, A. (eds.) AIED 2011. LNCS, vol. 6738, pp. 426–428. Springer, Heidelberg (2011)
4. Butko, N., Theocharous, G., Philipose, M., Movellan, J.: Automated facial affect analysis for one-on-one tutoring applications. In: 2011 IEEE International Conference on Automatic Face Gesture Recognition and Workshop (fg 2011), pp. 382–387 (March 2011)
5. Cu, J., Latorre, A., Solomon, K.Y., Tensuan, P.: SAM-D2: Modeling Spontaneous Affect. Unpublished Undergraduate thesis. DeLasalle University
6. D'Mello, S., Graesser, A.: Multimodal semi- automated affect detection from conversational cues, gross body language, and facial features. User Modeling and User-Adapted Interaction 20, 147–187 (2010), http://dx.doi.org/10.1007/s11257-010-9074-4, doi:10.1007/s11257-010-9074-4
7. Ekman, P.: Darwin, deception, and facial expression (2003)
8. Ekman, P., Friesen, W.: Detecting deception from the body or face. Journal of Personality and Social Psychology 29, 288–298 (1974)
9. Gupta, P., da Vitoria Lobo, N., Laviola, J.: Markerless tracking using polar correlation of camera optical flow. In: 2010 IEEE Virtual Reality Conference (VR), pp. 223–226 (2010)
10. Kapoor, A., Picard, R.W.: Multimodal affect recognition in learning environments. In: ACM Multimedia Conference, pp. 677–682 (2005)
11. Kort, B., Reilly, R., Picard, R.: An affective model of interplay between emotions and learning: reengineering educational pedagogy-building a learning companion. In: Proceedings of the IEEE International Conference on Advanced Learning Technologies, pp. 43–46 (2001)
12. Kleinsmith, A., Bianchi-Berthouze, N.: Affectivebody expression perception and recognition: A survey, vol. PP, p. 1 (2012)
13. Li, J., Oussalah, M.: Automatic face emotion recognition system. In: 2010 IEEE 9th International Conference on Cybernetic Intelligent Systems (CIS), pp. 1–6 (September 2010)
14. Nicolaou, M., Gunes, H., Pantic, M.: Continuous prediction of spontaneous affect from multiple cues and modalities in valence and arousal space. IEEE Transactions on Affective Computing PP(99), 1 (2011)
15. Picard, R.W.: Affective computing (1997)

Mobile Indoor Positioning Using Wi-fi Localization and Image Processing

Jeleen Chua Ching, Carolyn Domingo, Kyla Iglesia, Courtney Ngo, and Nellie Chua

De La Salle University, College of Computer Studies,
2401 Taft Avenue Manila 1004, Philippines
courtneyngo@gmail.com

Abstract. At present, there has been an increased interest in indoor positioning systems that propelled researchers to come up with various solutions. A number of these research either use Wi-fi Localization or image processing; each having problems with either accuracy or speed. In this paper, we propose a framework which will use both Wi-fi Localization and image processing to address this problem. The framework used Wi-fi Localization to calculate the estimated position of the user then refine the accuracy through image processing. The designed framework was able to surpass the performance of Wi-fi Localization algorithms in accuracy, and image processing in speed. Different techniques were also applied that improved accuracy and made the system calculate for the location faster.

Keywords: Wi-fi Localization, Image Processing, SIFT Algorithm.

1 Introduction

Numerous mobile applications currently make use of positioning technologies such as Global Positioning System (GPS) in determining the position of the user. These positioning technologies are fairly reliable in an outdoor setting, but are unreliable in an indoor setting due to physical obstructions. GPS, for instance, cannot provide a reliable position indoors because it requires direct line of sight from satellites. There are other approaches that can be used to address this such as Wi-fi Localization and image processing. Although Wi-fi Localization can be more accurate in indoor positioning than GPS, it still provides a large margin of error and are not accurate enough. Inside a building, a few meters of discrepancy may already point to a different room. Image processing can find an image from the database that will match the image captured by the user. This may work well but it requires high computing power and a large storage space, and speed will be compromised. Thus, the researchers proposed the use of both Wi-fi Localization and image processing. The two technologies when combined is intended to address the problems and limitations of each approach.

S. Nishizaki et al. (Eds.): WCTP 2012, PICT 7, pp. 242–256, 2013.

2   Review of Related Literature

2.1   Wi-fi Localization

Wi-fi Trilateration. Spherical Trilateration makes use of at least three routers to calculate the location of the user. It needs the position of the routers and distances between the device and the routers. The distance would be used to generate a circle around each router, and the estimated position of the user will be where the circles intersect [1]. Another kind of Trilateration is the Gauss-Newton spherical Trilateration. Based on the same constraints as spherical Trilateration, the only difference is that it uses a random initial position to calculate the final position. It then increases the value of the initial position until it becomes acceptable to the system. These two Trilateration techniques still have a large distance error. The former has a mean error of 9.7 meters and a maximum error of 23.48 meters while the latter has a mean error of 6.26 meters and a maximum error of 22.13 meters [5].

Wi-fi Fingerprinting. Another method to solve these problems is by using Fingerprinting. The Fingerprinting method has two phases: an online phase, and an offline phase. The signal strength of the various access points are collected at various reference points and stored in a database for the first phase. These signal information includes the interference and multipath patterns of the Wi-fi signal which will help in identifying two spatially close points. In the next phase, the signals obtained from the location of the user will be compared and matched with the training data in the database. A location estimation algorithm will be used to find the best match, such as the k-Nearest Neighbour algorithm. The position of the best match will be the estimated location of the user [6].

2.2   Gyroscope

A gyroscope is a device used to calculate the orientation of a device. The type used in mobile phones is the mechanical gyroscope or Micro-electromechanical (MEMS) gyroscope, which was created with the goal of creating smaller and more sensitive devices. The operating principle used in this gyroscope is that a vibrating object tends to continue vibrating in the same plane as its support rotates. Today, the MEMS gyroscope is found in electronic devices in various devices [8]. By calculating the change in angle of the carrying device to a fixed reference point, the yaw, pitch, and roll of the phone can be determined. The data gathered from the gyroscope will be used to minimize the candidate images when searching the database [4].

2.3   Image Processing

Finding similarities in images involves 3 basic steps. The first step involves finding unique and robust points in an image. To be considered robust, they must be able to withstand different changes which may be in terms of 3D perspective, illumination, rotation, and scale and it must be found in other images of the same scenery [1]. After finding these points, descriptors are made for each point through the creation of feature vectors to describe the neighbouring pixels surrounding it. The last step is the comparison of the features with the features of other images, usually through measurement of the distance of feature vectors [3]. Speeded Up Robust Features (SURF) is a point detector and descriptor created by Herbert Bay, Tinne Tuytelaars, and Luc Van Gool. It focuses on primarily achieving scale and rotation invariance and depends on the creation of reliable descriptors to address changes in perspective or skew. This enables the algorithm to be faster than other computationally heavy algorithms due to high dimensionality such as SIFT [3]. The SURF algorithm uses an approximated Hessian matrix in its detector, coined by the 3 authors as "Fast-Hessian" [3], which uses integral images as a way to shorten computation time [2]. Computing for the Hessian matrix of a point is defined as

$$H(x,\sigma) = \begin{bmatrix} L_{xx}(x,\sigma) & L_{xy}(x,\sigma) \\ L_{xy}(x,\sigma) & L_{yy}(x,\sigma) \end{bmatrix}. \tag{1}$$

where x is a point (x,y) in an image, σ is the scale, and L_{xx}, L_{xy}, and L_{yy} represent the convolution of the Gaussian second order derivative . The location and scale is chosen through finding the determinant of the Hessian matrix. A box filter is then used to approximate the Gaussian second order derivative in order to avoid dealing with the distortion of signals caused by using a Gaussian filter. This, along with the use of integral images, shortens computation time even more [3]. In order to achieve invariance, points must be examined through the different image scales (or the scale space). Unlike SIFT, SURF does not make use of an image pyramid (i.e. down sampling the images) in order to do this. Instead, it increases the filter size and applies the filter to the original image [3]. Non maximum suppression is then used on a $3x3x3$ pixel neighbourhood and the maxima is then obtained [2]. For the creation of the descriptor, the orientation is determined first through the use of Haar wavelet in the circle around the point. After finding the orientation, a square area that includes the orientation is used in order to obtain the descriptor itself [3].

2.4   Related Systems

There has been a similar study on creating a hybrid approach to estimate indoor location using smart phones with a built-in Wi-fi and camera. In [9], the proponents made use of two-dimensional markers which included location information throughout the test environment. The location of the user is then acquired by processing the markers in the picture taken by the camera. On the case that no markers are present in the image, the system would make use of the radio wave strength among nearby radio wave base stations. To eliminate the need for markers, this research proposes a different hybrid approach for indoor location.

RADAR is a system created by Bahl and Padmanabhan which makes use of radio frequency (RF) to locate users. It was created in the purpose of addressing the problems of Infrared (IR) wireless networks in locating users such as its short range and its inability to send data. Through triangulation using RSS, RADAR is able to obtain the users location. The k-Nearest Neighbour algorithm is used to compare among locations and find the best match for a particular signal strength detected.

There are two phases involved in RADAR: the offline phase and the real-time phase. In the offline phase, empirical data, specifically the coordinates of the user using signal strength data and the timestamp, are obtained. The second phase involves creating a signal propagation model which uses computer-generated signal strength data which is similar to what was obtained during the empirical data collection. The Wall Attenuation Factor (WAF) and Floor Attenuation Factor (FAF) propagation models were also used in order to account for signal attenuation due to building floors and walls [10]. RADAR is able to achieve an accuracy of 3 5 meters, with the precision of 50% approximately within 2.5 meters and 90% approximately within 5.9 meters.

Experiments show that the orientation and number of nearest neighbours negatively affect accuracy. Bahl and Padmanabhan [10] stated that the signal strength of a particular location changes along with the users orientation because the user can become an obstruction in some orientations. As for the number of nearest neighbours, having a large k introduces the chance of points far from the expected location to be included which decreases accuracy. The proposed framework aims to be able to accurately locate the user despite changes in the orientation.

The paper [12] presents an accurate and scalable system for determining the user location in an 802.11 wireless LAN (WLAN) framework. Like most RF-based System, there is an offline training phase and an online location determination phase. The noisy characteristics of the wireless channel, however, caused the samples measured in the online phase to deviate significantly from those stored in the radio map created during the training phase, limiting accuracy. The Joint Clustering technique presented in this paper uses a probability distributions of the signal strength to nd the most probable user location given the observed signal strength values. This enhances accuracy and tackles the noisy nature of the wireless channels. The Joint Clustering technique also uses clustering of map locations to reduce the computational requirements, increasing the scalability of the system. A cluster would be a set of locations sharing a common set of access points.

After evaluating the system in an indoor space spanning a 20,000 square foot area, results show a 90% accuracy to within 7 feet with very low computational requirements. Their system is based solely on RF signals which as mentioned earlier can have a lot of noise. The researchers proposes an alternative approach by using a hybrid of Wi-Fi localization and Image processing to mitigate the problems of using solely RF signals.

Horus is an RF-based location determination system that makes use of received signal strengths to infer the user location through probabilistic inference algorithms. To further enhance the accuracy, the paper suggest a model to identify the noisy characteristics of the wireless channel and develop techniques to handle them. Their approach is to treat the samples collected from an access point at a given location as a time series which is a set of observations generated sequentially in time. An autoregressive model would then be used to analyze the stochastic time series that would then be incorporated to the current system. Compared to original Horus system, the modified algorithm show significantly improved performance of about 2 feet, resulting in an average accuracy of about 2.15 feet.

The computational overhead of the modified technique is minimal, and the technique presented in the paper are general and can be applied to other probabilistic WLAN location determination techniques to enhance accuracy.

3   System Description

The proposed system involves a client and a server. The client will collect router information and images from the environment, and send these to the server along with the gyroscope orientation of the device. The two-phase framework will process the router information and orientation first with Wi-fi Localization to get a rough estimate of the position of the device. The second phase includes image processing which will refine the search using the image taken and the rough estimate of the first phase. The resulting position will be sent to the device after.

3.1   System Architecture

There are seven modules in the system. Three modules will reside on the client side: Wi-fi Scan, Gyroscope and Camera. The other four will reside on the server-side: Wi-fi Localization, Database Search, Image Feature Extraction, and Image Feature Comparison. The Wi-fi Scan Module is responsible for detecting the routers in the environment, the Gyroscope Module gives the current orientation of the device, and the Camera Module takes a picture being viewed by the mobile device. All of the information taken with these three modules will be sent to the server side. The Wi-fi Localization Module has two different algorithms to choose from: Wi-fi Trilateration or Wi-fi Fingerprinting; its basic purpose is to get a rough estimate of the position of the mobile device given RSSIs. The Database Search Module will be scouring the database for candidate images based on the Gyroscope orientation and rough position of the device, and it would also be used to return a message to the client later on. The Image Feature Extraction Module extracts the features from the picture taken with the mobile device, and will feed its results to the Image Feature Comparison Module. The Image Feature Comparison Module will compare the resulting features extracted from the said picture, and compare it with the features of the candidate images taken from the Database Search Module. The database ID of the candidate image with the highest matching rate will be used to find for position, which will be sent back to the client.

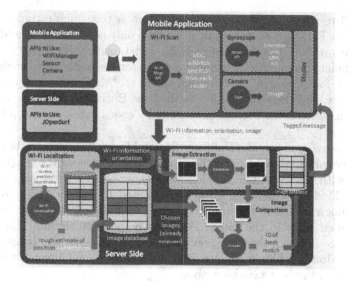

Fig. 1. System Architecture

3.2   Client Side

Wi-fi Scan Module. The Wi-fi Scan Module uses WifiManager API to scan results from detected routers, it gathers the MAC address of each router and its RSSI to the device.

Gyroscope Module. The Gyroscope Module uses the Sensor API to know the yaw, pitch, and roll of the device. The result of both modules will be put in a formatted string, which will be sent to the server.

Camera Module. The Camera Module uses the Camera API to capture the image being viewed by the camera of the device. It will store it in the external storage of the device, and then send it to the server right after the aforementioned formatted string is sent.

3.3   Server Side

Wi-fi Localization Module. The main task of this module is to get a rough position of the mobile device using either Wi-fi Trilateration or Wi-fi Fingerprinting. For Wi-fi Trilateration, it would only look for certain routers which positions are fixed and familiar to the system. From the shortened list, it would search for the lowest RSSI (best signal) to know which area/room the mobile device may be in. A set of correction factors is coupled with the resulting best signal. The correction factors will then be applied to each RSSI to compensate for expected permanent physical attenuation, and these RSSI values would undergo conversion to distance(cm). The set of distances together with the corresponding position of the router will be the input to the Wi-fi Trilateration algorithm.

This would result to an (x,y) coordinate, which is the rough estimate of Wi-fi Trilateration. For Wi-fi Fingerprinting, it would only need the MAC address and RSSI of each router. The server already has a database of RSSI entries for each reference point. From here, it would calculate for the k-Nearest Neighbour. It would check for entries in the database where the MAC addresses in the scan are not empty, and then sort the list by the least Euclidean distance. The rough position calculated is the average of the x and y coordinates of the top k entries.

Database Search Module. The Database Search Module connects to the database of the server and then searches for the candidate images using the rough position estimated with the Wi-fi Localization Module, and the orientation taken from Gyroscope Module. Candidate images are limited to images within 15 meters from the rough position. The resulting set of candidate features of the images would be used later in the Image Feature Comparison Module. These set is already in the form of features previously extracted with the Image Feature Extraction Module. This module will be used later to get the position of the candidate with the best matching rate.

Image Feature Extraction Module. The Image Feature Extraction Module uses JOpenSurf, a Java implementation of SURF, to extract features from the picture captured by the client. The number of octaves used in the SURF extraction is currently 5. The extracted features would be used in the Image Feature Comparison Module together with the set of candidate features of the images.

Image Feature Comparison Module. The Image Feature Comparison Module also uses JOpenSurf. It finds matches between the keypoints of the captured image and the features of each candidate image. In comparison, a filtering algorithm is used to ignore images that have keypoints greater than or less than 50% of the number of keypoints of the captured image.

4   Results and Discussion

For this research, tests were divided into framework effectivity and initial tests, where the framework effectivity will present how the combination of Wi-fi localization combined with image processing can improve the overall performance of the system while the initial testing will contain the tests that will help determine what environment configuration will bring the best result. It will include the following: comparison between Fingerprinting and Trilateration, framework initial tests, comparison between low resolution images and low resolution keypoints, and the use of different techniques. The criteria for testing the framework involves speed and accuracy. The target accuracy is achieving a position that has less than or equal to a $\sqrt{2}$ meter Euclidean error from the actual point. As for speed, the framework must perform faster than using image processing alone. These tests were done using a mobile device and a server. The mobile device used was a Samsung Galaxy S2 which has an 8 Megapixel camera, a Gyroscope,

Wi-fi connectivity, a dual core 1.2 GHz CPU, and is running on Android OS v 2.3.3 (Gingerbread). The server has 4 GB of RAM, an Intel Core 2 Duo 2.66 GHz CPU, and is running on a Windows 7 Ultimate 64-bit operating system. Aside from the tests mentioned, additional simulations were done which involved increasing the position padding value. This was to determine how well the Wi-fi Localization algorithm contributes to the performance of the whole system in terms of speed and accuracy. A simulating program was created to automatically test low-resolution images, masking images, masked low-resolution images, and padding to see their effects to the framework.

4.1 Framework Effectivity

Image Processing and Fingerprinting. Initial tests have been done to the two main algorithms, Wi-fi Localization and image processing. Both of these algorithms must pass a criteria in which the framework of our indoor positioning system depends upon. At this phase, the Wi-fi Localization algorithm must at least turn up with an average discrepancy less than the actual width and length of the area, and the image processing algorithm must at least be able to finish faster than a whole database image search. The environment of this initial testing was under the condition where the environment had minimal obstructions, and the image may or may not be feature-rich.

4.2 Image Processing and Fingerprinting

For image processing, we first ran the system to check all pictures in the database. Though it came up with the correct result, it took approximately 509 seconds to finish. Adding a pad of 1500 centimeters in the coordinate computed by Fingerprinting showed an apparent advantage, finishing at 81 seconds and giving out the correct coordinate.

Table 1. Comparison between Using Image Processing and Image Processing with Wi-Fi Fingerprinting

X	Y	Predicted	Image Processing Total Time (seconds)	Image Processing with Fingerprinting Total Time (seconds)
1700	1900	1700, 1900	517	73
2200	1200	2200, 1200	263	33
2400	1100	2400, 1100	746	137
Sum			1526	243
Average			508.67	81

Although the Wi-fi Fingerprinting technique is a faster method, there is still a need for better accuracy. For this, the research will employ image processing which is rather computation heavy but would be faster than normally since the Wi-fi Fingerprinting can provide a rough estimate to narrow down the database image searching.

Table 2. Comparison between Wi-fi Fingerprinting and Wi-fi Fingerprinting with Image Processing

	Wi-fi Fingerprinting	Wi-fi Fingerprinting with Image Processing
Average Euclidean Error (cm)	694.08	163.82
Time elapsed after receiving client data (ms)	37.66	445.50

To see the effectivity of adding a second phase, a test was made on Wi-fi Fingerprinting alone and Wi-fi Fingerprinting with image processing, measuring the effect on accuracy and speed for evaluation. From the table above, it has shown a 76.3978% improvement in accuracy but a 1209.6226% drop in speed. Nevertheless, the goal of getting a more accurate measure with the proposed two-phase framework was achieved.

Position Padding Changes. A simulation of tests were done which involved changing the position padding value. The purpose of these tests were to see how improving Wi- Fingerprinting would affect the performance of the whole framework in terms of speed and accuracy. The range of position padding values used were from 1 meter to 30 meters. The researchers conducted a simulation of tests using the test data log, which consisted of 32 test items. Instead of conducting Wi- Fingerprinting, the Wi- Localization results came from generating a pair of coordinates at random depending on the padding error given. The padding error ranges from 1 meter to 30 meters, up to twice the original padding. Suppose this batch of simulated tests had a padding error of 5 meters, the random coordinates to be use the random coordinates to be used for database search must be within 5 meters of the actual position. However, 32 items for each batch is a very small data set to infer from. To generate more data, each batch would simulate the initial 32 items for 10 iterations, resulting to 320 test items for each of the 30 batches. After all the tests, the correlation of the position padding value to Euclidean error and the correlation of the position padding value to the total time was computed. The correlation of the position padding value and the Euclidean error was 0.981, while the correlation of the position padding value and the total time was 0.987. Both padding and Euclidean error, and padding and total time are positively correlated. The positive correlation indicates that as the position padding value increases, both the Euclidean error and total time increases. Images of farther areas which can be located in different rooms in the environment are included in comparison which contributes to the increase of error and increase in comparison time.

Both padding and Euclidean error, and padding and total time are positively correlated. The positive correlation indicates that as the position padding value increases, both the Euclidean error and total time increases. Images of farther areas which can located in different rooms in the environment are included in comparison which contributes to the increase of error and increase in comparison time.

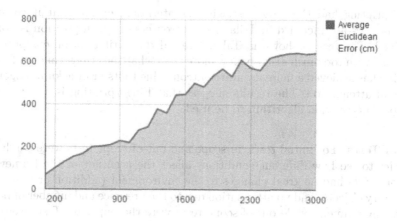

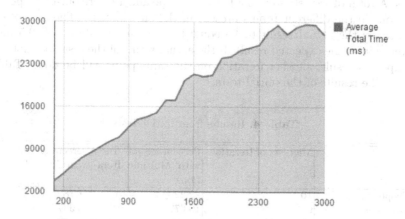

Fig. 2. Graphs showing the relationship between (top) Padding and Average Euclidean Error, and (bottom) Padding and Total time

4.3   Initial Tests

Wi-fi Fingerprinting and Wi-fi Trilateration. Wi-fi Trilateration, the first Localization algorithm, was tested using data taken from reference points, meaning that the mobile device was positioned at the marked coordinates. Wi-fi

Table 3. Comparison between Wi-Fi Fingerprinting and Wi-Fi Trilateration

	Average Error of X (m)	Average Error of Y (m)
Trilateration	14.36	72.88
Fingerprinting	3.76	3.94

Fingerprinting had the same test with Trilateration, except that it considers other routers not included in Trilateration (routers of which position is not determined. The results shown in Table 1 reveal that Trilateration computes for the point with too much discrepancy, more than what the criteria allows. Fingerprinting has achieved a more accurate outcome due to its prior acknowledgement of signal attenuation. The results suggest that Fingerprinting is a much more suitable Localization algorithm to be used.

Initial Tests. For initial tests, in-scope and out-of-scope tests were conducted in order to see how different conditions affect the accuracy of the framework. In-scope tests had no great changes to the environment (addition or removal of temporary objects) and no attenuation due to the presence and number of people in the environment while out-of-scope tests were the opposite. Tests were also done using distinct (unique and easily distinguishable) and indistinct images, and were done on both exact points (marked in the environment) and inexact points. A total of 48 tests were made (36 in-scope and 12 were out-of-scope) and were done in the different rooms present in the environment. The goal of each test was for the actual result to be equal to or to have only a +/- 0.5 meter difference from the expected result. If the actual result of the test does not meet the expected result, another iteration on the same point will be done. Table 3 presents the results of the initial tests.

Table 4. Results from Initial Tests

	Max No. of Iterations	Percentage of Tests with Multiple Iterations	Percentage of Tests with Single Iteration
In-Scope	10	22.22%	77.78%
Out-of-Scope	6	16.67%	83.33%
Both	10	20.83%	79.17%
Average	8.67		

Some tests needed multiple iterations in order to achieve the target accuracy because some captured images in the environment were indistinct. These indistinct images lacked features that could have made them more unique than other similar images. Also, a great change in the scene affected the extracted keypoints. Keypoints that were originally present in the image database could have been removed because of the addition or removal of objects which were originally present. Lastly, attenuation also affected the accuracy. There was greater attenuation during these tests because there were more people and other sources of attenuation present in the environment during testing compared to data gathering.

Low Resolution Images and Low Resolution SURF Keypoints. For the next set of tests, the resolution of the images were reduced from 3264 x 2448 pixels to 2048 x 1536 and 640 x 480 pixels. The average file size of these low resolution images was compared to the average file size of their SURF keypoints in order to see if there was a great difference between the two. It is important to determine which of the two, the raw low resolution images or its keypoints, would be faster to send from the device to the server. The data presented in Table 4 indicates that the difference is negligible. Therefore, sending either SURF keypoints or images from the phone to the server will not drastically affect the speed of the system.

Table 5. Low Resolution Images and Low Resolution SURF keypoints

	Images	SURF points
Number of Files	378	378
Cumulative File Size (bytes)	21,489,492	19,629,663
Average File Size (bytes)	56,850.00	51,930.33

Comparison of Techniques. Different techniques were tested on the framework to see how each affect and possibly improve the accuracy and speed of the system. These techniques involve decreasing the resolution of the images from 3264 x 2448 pixels to 2048 x 1536 and 640 x 480. The images from both the original resolution and the 640 x 480 resolution underwent additional processing including a masking process which removed keypoints of objects that can be easily changed or removed. The techniques that used masking did not use the algorithm that filters images based on the number of keypoints. Since masking removes keypoints, it is possible for an image in the database to be the best match for the captured image even if there is a great difference in the number of keypoints. The techniques shown in Table 5 are as follows: (A) unmasked original resolution (3264 x 2448 pixels) images, (B) unmasked images with a resolution of 2048 x 1536 pixels, (C) unmasked low resolution (640 x 480 pixels) images,(D) masked original resolution images, and (E) masked low resolution images.

Table 6. Summary of Results

	Technique A	Technique B	Technique C	Technique D	Technique E
Average Euclidean Error (cm)	229.50	204.51	163.82	212.27	190.82
Average Comparison Time (ms)	4539.91	1853.28	242.97	6183.81	394.91
Extraction Time (ms)	2962.41	1206.09	164.88	2971.47	192.63
Average Total Time (ms)	7359.97	3097.03	445.5	9192.94	625.79

Based on the results of technique A (using unmasked original resolution images) and technique D (using masked original resolution images), it can be concluded that masking improves accuracy, but at the expense of time. A possible reason why Technique D had a greater time was because masking did not use the filtering algorithm and thus, more candidate images were used in comparison. Although masking decreased the Euclidean error of techniques which used original resolution images, it worsened error in low resolution images, as indicated by the results of technique C and technique E. Decreasing the resolution of images from its original resolution causes loss of detail which removes keypoints, and masking these images further reduces the number of keypoints. This makes comparison and finding a best match problematic.

The results suggest that as the image resolution decreases, the average Euclidean error and total time decreases as well. Higher resolutions introduce more noise as compared to lower resolutions, which could reduce the chances of unreliable keypoints to be extracted and thus, reduce Euclidean error. Among all the techniques used, using low resolution images gave the best results with only 1.64 meters in average Euclidean error.

The proposed framework is able to perform with an accuracy of 1.64 meters and precision of 94% less than 5 meters, 64% less than 1 meter, and 32% less than 0.5 meters.

5 Conclusion

The researchers were able to meet the objective of creating a two-phase framework for indoor positioning using Wi-fi Localization and image processing. Through testing, the group discovered that Fingerprinting outperforms Trilateration. Trilateration fails when only 2 routers are detected and it has a larger average and maximum error compared to Fingerprinting. Further research led the group to choose the SURF algorithm over SIFT, and other image manipulation techniques. According to literature, SURF is a faster algorithm than SIFT, and its speed did not greatly compromise accuracy. The researchers were also able to determine a criteria for evaluating the framework. Our criteria for the framework involves the target accuracy defined by achieving a position that has a $\sqrt{2}$ meter Euclidean error from the actual point, and speed achieved by outperforming image processing alone. The framework was able to meet the criterion for speed. Initial tests on the framework revealed the different limitations of the framework involving distinctness of images, changes in the environment, and attenuation. The results of the next simulations to determine possible improvements on the framework in terms of speed and accuracy led the researchers to conclude that leaving the algorithms to the server rather than the client would be more suitable. Results of other simulations which involved masking and lower resolution images indicate that masking decreases error, at the cost of speed. Masking low resolution images, on the other hand, generated a greater error. Because unmasked low resolution images generated the lowest Euclidean error, this led to the conclusion that using unmasked low resolution images generated the best, most accurate results. The results of simulations, those involving padding, indicate that the system is able to generate much more accurate positions at a faster rate when the Fingerprinting error decreases. Determining suitable algorithms and the results from the tests conducted enabled the researchers to successfully meet the objective of creating a framework for indoor positioning.

6 Recommendations

Future study for the system can include developing a faster algorithm for Wifi Localization and image feature comparison without compromising accuracy. Finding a method to address varied elevation of the environment, non-lateral orientation, and image amplification can also be done. The proposed indoor positioning framework can play an important role in the development of augmented reality systems and other mobile applications. With proper indoor positioning, users will be able to receive much more useful and necessary information from the system. Navigation systems in establishments such as malls and museums, for example, can present information to users upon going to a particular position. Advertisements in malls can also be shown to users in the same manner. Another example are scavenger hunt type of applications which can place digitally rendered items on positions provided by the application.

References

1. Cook, B., Buckberry, G., Scowcroft, I., Mitchell, J., Allen, T.: Indoor Location Using Trilateration Characteristics. In: London Communications Symposium (2005)
2. Bay, H., Ess, A., Tuytelaars, T., Van Gool, L.: Speeded-up robust features (SURF). In: Comput. Vis. Image Underst., pp. 346–359. Elsevier Science Inc., New York (2008)
3. Bay, H., Tuytelaars, T., Van Gool, L.: SURF: Speeded up robust features. In: Leonardis, A., Bischof, H., Pinz, A. (eds.) ECCV 2006, Part I. LNCS, vol. 3951, pp. 404–417. Springer, Heidelberg (2006)
4. Burg, A., Meruani, A., Bob, S., Wickman, M.: Mems gyroscopes and their applications
5. Herranz, F., Ocana, M., Bergasa, L., Sotelo, M., Llorca, D., Parra, I., Llamazares, A., Fernandez, C.: Studying on WiFi range-only sensor and its application to localization and mapping systems. In: IEEE ICRA 2010 Workshops, pp. 115–120 (2010)
6. Mok, E., Retscher, G.: Location determination using WiFi fingerprinting versus WiFi trilateration. Journal of Location Based Services Archive, 145–159 (2007)
7. Navarro, P.: Quan: Wi-Fi Localization using rssi fingerprinting (2010)
8. Novik, K.E.: U.S. Department of Commerce. In: Proceedings of the International Symposium on Advanced Radio Technologies, Colorado (2006)
9. Hattori, K., Kimura, R., Nakajima, N., Fujii, T., Kado, Y., Zhang, B., Hazugawa, T., Takadama, K.: Hybrid Indoor Location Estimation System Using Image Processing and WiFi Strength. In: Proceedings of the 2009 International Conference on Wireless Networks and Information Systems, pp. 406–411. IEEE Computer Society, Washington (2009)

10. Bahl, P., Padmanabhan, V.N.: RADAR: An in-building RF-based user location and tracking system. In: Proc. IEEE INFOCOM, vol. 2, pp. 775–784 (2000)
11. Liu, H., Darabi, H., Banerjee, P., Liu, J.: Survey of wireless indoor positioning techniques and systems. IEEE Transactions on Systems Man and Cybernetics Part C, 1067–1080 (2007)
12. Youssef, M., Agrawala, A., Udaya Shankar, A.: WLAN location determination via clustering and probability distributions. In: IEEE Int. Conf. Pervasive Comput. Commun., pp. 143–151 (March 2003)
13. Youssef, M., Agrawala, A.K.: Handling samples correlation in the Horus system. In: IEEE INFOCOM 2004, Hong Kong, vol. 2, pp. 1023–1031 (March 2004)

Automated Vehicle Entrance Monitor Using Pattern Recognition via Haar Classifiers

Jordan Aiko Deja, Darrell Talavera, Leonard Pancho, and Erika Svetlana Nase

College of Information Technology, Malayan Colleges Laguna, Philippines
japdeja@mcl.edu.ph, {darrell.talavera,
leonard.pancho,erika.nase}@live.mcl.edu.ph

Abstract. The trend on the application of Digital Image Processing has paved for a future consists of an automated workforce where the goal is to minimize the number of errors that is induced by continuous hours of working. This setup is followed in institutions where access to the premises of a vehicle is determined by the detection of an authorized sticker. The study developed a prototype, whose effectiveness, accuracy and speed in automatically detecting the car sticker was tested in real-time. Research results show that the technology can be integrated and can perform better with more fine-tuning.

Keywords: Haar-Classifiers, OpenCV, Machine Vision, Real-time Pattern Recognition.

1 Introduction

Nowadays, technology had continued to expand through different ways or methods that speed up daily processes. Automation of manual processes can be considered as a renowned way of exhibiting the present state of technology. Through this technique, accuracy would be maintained.

Institutions in the Philippine setting employ a security strategy where only authorized vehicles can be allowed to enter private premises. These vehicles are recognized with the installation of a secure car sticker. Vehicles attempting to access the premises will only be granted access if a security guard-on-duty recognizes such sticker installed on the vehicle. On average security guards work 8-hour shifts whose duties involve strenuous activities. As such the possibility of misplaced decisions in granting access to some unauthorized vehicles is not far from happening. Possible causes that might add up to the difficulty in making these decisions might be induced by poor lighting, or impaired vision or even tiredness. With the understanding that computers, unlike humans, who do not get tired and whose performance do not deteriorate over a period time, it has been proposed in this study that the performance of such computers be measured in this similar task.

Rather related studies are being deployed in the setting which includes the use of machine vision on computers to detect the ripeness and redness of harvested tomatoes [1]. Image Processing technologies and Machine Learning techniques are employed by the researchers and recent results have proven to be significant [1]. The success of the study in [1] resulted to a 97% recognition accuracy. The goal of the study at hand is to

S. Nishizaki et al. (Eds.): WCTP 2012, PICT 7, pp. 257–265, 2013.

present a similar approach, by presenting an automated solution to this typical manual process. This paper discusses how automating the sticker-recognition process might be effective, putting up groundwork to future works that would merge with Robotics (especially on the part of controlling the bar used in gates in private premises).

The software would primarily integrate a visual feature through the use of a web camera. The software will detect the Malayan Colleges Laguna (MCL) car sticker through the video feed. In the latter portions of this paper, it is presented that on a right amount of images used for training the classifier, the detection of MCL car sticker was possible. It is shown that the effectiveness of the detection of the template is based on the training images used.

2   Research Design

This study will focus on detecting MCL car stickers to automatically permit the vehicle in entering the premises of MCL. The detection of the car sticker is through the use of pattern recognition via Haar Classifier. The algorithm for detecting an object is already designed and the proponents just generated a Haar classifier[2] from an open source tool for computer vision called OpenCV for detecting the MCL car sticker.

Since, the sticker that was used in this study is the newest revision of MCL car sticker, thus, any old modification of MCL sticker will not be considered and recognized by the software. A web camera will be used for capturing the video. Also, the sound that will be recorded during the process will be rejected. Due to the image training at a specific distance only, the detection of the car sticker will not be effective if its distance from the web camera is too far. Figure 2 describes the combined architecture and methodology design of the study.

The main hardware requirement for the implementation of this study is a web camera. It will be used to get a video feed so the MCL car sticker will be detected. The specific individual activities needed to be done for the study are the following: gathering of positive images having the object (in this case, it is the MCL car sticker), negative images without the object and test images that also contain the object but different from the positive images. These will be done through capturing videos and each of these frames will be saved through another tool called Positive Builder. Another activity to be done is to "train the computer" using the positive and negative images. Lastly, test images will be used to test the Haar classifier using the tool Open CV.

This system needs to go through certain important method which are first the collection of image, second is the training process using Haar-Training method, next is coding implementation and last is to match the object with the dataset which in this case was the cascaded image threshold which is contained in the xml file or the purpose to determine whether the desired object is exist or not[2]. Object detection system is given an image patch of known size or a feature and is to decide whether this features stemmed from an object, or a non object. For the purpose to get a reasonable accuracy of object detection performance, the Haar- classifier is applied to this system. Haar-Classifier encodes the existence of oriented contrasts between regions in the image. A set of these features can be used to encode the contrasts exhibited by an object. The detection technique is based on the idea of the wavelet template that

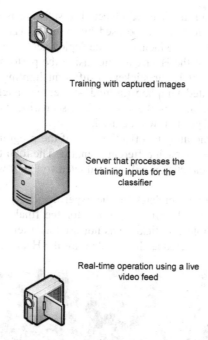

Training with captured images

Server that processes the
training inputs for the
classifier

Real-time operation using a live
video feed

Fig. 2. Implemented research setup

defines the shape of an object in terms of a subset of the wavelet coefficients of the image. Haar-like features are so called because they share an intuitive similarity with the Haar wavelets. Historically, for the task of object recognition, working with only image intensities (i.e. the RGB pixel values at each and every pixel of image) made the task computationally expensive. This feature set considers rectangular regions of the image and sums up the pixels in this region. The value this obtained is used to categorize images.

3   Methodology

The initial part of the experiment is gathering positive, negative, and test images. The collection of positive images contains the object you want to detect; collection of negative images are images that does not contain the object, and the collection of test images are for testing purposes only. Positive images can be easily created using a tool called Positive Builder. Also, to create a vector file that will be used for the training, the proponents need to create a text file with the positive images' path and coordinates of the object to detect. It will then be compiled into a vector file using the createsamples.exe. Another text file will be created for the negative images' path. The vector file and the text file of negative images will be used to train the Haar classifiers [2] that will output extensible markup language (.XML) file; the duration of the training will depend on the number of images that will be used. Finally, .XML file will be used as storage of information of the images compiled in the previous phase and will be used by the program to recognize the MCL sticker in the video stream. If the program detects the MCL sticker, it will display a circle around the detected object. The software is needed to have a deployment or installation before it will be used by the user. The software requirements needed to run the program will be included in the deployment.

The proponents will test the Haar classifier using the performance.exe in OpenCV to recognize the correct MCL car sticker in different lighting setups. Also, the researchers will test the tilted template (around 45 degrees angle) that may possibly occur when already implementing the system. The scenarios stated above are equally distributed in 100 test images that will be used.

The researchers will monitor the effectiveness of the program for the detection of the MCL car sticker tested in each different scenario. The total time of processing the image and validating the car sticker will be evaluated so that it will be efficient for the future use.

The proponents used two templates for the experiment. For the first to fourth trial, researchers used the (a) whole MCL logo and for the final set is the (b) MCL car sticker. However, the whole car sticker was not used as a template. The (c) logo and the school name were considered as the template for the Haar training.

Fig. 1. From L to R: MCL logo, MCL car sticker, and the template used for final trial

The researchers collected positive images, negative images and test images[4]. See Table 3 for the summary of images collected in the five trials. In trial 1, the number of positive images used was 150, for negative images there are 150, and the test images there were 100. Total images used were 300 and all of the images were subjected to create region of interest points by cropping each images in the positive builder program. After 5 hours of training, we tried to use the .XML file of the first trial.

Second trial was just a continued first trial, where the 350 positive and 350 negative images are added to the set of positive and negative images. Therefore, the total number of images of the second trial is 1000. Then the second trial goes to the same process as first trial does. Trial 2 consumed 12 hours before the .XML file was created.

In trial number 3, the positive and negative images used were from the previous trials but 250 positive and 250 negative images were added in the training. The training occurred for 24 hours.

Here is the sample of images used in the first to third trial. From left to right, (a) positive image, (b) negative image, and (c) test image.

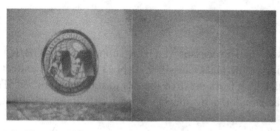

Fig. 2. First to Third Trial Images: (a) Sample positive image, (b) Sample negative image (c) Sample test image

In the fourth trial, the images used were also from the previous trials and also added 500 more images. 250 positive images were used but the logo template is different from the previous trial. The researchers used the logo from the MCL ID to add effectiveness in detecting the logo. Another 250 negative images were also added.

Fig. 3. Fourth Trial Images: (a) Sample positive image, (b) Sample negative image (c) Sample test image

On the final trial of the training, the proponents already used 2000 images for each positive and negative. And, the template was different from trial one to four. The training duration of this trial is 122 hours which is approximately 300% longer than the duration of fourth trial (See Fig 4). It is because the computer that was used in the training is experiencing memory insufficiency thus, it takes the computer to train the classifier much longer than the expected time which is twice the duration of fourth trial.

Fig. 4. Sample positive image (b) Sample negative image (c) Sample test image

4   Research Results

A Netbook with an Atom-based Intel Processor (1 GHz) supported by 1GB RAM memory was used for the training of the Haar Classifier. Table 1 shows data collected by the researchers through trial 1 to 5 from positive images to training duration. The training duration is the time accumulated by the OpenCV haartraining.exe for the creation of XML file.

Table 1. Training duration for each trial

Trial Number	Positive	Negative	Total Training Images	Test Images	Training Duration
1	150	150	300	100	5 hours
2	500	500	1000	100	12 hours
3	750	750	1500	100	24 hours
4	1000	1000	2000	100	35 hours
5	2000	2000	4000	100	122 hours

Note that the test images are different from the evaluation. Test images are used to test if the Haar training was successful in detecting the template and was done in the command line while in evaluation; the proponents used the generated XML file for the software to detect the template in a video feed.

Table 2. Result of Haar Classifier Using Performance.exe

	Trial 1	Trial 2	Trial 3	Trial 4	Trial 5
Hits	9	79	41	91	87
Missed	91	21	59	9	13
Total	100	100	100	100	100

After the training has finished, the researchers tested the Haar classifier (XML file) using the performance.exe of OpenCV. The summary of the result of testing is shown in Table 3. "Hits" means the classifier detected the object in the image while the "Missed" means the classifier was not able to detect the object in the image.

After the researchers have tested the classifier, they evaluated the time the software can detect the template in the video feed given that the position of the template in the video was stable. See Figure 6 for the sample of what the proponents can see if the car sticker was detected.

The farthest distance that the trial 2 can detect the template in the video feed is 16.51cm. Trial 3 can detect the template as far as 28cm while in trial 4 it can detect within the distance 31cm. Trial 5 can detect the template with the farthest distance of 17.5cm (see Figure 7) since the template is different from the trial 1 to 4 and it is more detailed so the computer cannot "see" clearly at far.

Fig. 5. Screenshot of Detected Car Sticker in the Video Feed

Fig. 6. (a) Front view and the (b) side view of the farthest distance the trial 5 can detect

Below are the results of evaluating the Haar classifier in three scenarios. First scenario was at normal lighting and projection. The second scenario was at different projecttion (the sticker was tilted vertically). Lastly, the template was at bright lighting.

Table 3. Result of Testing the Object in Normal Scenario in Milliseconds

	Trial 1	Trial 2	Trial 3	Trial 4	Trial 5
Set 1	Fail	Fail	900	1,000	620
Set 2	Fail	7,900	1,100	400	780
Set 3	Fail	22,400	700	700	970

Table 4. Result of Testing the Object When It Is Tilted Vertically in Milliseconds

	Trial 1	Trial 2	Trial 3	Trial 4	Trial 5
Set 1	Fail	Fail	Fail	Fail	1,170
Set 2	Fail	Fail	Fail	4,000	6,720
Set 3	Fail	Fail	Fail	Fail	2,360

Table 5. Result of Testing the Object with Bright Background in Milliseconds

	Trial 1	Trial 2	Trial 3	Trial 4	Trial 5
Set 1	Fail	3,000	7,300	4,200	5,000
Set 2	Fail	16,000	7,900	5,500	4,000
Set 3	Fail	20,000	6,600	4,900	7,120

During the testing and evaluation in trial 1, the software cannot detect the MCL logo due to small number of training images. Compared to trial 2, which increased the number of total positive images used up to 500; it detects the MCL logo within the time of 7,900 to 22,400 milliseconds within the same distance as it was trained. Trial 1 and 2 has failed to detect the logo when it was tilted vertically. But, trial 2 can detect the logo if it is in bright background with the time of 3000 to 20000 milliseconds.

In trial 2, the number of sets that the template was detected increases compared from trial 1 but there is a greater number of fail detection than the successful detection of the template. The reason behind the detection failure is also the number of training images. The over-all time elapsed in detecting the template is not much fast.

In trial 3, the number of sets that the template was identified raised in contrast from trial 2. The general time elapsed in detecting the template decreased compared from trial 2.

In trial 4, the number of successful detection was greater than the number of fail detection. As a whole, the time elapsed in detecting the template has a little bit difference from the trial 3. The average detection of trial 3 is 900 milliseconds while trial 4 has an average of 700 milliseconds. Also, in this trial, the software can already detect a tilted template, but still, it's not effective since only 1 out of 3 tries was successful.

In the last trial, which is the template that is different from the previous four trials and has the most number of trained images, almost all of the sets were detected. The result shows that the detection of the template became more effective since it used 4000 images. However, in Table 4.5, the result shows that in bright background, it is faster to detect the template in trial 4 than trial 5, it is because the car sticker has a glossy surface, and it forms small light reflections.

The variation of the scenarios is considered as a factor in the results achieved because the training of images was focused on a normal scenario. Another reason is if the object is not in its normal scenario, some of its details were not clear for the web camera to see. Thus, there is a greater number of fail detection than the successful detection of the template.

Overall result of the experiment, the data gathered during the evaluation of the Haar classifier shows that if the template is too far from the camera, the software cannot detect the car sticker. It is because the computer cannot "see" clearly the template anymore and the greater the number of training images, the better the detection of the template is.

5   Conclusion

The researchers were able to develop the prototype software that can be integrated with robotic equipment, so it could control the authorization of vehicles entering the premises. The integration with this equipment was not included in this study because it is believed to be out of scope in terms of the skills and capabilities of the researchers. Furthermore, the study is expected to be qualified as groundwork, should a robotic equipment that interacts with the developed prototype in this study, will be integrated. The software prototype on an average can be detected in a period of 3 to 6 seconds. Though performance-wise this may be slow as compared to the average time it takes for a typical security guard to detect the sticker, the results promote an opportunity for improvement. By the 5th trial (which is believed to be the most trained prototype), it has a 100% recognition accuracy rate. Should the quality settings of the camera, the number of training images be increased and the processing power of the learning computer unit would be improved, it is expected that the real-time detection of the sticker, would speed up in terms of time. Nevertheless, the study performed is only at its initial stages and may produce promising results in the future should the necessary enhancements be applied.

Acknowledgements. The authors of this study would like to thank Flordeliza Fernandez and Agnes Jade Aclan for the support they have provided especially in the training and provision of the appropriate tools used for Digital Image Processing.

References

1. De Grano, A.V., Pabico, J.P.: Automating the Classification of Tomato (Lycopersicon esculentum) Maturity Using Image Analysis and Neural Networks. Transactions of the National Academy of Science and Technology of the Philippines 29(1), 131–132 (2007)
2. Chang, C., Yulin, W.: Face Detection Based on Extended Haar-like Features. In: Transactions of the 2nd International Conference on Mechanical and Electronics Engineering (ICMEE 2010), pp. 442–445 (2010)
3. Huiyong, Y., Wei, L., Kun, X., Jian, Q., Xin, W., Lingwang, G., Zuorui, S.: Research on Insect Identification Based on Pattern Recognition Technology. In: 2010 Sixth International Conference on Natural Computation (ICNC 2010), pp. 1–3 (2010)
4. Barnes, D.: OpenCVHaar Cascade Training Positive/Negative Image (2011), Retrieved from Google Code: `https://code.google.com/p/opencv-haar-cascade-positive-image-builder/` (Date accessed: March 03, 2012)

A Study on Activity Predictive Modeling
for Prompt and Delayed Services in Smart Space

Danaipat Sodkomkham[*], Roberto Legaspi, Satoshi Kurihara, and Masayuki Numao

The Institute of Scientific and Industrial Research, Osaka University
8-1 Mihogaoka, Ibaraki, Osaka, 567-0047, Japan
{danaipat,roberto,kurihara,numao}@ai.sanken.osaka-u.ac.jp

Abstract. Knowledge about position and activities of the participants is commonly used in location-based services and applications in smart environment, which need to know an approximated location of the users to provide a proper service. Assistive services provided in the smart space can be divided into two categories: prompt services and delayed services. Prompt services can be served instantly and require no preparation time. On the other hand, delayed services need some time to prepare and be ready. When users are moving in an environment doing tasks, knowledge of the next location or destination of those movements can be used to assist the system to give more accurate system responses for the prompt services that can be served right away when the user arrives in his/her destination. These services require the following knowledge to operate: 1) a predicted location of the users or a plausible destination, and 2) a predicted time of arrival. For these two requirements, a predictive temporal sequential pattern mining algorithm is proposed in this paper, which is a method aimed at predicting the next location of a moving object from its temporal and spatial context. The prediction uses previously extracted temporal sequential patterns, which represent behaviors of moving objects as sequences of locations frequently visited within a certain speed. A decision tree based classifier is trained from the temporal sequential patterns and used as a predictor for the next location that is most probable location to be visited within the movement sequence. Moreover, A preliminary study on a predictive model for delayed services is also discussed in this paper. Finally, a performance evaluation of the methods tested over a real dataset is presented.

Keywords: Smart environment, Temporal sequential pattern Mining, Human movement pattern mining, Human activity pattern mining.

1 Introduction

In the design of a smart environment or intelligent space, knowledge about human behavioral patterns is important when designing services, and applications that could capture user patterns, predict their needs, and provide a proper system response are

---

[*] Corresponding author.

S. Nishizaki et al. (Eds.): WCTP 2012, PICT 7, pp. 266–278, 2013.

desirable. Particularly, a model of user mobility pattern in the space is required by various kinds of services. Hence, we focus on developing a system that can learn user movement patterns and make a prediction about the future location, where a user is heading including their target activity and the time of his/her arrival to that location.

Assistive services provided in the smart environment can be classified into two categories. First, prompt services are services that can be provided instantly and require very short preparing time. On the other hand, delayed services require more time to prepare and be ready to use. For example, automatic light is an example of a prompt service, but air conditioner is a delayed service. In this paper, we design predictive models especially for both prompt services and delayed services.

Main feature of prompt services is that they can be activated immediately without preparing period. The lights can be turned on instantly after the system recognizes current movement pattern and predicts that user is heading to someplace, where plausibly need more lights. The proposed method for prompt services is composed of two parts. First, the temporal sequential pattern mining algorithm is specially designed to extract the movement patterns that occur at a certain frequency. The movement patterns are represented by a sequence of locations and time interval between two locations in a sequence. This time interval factor is used to indicate the movement speed of an object. Second, the predictive method uses previously extracted movement patterns to train its classifier to predict the next user location.

The three main steps of our approach are as follows:

Movement Detection. In our experiment we use the infrared (IR) sensors that are basically used for distance measurement to detect a state, where there an object is at a certain location in the experimental space. Afterwards, sequences of movements are generated by concatenating these events together. A concise representation of these movement sequences consists of two components: 1) a location id, and 2) timestamp of the visit. Finally, each occurring sequence for each movement is logged into the sequence database. Note that our sensors itself cannot distinguish different people. Furthermore, if two or more users move at the same moment, those movements will be detected as one mixed movement sequence. However, IR sensors are unobtrusive, i.e. they need not be attached on subjects and permit the subjects to move more naturally in the space. Because of this advantage, we prefer the IR sensors to cameras or RFID tags.

Temporal Sequential Pattern Extraction. From the sequence database acquired previously, the temporal sequential pattern mining algorithm is executed to extract frequent patterns of movements with a typical movement speed. The method we use in this step is modified from the well-known PrefixSpan algorithm [7] to support temporal context that indicates movement speed. The algorithm is explained more in detail in section 3.

Predictive model construction: Given the temporal sequential patterns, we trained a decision tree-based classifier, called C4.5 [14], to identify to which in a set of 'next location' classes will any new observed movement sequence belong. To test the classifier, given a new movement sequence S, we used the prediction tree to predict the next location of S.

The performance of the method is evaluated against a real dataset over 5 weekdays of movement sequences collected by the IR sensors network installed on workspaces, hallway, and tearoom in our laboratory. An experiment is designed to evaluate the prediction accuracy of the method. We use the predictor trained from previous 24 hours of movement patterns to predict the next location of a new movement in real time to evaluate the prediction accuracy. The results of the experiment show that considering temporal context helps us achieve higher accuracy and efficiency.

Lastly, we propose an initial design of a predictive model for delayed services. Some services, such as air conditioner, coffee brewer, and electric water boiler, need some preparing time. The previous method, which can predict ahead at most 30 seconds, will not work on this kind of delayed services. In this study, time context, temperature, brightness of the room and historical usage data of those devices are used for the predictive model. The model is designed to predict at a certain period of day where and which facilities are being used.

The rest of the paper is organized as follows. Section 2 reviews related works. Section 3 describes the temporal sequential pattern extraction in detail. Section 4 explains how to build a predictive model for prompt services from sequential patterns extracted from the previous step, and discusses how to predict the next location for a newly observed movement using the predictive model. The design of predictive model fro delayed services is discussed in Section5 and the experimental results and performance studies of both approaches are presented in section 6, and we conclude our study in section 7.

2   Related Work

In this section we summarize some relevant studies related to the sequential patterns mining and location prediction.

2.1   Sequential Pattern Mining

Sequential pattern mining algorithms are designed to extract a set of frequently observed orders or subsequences as patterns from a dataset called sequence database. The sequence database consists of ordered events or items that are annotated with or without time. There are many researchers who study on sequential pattern mining [1, 5, 7, 9]. Since its performance is the best compared to GSP [5] and SPADE [9], the PrefixSpan [7] algorithm has been applied to many problem areas and has been provided with various extensions and modifications in different directions. However, sequential patterns extracted from PrefixSpan do not provide knowledge about the time span between items in the sequential patterns that could further support decision making. One solution to this problem has been introduced in [2], that is time-interval sequential mining algorithm was designed to extract not only frequently occurring sequences but also the time intervals between successive items. Time intervals have been commonly presented as having definite range [3, 7]. This, however, has led to strict or rigid boundary problems. When a time interval is near the boundary of two adjacent predetermined ranges, the tendency is to either ignore or overemphasize the interval [8]. Hence, in our previous study [13] we modified the Apriori algorithm to deal with the sharp boundary problem given multiple time intervals using clustering wherein we can specify a number of clusters. We used in particular the k-means method. However, its performance is not good enough to create a promising predictive model, which becomes the target of this work. Hence, we would like to modify to better predict the future location of a moving human subject. The proposed method in this paper also uses a concept of cluster analysis that deal with multiple time interval values with clusters of time intervals that are similar to each other within the same cluster as opposed with to those in other clusters. We integrate this idea into the PrefixSpan algorithm to create a temporal sequential pattern mining algorithm that allows multiple time interval sequential patterns with more flexible time interval data.

2.2   Trajectory Pattern Mining and Location Predictor

The concept of clustering sequential patterns and time intervals was also discussed in [11]. A trajectory pattern mining algorithm was a PrefixSpan based sequential pattern mining algorithm specially designed for spatio-temporal datasets. The algorithm finds a set of sequences of regions that are frequently visited by subjects and with typical time intervals. In [11], the time intervals are clustered using a density based clustering algorithm to handle multiple-time intervals between regions. Consequently, the sequential patterns found from the trajectory pattern mining algorithm are used to build a predictive model called "WhereNext" [12]. WhereNext is a location predictor that uses trajectory patterns as predictive rules. The data structure called the prediction tree [12], is constructed using all sequential patterns. The nodes of the tree are regions and edges representing represent typical time intervals between two successive regions. A matching method needs to be defined afterwards to match test instances with the model and make a prediction.

The difference between the work of WhereNext and ours is that we use a decision tree learning algorithm to build a prediction tree instead of manually selecting attributes for every node and edge of the tree without knowing how well each attribute separates the training examples according to their target class (i.e., the region where the subject is moving towards). This obviously makes our prediction tree smaller and predict faster. However, the time complexity for constructing the prediction is much higher in our case. We use a decision tree learning algorithm called C4.5, which has a time complexity of $O(mn^2)$, where m is the size of the training examples and n is the number of attributes. On the other hand, the time complexity of the prediction tree construction phase in WhereNext is $O(lm)$, where l is an average length of the patterns and m is the size of the training examples.

3   Temporal Sequential Pattern Mining

In this section we propose the Temporal Sequential Pattern Mining (TSPM), an algorithm that modifies PrefixSpan to determine temporal context in the candidates pruning step. The PrefixSpan algorithm adopts a pattern-growth approach to sequential pattern mining as developed by Pei et al. [7]. PrefixSpan is a divide-and-conquer algorithm that extracts subsequences that appear in a dataset with frequency no less than a user-specified threshold. The first scan finds length-1 sequential pattern that satisfy the minimum threshold. Each sequential pattern is treated as a prefix and used to project over the dataset to find longer sequential patterns. This process recurs until there is no more prefix left to be projected. The summary of the PrefixSpan can be found in Algorithm 1.

Basically, our movement sequences have temporal annotations that imply speeds. For example, (A, 10:20)→(B, 10:22)→(C, 10:30) can be read as "a person takes 2 minutes from location labeled 'A' to location 'B' and 8 minutes from 'B' to 'C'. The main idea of our approach that we add into PrefixSpan is that if we could extract typical movement speeds of particular movement patterns, the algorithm can ignore all projected candidate sequences that happened in infrequent speeds. An example of such cases is when it is 10 kilometers between 'B' and 'C'. It is obviously impossible for a person to walk from 'B' to 'C' within 10 minutes, which is also uncommon time interval found in movements between 'B' and 'C'. Hence, the algorithm can reject this sequence pattern because it is not practical.

TSPM uses a clustering algorithm called k-means clustering [15] to assign in groups time intervals appearing in each of the sequences. The clustering step is put into the PrefixSpan in frequency checking step. The projected sequential patterns that have lower frequency than the user-specified threshold, i.e., minimum support, will be discarded and the patterns that were clustered into a group that has smaller size than a minimum threshold, i.e., the minimum t-support, will be discarded as well. The pseudo-code of TSPM is presented in Algorithm 2.

Algorithm 1: PrefixSpan
Input: A dataset *seqDB* of sequences database, and a minimum support *minSupp*
Output: A set of sequential patterns

```
  L=1;
  Prefix_{L=1} = findLengthL1patterns(seqDB)
  while Prefix_{L} • 0 do
        Prefix_{L+1} = {}
        for each Prefix in Prefix_{L} do
                if Prefix.support > minSupp do
                        output(Prefix)
                        ProjectedSequence = project(seqDB, Prefix)
                        Prefix_{L+1}.add(ProjectedSequence)
                end if
        end for
        L++
  end while
```

Algorithm 2: Temporal Sequential patterns Mining (TSPM)
Input: A dataset *seqDB* of sequences database, and a minimum support *minSupp,* a minimum temporal support *minTSupp,* and number of time interval clusters *k* for k-means clustering algorithm.

```
Output: A set of temporal sequential patterns in a set of couples (se-
quence pattern, time interval cluster)
    L=1;
    Prefix_{L=1} = findLengthL1patterns(seqDB)
    while Prefix_L • 0 do
        Prefix_{L+1} = {}
        for each Prefix in Prefix_L do
          if Prefix.support > minSupp do
                timeintervalClusters = k-means(Prefix.timeinterval, k)
                  for each cluster in timeintervalClusters do
                      if cluster.size > minTSupp do
                          output(Prefix, cluster)
                          ProjectedSequence = project(seqDB, Prefix)
                          Prefix_{L+1}.add(ProjectedSequence)
                      end if
                  end for
          end if
        end for
        L++
end while
```

4   Predictive Model Building for Prompt Services

Our approach uses a decision tree-based classifier to identify to which of a set of 'next location' classes will a newly observed movement sequence belong. All sequential patterns extracted from TSPM algorithm are used as training examples for the classifier. Based on our hypothesis that temporal context will help the predictor achieve more accuracy, the time intervals and timestamps of each of sequence are also included as classifier features. The time interval attributes are time durations that subjects usually take from one location to another, and the timestamps are basically time logs that indicate frequent periods (in hour unit of time) of the day that these patterns were observed frequently. Consequently, features of the model include 1) length n-1 sequences of locations, when n is the length of the patterns, time intervals, and timestamps. Finally, the last element of the sequences is treated as a target class that we want the classifier to identify.

Given a prediction tree previously built T and a new movement sequence S, T classifies S to a target class that has the highest possibility to be the next location of S. Classification uses previously visited locations, time intervals between successive locations in the sequence, and an observed timestamp of the sequence for its decision.

5   Predictive Model for Delayed Services

As mentioned earlier, the predictive model that adopts TSPM (or any other sequential pattern mining algorithms with human movement pattern) cannot be used to predict long future. However, in the experiment, TSPM has shown us that time context gives some insight about which locations or devices will be used, and it helps improve the accuracy of the prediction. We also have gained some knowledge from that experiment that some activities, services, or devices are used habitually and related to some environmental context. For instance, Microwave oven is used only during 10:00 to 14:00. Air conditioners in the meeting room are turned on only when the temperature is over 30 degree Celsius on Tuesday or Wednesday during 13:00 to 16:00 in summer. These observation leads to our initial study on how accurate that a predictive model, which is built only from the environmental context, can perform.

The predictive model for delayed services in this study uses Naïve Bayes classifier to identify whether or not a certain facility is being used at a certain state of given contexts. The model is designed to learn activities patterns by observing a correlation between user activities and environmental contexts. To capture activities in the experimental space, we basically interpret a certain activity by detecting usage of particular facilities. For example, a microwave oven and a refrigerator are correlated to food-related activities.

The environmental contexts in our project cover three aspects. Firstly, we use time periods of a day as a temporal context because time is highly associated with some activities that habitually happened. Secondly, the temperature of the room is a proper guide to tell when to switch on the air conditioners. Lastly, brightness of the room can imply need of lights in a room, if user is going to use the room.

However, during period of the experiment the air conditioners in the experimental room had never been used because of university energy saving policy. Hence, in this work, we have to omit the temperature context. So, features vector is composed of:

- 1st -24th attribute are periods of time attributes (0.00-1.00 to 23.00-0.00) when some facilities were being used at that time period.
- 25th attribute is the lighting context (dim or bright)
- 26th attribute is the facilities name being used at that time (Oven, Fridge, and Light)

More discussion about the performance evaluation on this approach is presented in the next section.

6   Experiment Results

6.1   Performance Evaluations on the TSPM Based Predictive Model for Prompt Services

We have conducted a performance study to evaluate the accuracy of the proposed method and compared this with the predictive model built by PrefixSpan without any temporal context. As in [13], dataset from IR sensors are used. 60 IR sensors have been installed extensively in our experimental space as shows in Figures 1 and 2. The experimental spaces are composed of three main sections. First, the student room, where we installed a total of 36 sensors in student workspace cubicles to detect movement from students. Second, we installed 20 sensors in the tearoom, to detect usage of the printer (and copy machine), teapot, refrigerator, microwave oven, kitchen sink, couches, and TV. This space is shared among students and all laboratory members. Sensors were installed to detect movements and also to know the usage of facilities provided in the room. Finally, we installed 4 sensors along the hallway, between the student room and the tearoom to observe movements between these two sections. Mobility data used in the experiments are collected from approximately 10 subjects coming to work in the space. Data were collected 24 hours a day for 5 days from Monday to Friday.

Fig. 1. IR Sensors are used to detect movement in the experimental space

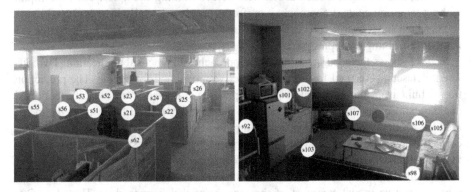

Fig. 2. IR sensors installed in the student room and the tearoom

All experiments were performed on a 2.67GHz Intel Xeon PC with 6GB of main memory. All the algorithms are implemented in Java. The parameters of TSPM are set as follows. The minimum support is set to 10 occurrences/day and the minimum t-support is 2 data points/day. Specifically, the minimum support used in this setting is an absolute support value, which is basically a frequency. This means that we concern ourselves with only movements that happened at least 10 times in one day. In the same way, the minimum t-support indicates the size of a time interval cluster. This simply means that a certain path must be traversed with the similar speed at least twice. Similarity in this case is determined by the time interval cluster, where each sequential pattern was assigned to during the clustering step. Lastly, parameter k in the k-means is set to 5.

Table 1 compares the precision of two approaches. First, the proposed method, which is concerned with the temporal context of the movement patterns, and the second approach that ignores all temporal contexts. The precision of class A is the ratio of correctly predicted results (user eventually visited a predicted location A) and number of times that a user actually visited location A. Table 1 shows the weight average precision of all predicting locations (from every rooms in the experimental space). A 10-fold cross validation was used in every dataset. The result clearly shows that temporal context helps the model predict more accurately.

Figure 3 shows the performance of TSPM and PrefixSpan in detail. F-measure is used to evaluate the prediction accuracy. The 5-day dataset is used to train and test the model using 10-fold cross validation. The column labeled as class is the next location that we want the algorithm to predict. For example, s92 is a sensor ID representing an area around a teapot; s95 is at the door, and s35 is at a student's desk. The results show the accuracy of two approaches using the F-measure. We use F1-score for the F-measure. The performance of TSPM is higher than the PrefixSpan for all classes and shows no problem with the unbalanced dataset. As showed in Figure 4, size of the training data of different classes are not equal. The smallest class is s34 with only 11 examples, and the biggest class is s92 with 30,118 training examples. The PrefixSpan without temporal context approach, it performs poorly especially in s13, s26, s34, and s103, which have nearly zero in both precision and recall rate (See Figure 4 for the number of training examples per class).

Table 2 shows the accuracy of the two approaches in practical application. Two different datasets were used for training predictive model and for testing. The prediction tree is built from the temporal sequential patterns extracted from a dataset collected one day before, and the dataset collected in the next day is used for testing. We measured the accuracy of two approaches by a ratio of correctly predicted location and incorrectly predicted location. The TSPM approach also gives slightly better accuracy than the PrefixSpan. However, a low performance in both approaches on Thursday indicates that a predictive model built on a small dataset on Wednesday cannot produce a representative model for most of the movement that happened on Thursday. One plausible solution for this problem is increasing the size of the datasets.

Table 1. Precision of TSPM and PrefixSpan w/o temporal context

Dataset	Mon	Tue	Wed	Thu	Fri	All5days
TSPM	93.2%	88.7%	91.0%	97.0%	90.4%	95.7%
PrefixSpan	93.2%	75.4%	63.3%	69.8%	63.3%	68.9%
Size of the dataset	2,965	2,399	4,217	46,103	1,631	57,315

Table 2. Accuracy comparison between TSPM and PrefixSpan in a Next Day prediction experiment

Training	Mon	Tue	Wed	Thu
Prediction	Tue	Wed	Thu	Fri
TSPM	78.1%	49.4%	23.1%	58.0%
PrefixSpan	78.1%	44.9%	22.1%	48.3%

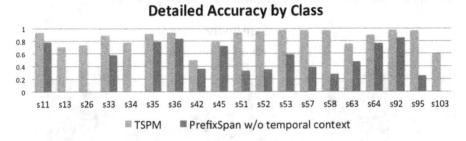

Fig. 3. Detailed Accuracy by Class measured by F-measure (F1-score)

Fig. 4. Size of a training set per class

6.2 Performance Evaluations on the Environmental Contexts Based Predictive Model for Delayed Services

This experiment uses different dataset. In this part, we setup sensors to detect usage of the microwave oven, refrigerator, and the light switches in the tearoom. A precision light sensor was also installed in the tearoom to measure brightness of the room during the experiment. As mentioned earlier in section 5, the model is built from Naïve Bayes classifier. We assume that all features are independent to each other. The time context and the lighting context are used as features and a facility (oven, fridge, light s), which is being used at that moment, is the target class of the classifier.

We use 2 weeks of data and the model is built and validated by 10 fold cross-validation. Figure 5 shows the sensitivity (i.e., True positive rate) or the ability of the model to identify the positive results. The sensitivity shows the proportion of actual positives, which are a certain device, is correctly identified as being or not being used at a given context. Graph shows that the prediction sensitivity of 'Used' (a certain device is correctly identified as being used at a given context) in 'Oven' and 'Light' cases are quite low because the dataset is skewed toward the 'Fridge' case, which is the most frequently used facilities during the experiment (usage share is shown in Figure 6), but the sensitivity is high in the 'Fridge' case. On the other hand, the sensitivity of 'Not used' (a certain device is correctly identified as not being used at a given context) is fairly high in 'Oven' and 'Light' cases, but quite low in the 'Fridge' case. Table 3 also shows detailed performance evaluation per device.

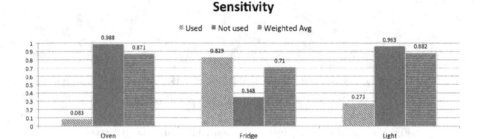

Fig. 5. Sensitivity (True positive rate)

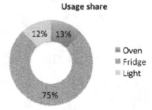

Fig. 6. Usage share of facilities in the tearoom (experimental space)

Table 3. Performance evaluation per device

Device	TP rate	FP rate	Precision	Recall	F-score	Class
	0.083	0.012	0.5	0.083	0.143	Used
Oven	0.988	0.917	0.879	0.988	0.93	Not used
	0.871	0.8	0.83	0.871	0.829	Weight Avg
	0.829	0.652	0.795	0.829	0.811	Used
Fridge	0.348	0.171	0.4	0.348	0.372	Not used
	0.71	0.533	0.679	0.71	0.703	Weight Avg
	0.273	0.037	0.5	0.273	0.353	Used
Light	0.963	0.727	0.908	0.963	0.935	Not used
	0.882	0.646	0.86	0.882	0.866	Weight Avg

7   Conclusion

To design a service or application in a smart environment research, we need to model human movement behavioral patterns where the knowledge of future user movement is vital. We designed two predictive models for two kinds of assistive services found in the smart space that we have differentiated them by services preparing time. The prompt services are type of services that can be activated and serve to users instantly without delay, but the delayed services require more time to be ready to be served.

Firstly, to create a model for the prompt services, we designed a sequential patterns mining algorithm, called Temporal Sequential patterns Mining algorithm (TSPM) especially for analyzing human movement patterns. Afterwards, the movement patterns are used to build a predictive model that was constructed by C4.5. A future location is treated as a class where a newly observed movement sequence will be classified. One hypothesis that we made in this work is that temporal context of such movement patterns (e.g., time span between two locations, typical time of the day that a certain pattern appears) could help the predictor perform with smaller error. TSPM is designed to handle multiple time intervals that appear in sequential patterns by employing k-means clustering, a specific number of clusters k can explicitly stated. Different from [13], we developed TSPM based on the PrefixSpan algorithm as an alternative to an Apriori algorithm because of its speed and less memory complexity problems. Lastly, the prediction part of our approach is different from [12] since we used decision tree learning to build the prediction tree, which basically greedy selects an attribute that is most useful for classifying examples to create a decision node first. This gives a more compact and smaller tree than a prediction tree directly constructed by the sequential patterns.

We implemented and tested our approach against the prediction method that do not use temporal contexts. The results show that the temporal context of movement speeds and time helps the predictor to achieve higher performance by 27% in the average.

We also conducted an experiment that simulated a real application, where movement patterns are mined and the predictive model is built beforehand offline, and used to predict the future location for a movement in the succeeding day. The result also shows an improvement in prediction accuracy by 2% on the average, whereas the test cases with lowest and highest improvement got 0% and 10% respectively. The reason behind this situation is that the number of users cannot be fixed; therefore we got a different number of movement patterns on different training days. Too small number of training examples could lead to problems because they cannot be used to generate a representative model for all cases.

However, the predictive model built from sequential patterns of human movement cannot predict events in long future. Then, we have proposed the initial studies on using the environmental contexts (i.e., time, temperature, and brightness) to build the predictive model that models user's activities habitual patterns in the smart space. We used Naïve Bayes in this study and we also evaluated the performance of the model with real dataset. The results show that time context and lighting context are correlated with some activities. Nonetheless, the prediction accuracy is still low because of the skewed dataset problem. In future work, we plan to study on more contextual information and design a new model that can tolerate skewed dataset to achieve higher performance.

References

1. Agrawal, R., Srikant, R.: Mining sequential patterns. In: Proc. 11th International Conference on Data Engineering, pp. 3–14 (1995)
2. Chen, Y.L., Chiang, M.C., Ko, M.T.: Discovering time-interval sequential patterns in sequence databases. Expert Systems with Applications 25, 343–354 (2003)
3. Chen, Y.L., Huang, T.C.K.: Discovering fuzzy time-interval sequential patterns in sequence databases. IEEE Transactions on Systems, Man, and Cybernetics-Part B: Cybernetics 35(5), 959–972 (2005)
4. Hafez, A.: Association of dependency between time series. In: Proc. SPIE, vol. 4384. SPIE Aerosense (2001)
5. Hirate, Y., Yamana, H.: Generalized sequential pattern mining with item intervals. Journal of Computers 1(3), 51–60 (2006)
6. Hu, Y.H., Hang, T.C.K., Yang, H.R., Chen, Y.L.: On mining multi-time-interval sequential patterns. Data and Knowledge Engineering 68, 1112–1127 (2009)
7. Pei, J., Han, J., Mortazavi-Asl, B., Wang, J., Pinto, H., Chen, Q., Dayal, U., Hsu, M.C.: Mining sequential patterns by pattern-growth: The PrefixSpan approach. IEEE Transactions on Knowledge and Data Engineering 16(11), 1424–1440 (2004)
8. Srikant, R., Agrawal, R.: Mining sequential patterns: Generalization and performance improvements. In: Apers, P.M.G., Bouzeghoub, M., Gardarin, G. (eds.) EDBT 1996. LNCS, vol. 1057, Springer, Heidelberg (1996)
9. Zaki, M.J.: Spade: An efficient algorithm for mining frequent sequences. Machine Learning 42(1-2), 31–60 (2004)
10. Jiawei, H., Hong, C., Dong, X., Xifeng, Y.: Frequent pattern mining: current status and future directions. Data Mining and Knowledge Discovery 15(1) (August 2007)
11. Giannotti, F., Nanni, M., Pinelli, F., Pedreschi, D.: Trajectory pattern mining. In: KDD 2007, pp. 330–339 (2007)
12. Monreale, A., Pinelli, F., Trasarti, R., Giannotti, F.: WhereNext: A location predictor on trajectory pattern mining. In: Proceedings of the 15th ACM SIGKDD International Conference on Knowledge Discovery and Data Mining, Paris, France, June 28-July 01 (2009)
13. Legaspi, R., Sodkomkham, D., Maruo, K., Fukui, K., Moriyama, K., Kurihara, S., Numao, M.: Time-Interval Clustering in Sequence Pattern Recognition as Tool for Behavior Modeling. In: Nishizaki, S.-y., Numao, M., Caro, J., Suarez, M.T. (eds.) WCTP 2011. PICT, vol. 5, pp. 174–186. Springer, Heidelberg (2012)
14. Quinlan, J.R.: C4.5: Programs for Machine Learning. Morgan Kaufmann Publishers (1993)
15. MacQueen, J.B.: Some Methods for classification and Analysis of Multivariate Observations. In: Proceedings of 5th Berkeley Symposium on Mathematical Statistics and Probability, pp. 281–297. University of California Press (1967) (retrieved April 07, 2009), MR0214227. Zbl 0214.46201

Plagiarism Detection Methods by Similarity
and Its Application to Source Codes of C Program

Hiroyuki Tominaga and Kazushi Ueta

Kagawa University
2217-20 Takamatsu, Kagawa 761-0396, Japan
tominaga@eng.kagawa-u.ac.jp

Abstract. We propose methods based on similarity for detecting plagiarism in reports submitted online. We adopt two approaches for the similarity judgment: a text based, minimum operations edit distance and a binary based, file compression ratio information distance. Token analysis and the removal of redundant information are implemented as preprocesses. We aim to realize an adequate judgment precision and cost performance. The tool works as an independent application and can be embedded as a plug-in module in various educational support systems. The detection tool is used to not only inform the teacher of fraudulent activity but also warn the student at the time uploading files. The tool is demonstrated through application to C program source codes. We calculate both the similarity of the original and assembled codes to detect superficial changes in the codes. We carried out experiments for short answer programs to introductory problems and longer application-level codes. We analyze the results and discuss anomalous codes.

Keywords: Detection methods for online-report plagiarism, Similarity of C program source codes, Edit and information distance.

1   Introduction

Recently, as computer environments in colleges have improved, teachers have started planning lessons in which every student uses a PC in the classroom; a student may use either a notebook PC or a mobile terminal during the lesson and may also be able to access the campus network from home. Various e-learning systems and learning management systems (LMSs) have been introduced and implemented in colleges. Teachers tend to utilize online teaching materials and tools to distribute problems and to receive report submissions. Online submission is highly convenient for teachers in that report files can be managed and marked automatically by a computer system. However, these online systems may also encourage and easily allow students to copy the work of others. For the integrity of educational institutions, it is essential that imprudent acts of plagiarism are dealt with appropriately [1][2][3].

We propose here methods to detect plagiarism by examining the similarity based on string distance. We have developed practical tools to be used as plug-in modules for various LMSs. We adopt two methods based on the edit and information distances

S. Nishizaki et al. (Eds.): WCTP 2012, PICT 7, pp. 279–294, 2013.
© Springer Japan 2013

to judge the similarity between submitted reports. We prepare several preprocesses according to the properties of target data, and in particular, we focus here on C programming exercises and evaluate the similarity of various source codes. Because perfect judgment is very difficult, the tools must offer support functions and a user-friendly interface, as well as providing a suitable balance between efficiency and precision.

2   Issues Surrounding Plagiarism

Copying from online sources is a very easy operation known as "copy and paste" that can be performed simply with a mouse; the process is called "copi-pe" in Japanese slang. Many students may perform this action on a daily basis. Sources of plagiarism in student reports can be separated in two categories. The first is that from web documents such as Wikipedia, which is often assisted by useful search engines, i.e., the so-called Google-Copy-Paste Syndrome [4]. The second is copying from other students.

To detect instances of web-document plagiarism, all documents on the web are searched for the original copy. This requires a crawler function or cooperation with existing search engines. However, the system can also easily gather content beforehand such as is done in the well-known PAIRwise [5] and TunItIn (iParadigms) [6][7] systems. In Japan, "Copypelna," developed by Dr. Sugimitsu and ANK Co., Ltd [8] has become widely used. Some research relating to the implementation of the former two systems has been published [9][10][11][12][13].

To detect instances of student-student copying, similarity among all pairs of student reports must be checked. This process is time consuming when there a large number of reports. Copypelna has a cross-checker function; however, it seems to be only applicable to Japanese documents. It is difficult to detect definitively instances of plagiarism using automatic methods and tools, and so it is necessary for the teacher to be the final judge. Moreover, the precision of the detection will depend on the properties of the target, for example, free style essays, an explanation of a technical term, a solution to a mathematical problem, or a computer program.

We have mainly taught information engineering and programming exercises, and plagiarism in this case is derived mainly from student-student copying as there few instances of entire code that can be copied. Our first target, therefore, is C source code in an introductory class, and we focus here on methods for detecting student-student copying.

3   Similarity Based on String Distance

3.1   LED Similarity by Edit Distance

A common method for measuring the similarity between strings is based on the edit distance. This distance is the minimum number of edit operations needed to transform one string to another. The simplest distance is the Levenshtein edit distance (LED) that adopts three basic operations for a letter: delete, insert, or exchange. The LED method is widely used in various application fields such as gene analysis in bio-informatics. For example, between the two strings "bcabc" and "abdd", there are two transformation processes using only basic operations as shown in the box below. The edit distance in transforming "bcabc" to "abdd" is 4, which is the minimum count.

a b d d	1 letter match	5 operations	as edit distance
b c a b c			
a b d d	2 letters match	4 operations	

Let |X| be a string length of X and X' be the string without the first entry. The LED between two strings of S and T is defined by

$$
LED(S,T) = \begin{cases} |S| & ; T = \emptyset \\ |T| & ; S = \emptyset \\ LED(S',T') & ; the\ same\ first\ letter \\ min \begin{cases} LED(S',T) \\ LED(S,T') \\ LED(S',T') \end{cases} & ; else \end{cases}
$$

The definition requires a large calculation time because of the redundant tree-like expression of the recursive call according to the string length. An efficient algorithm in this case is the dynamic programming method. In the working table shown in Fig. 1, the path with the minimum cost from the top left to bottom right is chosen. Starting at 0 in the top-left cell, the adjoining cells are calculated based on the minimum number of edit operations to change to the desired character. Finally, the value in the bottom-right cell is the distance.

	φ	b	c	a	b	c
φ	0	1	2	3	4	5
a	1	1	2	2	3	4
b	2	1	2	3	2	3
d	3	2	2	3	3	3
d	4	3	3	3	4	4

Fig. 1. The working table for the calculation of the LED

It is in fact not necessary to calculate the values of all the cells in the working table [14][15][16]. If you can estimate the upper bound u of the distance, you can just calculate the cells in the zone area with a width of u/2 along the diagonal line. The upper bound u is at most the length of the longer string. In the above table, the grey cells can be removed from the calculation. If you only need to detect pairs that have a small distance, then a threshold v can be set as the upper bound u, and cells in this narrower area can be calculated. Moreover, the calculation can be aborted when all the values on the frontier of the calculated cells exceed the threshold.

The LED gives the absolute difference value of strings and depends on the string length. In general, $|S| \trianglerighteq |T|$ implies $|S| - |T| \leq LED(S,T) \trianglelefteq |S|$, and specifically, $|T| \models \phi$ implies $LED(S,T) \dashv |S|$. Therefore, we define the similarity degree as follows:

$$SimLED(S,T) = 1 - \frac{LED(S,T)}{max\{|S|, |T|\}}$$

The LED is normalized in the interval [0,1] by the ratio to the string length. The value of SimLED is 1.0 in the case that the strings are exactly the same and 0.0 in the case of that the strings contain no matching letters.

3.2   NCD Similarity by Information Distance

Another method for detecting similarity is based on the information distance as expressed by the Kolmogorov complexity. The Kolmogorov complexity $K(S)$ is the minimum algorithm description length necessary to generate a string S, and it represents the complexity and randomness of the string, that is, the lower bound of the data compression size. The complexity is an abstract definition that does not depend on any specific program language, but in reality, you suppose a certain programming language or description. For example, in Ruby script, the string "01230123012230123" is described as print "0123" *4. Thus for the string T = "0123", you obtain the shorter description of print T *4. The relative Kolmogorov complexity $K(S|T)$ is the amount of a string S that can be reproduced by using the information included in a string T. As the inequality $K(S) \geq K(S|T)$ is valid in general, $K(S) - K(S|T)$ indicates the improvement in the data compression of S by using T, i.e., the information distance represents the overlapping information in T and S.

It is not actually possible to calculate the proper amount by searching all description methods. Therefore, we use the normalized compression distance (NCD), which is the effective approximation of the Kolmogorov complexity by a specific compression algorithm [17]. The NCD is also utilized for analysis in music and literature. The absolute amount $C(S)$ is the compression size of data S, and the relative amount $C(S|T)$ is $C(S|T) = C(TS) - C(S)$ using a concatenated string TS. The NCD also represents the absolute difference value and depends on the original data size. Therefore, we normalize the distance as follows:

$$NCD(S,T) = \frac{C(TS) - C(T)}{C(S)} \; ; C(S) \geq C(T)$$

The NCD is 1.0 in the case of completely different strings and 0.0 in the case of exactly the same strings. We define the NCD similarity as

$$SimNCD(S,T) = 1 - \frac{C(TS) - C(T)}{C(S)} = \frac{C(S) + C(T) - C(TS)}{max\{C(S), C(T)\}}$$

However, because actual compression tools add header information, this value may be less than 0.0 or slightly over 1.0 but this is not a significant problem in practice.

4   Plagiarism Detection Methods Based on Similarity Calculation

4.1   Our Approach and Proposed Methods

We have proposed methods based on similarity for detecting plagiarism in online reports [18]. We adopt here both the LED and NCD approaches for the similarity judgment, combining them according to the specific purpose, and preprocessing methods such as morphological analysis and redundant information removal are applied to the target. We aim to realize an adequate judgment precision and cost performance. The detection tool works as an independent application and can also be embedded as a plug-in module in various educational support systems. The detection tool not only exposes instances of fraudulent activity but also gives a warning to the student when they upload their files. Here, we apply the detection methods to C program source codes. We calculate the similarity for both the source codes and assembled codes. The similarity of the assembled codes detects superficial or cosmetic changes to the codes that are performed by students to conceal their plagiarism.

4.2   Consideration of Both Methods

Even though the dynamic programming method is used, the cost of calculating the LED similarity is proportional to the square of the string length. Practical, fast detection of plagiarism in reports with long sentences is therefore somewhat difficult. However, the comparison consists of checking whether each letter is equal and if the comparison unit is instead a larger object, then the object length and calculation time can be reduced.

In general, because students refer to their textbook and notebooks for the best answer, their reports are highly likely to contain similar partial phrases. The similarity value in this case is not 0.0 but exceeds a certain value. However, when one of the documents is very short, the LED is almost equal to the longer string length, and hence the similarity value is close to 0.0.

For the NCD approach, a document with a smaller size does not influence the compression size, and the NCD similarity in this case is also close to 0.0. To detect plagiarism, the similarity is thus an inadequate value because the distance is deceptively estimated as that of the lager document. To obtain a useful matching parameter, preprocessing is necessary to group the targets into a certain data size.

4.3   Preprocessing for C Source Codes

We propose the following preprocessing methods for the C source codes. All comments in the code are first deleted, and we adopt a token analysis to convert the code into a token sequence. We extract reserved words and user definition symbols to remove separator elements like parentheses and commas (although there are a large number of such elements in the code, they do not influence the originality of syntactically correct codes). For a problem which in partial code must be completed, the common code must be neglected. We constructed a simple parser to do this.

We also consider another approach by assembling. This preprocessing method translates the original text code into intermediate code in assembly language, and the given code is parsed into a token sequence. This method excludes the influence of cosmetic changes such as replacing a variable name or rewriting a "while" statement as a "for" statement. The data size of the token sequence is larger than that of the original code, and the token sequence may give more a precise similarity value for finding more savvy plagiarism.

These preprocessing methods are applied in two steps. In the first step, all the original code is checked for any instances of matching. In the second step, a subset of candidates for which plagiarism is suspected is selected and matching using the assembly intermediate is applied to these candidates.

5   Implementation of the Plagiarism Detection Tool

5.1   Implementation of the Tool

The preprocessing and main plagiarism detection process are implemented using Ruby script. We have also prepared the system interface to work as a plug-in module for web applications. Because of the long computation time of the LED similarity, the main Ruby script calls a C program for this calculation, and the GNU complier collection (GCC) is adopted as the assembler tool. For the NCD similarity, we use both the gzip and bzip2 compression tools. Compared to bzip2, gzip has a low compression ratio and a quick execution time. An algorithm with a high compression ratio will improve the precision of the similarity judgment, but we offer both compression tools to the user.

The LED and NCD similarity detection tools are designed as filter commands. The user can select both approaches and suitable preprocesses depending on the properties of the target, and the parameters of each module can be adjusted with flexibility. The user interface displays the judgment results and suspected pairs.

5.2   The GUI for the Independent Tool

The independent tool has a web-based graphical user interface (GUI). A screen shot of the input page is shown in Fig. 2(a). The target data can be input as a CSV format file of the file list or as a directory path. For the LED similarity, a tokenizer for the source code or intermediate code is selected and a similarity threshold is set. For the NCD similarity, a specific compression tool is selected. Short codes can be removed by a filter based on the number of strings. If the process is likely to be time consuming, the results can be sent later by e-mail. A screen shot of the results page is shown in Fig. 2(b). The matching pairs are ranked by the similarity and the distribution is shown at the top. When more details are necessary, the files can be compared side by side as in Fig. 2(c).

| (a) The input page | (b) The results page | (c) The details page |

Fig. 2. Screen shots of the independent tool

5.3   Plug-in Interface for WebBinder

We are developing the student portal site PASPort, which cooperates with several lesson support services and aggregates personal information from the campus site [19]. It offers main two services: document and schedule management. The document management service, WebBinder, supports students' report tasks (Fig. 3) and is an extension of an online storage service that provides a bridge between several teacher sites, the file sharing server, and a local student PC. Students can access the service both inside and outside the campus network. The system offers support functions according to the three phases of acquisition, working, and submission. For programming reports, the submission conditions in terms of selection of problems and the requirement of several product files are somewhat demanding. The proposed detection tool is used in two ways: to inform the teacher of illegal action by a student and to give the student a warning message before submission.

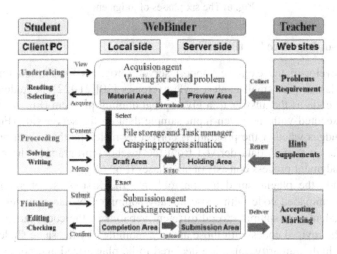

Fig. 3. The report document support system WebBinder

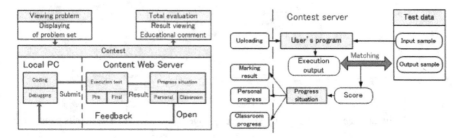

Fig. 4. The system configuration of tProgrEss   **Fig. 5.** The modules and data in code judgment

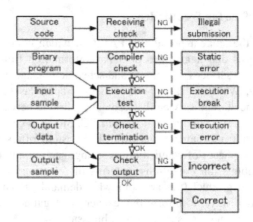

Fig. 6. The six phases of judgment

5.4   Plug-in Interface for tProgrEss

We have developed the exercise support server tProgrEss for introductory C programming classes with small contest style exercises [20][21][22]. A student uploads their source code as the answer to a given programming problem (Fig. 4), it is compiled and executed with the given input sample data on the server side (Fig. 5), and the server judges whether the program is correct by matching the execution result to the given output sample data derived from the teacher's program as the correct answer. The judgment process consists of six phases (Fig. 6).

We include the plagiarism detection tool into the first phase of submission. It matches the uploaded code with those already submitted by other students, and although a student may submit several different versions of the code, the system does not match the revised codes with those already submitted by the same student. Codes with a very high similarity value are declared to be plagiarized and not accepted. In the case of a rather high value, the system warns the student about possible plagiarism but proceeds to the next judgment phase. This code is flagged for later review by the teacher. The judgment result is then displayed on the student page. Because tProgrEss has been developed in Ruby, it is relatively easy to embed the detection tool, which is also implemented in Ruby. As the server is expected to deliver a real-time judgment, adequate preprocessing is important for the efficiency.

6   The First Experiment

6.1   Target Data

The target data for the first experiment were C program source codes submitted over the last 5 years in the third-year "Software Development Exercises" course at our information engineering college. One of the themes of the class is related to knowledge information processing. The problem related to the source codes considered here was to describe the student's original strategy for playing Poker. The function "strategy()" receives 5 cards and must return a selected card to be discarded. The student must implement the function from a prototype in the "Poker.c" file, and the main common process for the game is in located in another file ("PokerExec.c"). There may be programs that can be copied from web pages, and there is likely to be similarity in the answer programs as a result of students copying from each other. It is therefore a suitable target for testing the plagiarism detection tool. We considered 170 source codes and 14450 matching pairs. The average file size was 9546 bytes, the average number of lines without comments was 225, and the average number of letters was 4865. In the original codes, the average number of tokens of reserved words and user declaration symbols was 617 and that in the assembled intermediate code was about 3800, which is approximately six times greater. The average size of the gzip compressed files is 928 bytes, which is five times lower.

6.2   Processing Time

For the similarity calculation using the LED, the processing time for the parsing preprocess is less than several seconds, and the time for the main matching process is about 8 min. The processing time for matching one target to one of the other 169 targets is about 2 s. The time for a pair is less than 0.02 s. For the assembled codes, the assembling preprocess takes about 20 s, but the time for the main process is over 3 h. Because several codes contain some syntax errors, the number of assembled intermediate codes was reduced to 157. The time to compare one target and one of the others is about 2 min. These times are 36 times longer than those of the original codes because the file sizes are larger. To calculate the NCD similarity, the preprocessing time for the gzip compression is 2.0 s, and the main processing time is 23 s. The total processing time for one pair is 0.14 s. The machine used in the experiments reported here had an Intel Core2Duo 3.0 GHz CPU and 4 GB of DDR2 RAM running the 64-bit version of Linux CentOS 5.

6.3   Results of the Experiment

The correlation of the LED similarity is high to middling between the original source code and assembled intermediate code (Fig. 7). Reviewing the matching pairs with a high correlation, we judge 5 pairs to be apparent plagiarism. The programs in two of the 5 pairs were submitted by senior and junior students of the same laboratory.

In terms of efficiency, the LED similarity calculation takes longer than that of the NCD. For middle-level programming exercises, the NCD should be adopted as the main check and the LED should be used for more precise checking.

The correlation between the NCD and LED similarities was not particularly high. The pairs with high similarity values included instances of no plagiarism. We recognize this as being due to the existence of anomalous codes. These are codes that contain little revision of the given prototype and contain redundant, continuous "if" statements instead of a "while" statement. The former is close to being null code, while the latter is regarded as self-"copi-pe" with a high compression ratio because of the redundant code. These codes introduce noisy data into the plagiarism detection and should be removed. However, they need educational instruction. As these codes have irregular values of the ratio of the token number of the original to assembled codes, we can identify them in preprocessing.

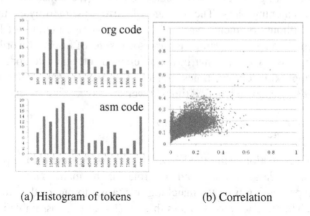

(a) Histogram of tokens   (b) Correlation

Fig. 7. The result for C source codes

6.4   Detection of Anomalous Codes

We attempt to detect anomalous codes by considering the file size and the ratio between the original source code and the compressed data file sizes after preprocessing. Fig. 8(a), (b), and (c) show histograms of the file size of the original source code, assembled intermediate code, and gzip compressed binary data, respectively. Fig. 9(a) and (b) show histograms of the file size ratio. For the assembled codes in Fig. 9(a), values between 3.0 and 7.0 are regarded as normal, but 9 codes with a value of less than 2.0 are incomplete and have some compiler errors. Codes with values of more than 9.0 contain redundant descriptions. These codes contain raw code that has not been optimized. For example, they may contain too many "if" statements in the form "if (hd[0]%13 == hd[1]%13 && hd[2]%13 == hd[3]%13) return 4;". These are low quality codes that must be re-factored through the nesting of "if" statements. For the compressed binary data in Fig. 9(b), values between 0.15 and 0.35 seem to be normal. Codes with values of less than 0.1 also contain redundant descriptions and self-correlation. These codes may be revised by replacing a function reference. Codes with values of more than 0.4 are too simple and have a very small size. These may be almost skeleton codes. Before detecting similar code, these anomalous codes must be identified and removed from the comparison.

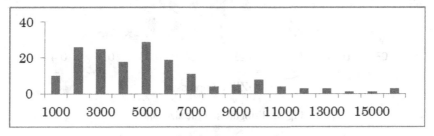

(a) Original source code

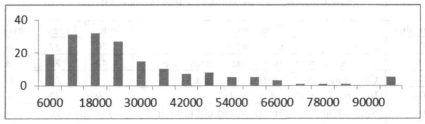

(b) Assembled intermediate code

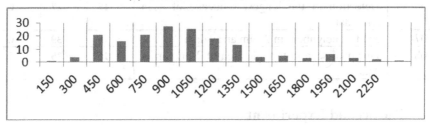

(c) GZIP compressed

Fig. 8. Histograms of the file size

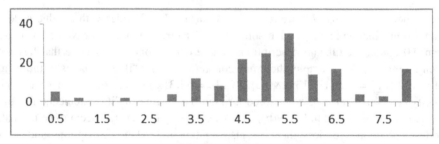

(a) Assembled

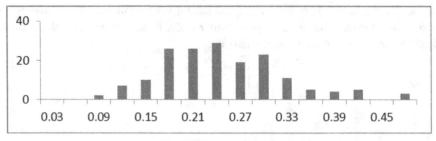

(b) GZIP compressed

Fig. 9. Histograms of file size ratio

Table 1. Problems for introductory C programming exercise

Number	Contents	Codes	Lines	Level
Q01	Figure output by ASCII art	70	27	B
Q02	The second maximum value in array data	47	29	B
Q03	Calculation of algorithm with radix 100	51	26	B
Q04	Reverse operation of array data	39	42	B
Q05	Sum of amounts in rows and columns in 2-dim array data	41	40	B
Q06	Detection of the longest sequence of 1 in array data	18	27	C
Q07	Sum of neighbors in 2-dim array data	16	39	C
Q08	Diagonal scan in triangle area of 2-dim array data	16	35	C

7   The Second Experiment

7.1   Target Data

The second experiment involved contest problems for the first-year introductory C language programming exercise via tProgrEss. The problems are listed in Table 1. Each answer can be implemented in about 30 lines, and the purpose of the problems is to check for an understanding of basic programming grammar. The students are free to select the problems, and as the latter problems are more difficult, only a few students attempted these. Compared with the codes considered in the first experiment, these codes are easier and smaller, and therefore, there is a tendency for the answer programs to be similar.

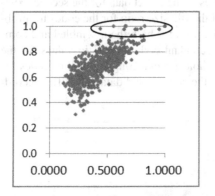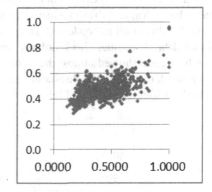

Fig. 10. The correlation for Q02 between the LED similarity of the source codes and (a) the LED similarity of the assembled codes and (b) the NCD similarity

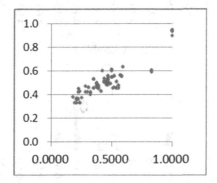

Fig. 11. The correlation for Q06 between the LED similarity of the source codes and (a) the LED similarity of the assembled codes and (b) the NCD similarity

7.2   Results of the Experiment

Because the number of pairs and the code size are rather small, the processing is relatively fast. The LED similarity calculation of the assembled codes takes less than 20 s. Similar pairs detected as plagiarism were found in all problems. We discuss problems Q02 and Q06 in more detail here. Q02 is easier but both are around 30 lines in length. Fig. 10 and 11 show the similarity correlations for Q02 and Q06, respectively. The circled area in Fig. 10(a) shows significant difference in both values, that is, although the similarity of the original codes was not high, that of the assembled codes was high. This is because the codes in the pair have only changed slightly and differ

only in terms of variable names. The similarity of the assembled codes is more efficient in detecting actions students have taken to conceal their plagiarism. In the short,
easy problem of Q02, the correlation coefficient between the original codes and the
compressed binary data is low (0.66), whereas in the more difficult Q06 problem, it is
very high (0.94).

We also attempted to detect anomalous codes in the target data for the second experiment. Fig. 12(a) and (b) show histograms of the file size ratio for the codes for problem Q06. Almost all the codes return normal values for both the assembled and compressed data. A manual check did not find redundant code. The ratio values for the
assembled code showed almost the same threshold for detecting anomalous codes for
all the problems, while the threshold ratio for the compressed data may differ according to the difficulty of the problem.

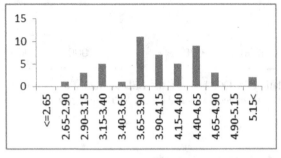

(a) Assembled

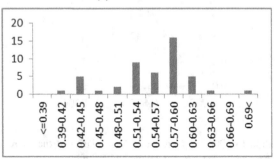

(b) GZIP compressed

Fig. 12. Histograms of file size ratio in Q06

8   Conclusion

We have proposed efficient methods based on similarity for detecting plagiarism in
reports. The purposes were to inform the teacher of the illegal actions of the student
and to warn the student before submission. We developed the detection tools and a
plug-in module prototype. We adopted both the LED similarity, based on the Levenshtein edit distance, and the NCD similarity, based on the information distance by

data compression. We mainly focused on C language source codes submitted as programming exercises and aimed to achieve an adequate balance between the precision and efficiency. We also prepared necessary preprocessing steps to reduce the processing time of the LED similarity by converting the codes into token sequences or by using assembled intermediate codes. For the NCD similarity, we adopted the gzip and bzip2 compression tools. The independent tool and plug-in module use a GUI and system interface, respectively.

We carried out experiments using C source codes. The LED similarity was successful in detecting apparent plagiarism and was relatively fast in comparing original source codes. However, it took much longer to calculate the similarity for the assembled intermediate codes, and so this method should be used only in the case of selected suspicious pairs. For the NCD similarity, the detection result includes noise because of anomalous codes. Our methods can also detect anomalous code through the file size ratio of the original source code without comments and its compressed binary data.

In future work, we will improve the precision of the method, develop more appropriate preprocesses, and further increase the efficiency of the implementation. The tools will find practical use in WebBinder and tProgrEss.

References

1. Maurer, H., Zaka, B.: Plagiarism - A Problem And How To Fight It. In: Proceedings of ED-MEDIA 2007, pp. 4451–4458 (2007)
2. Wang, Y.: University Student Online Plagiarism. International Journal on E-Learning 7(4), 743–757 (2008)
3. Suarez, J., Martin, A.: Internet Plagiarism: A Teacher's Combat Guide. Contemporary Issues in Technology and Teacher Education 1(4), 546–549 (2001)
4. Maurer, H., Kulathuramaiyer, N.: Coping With the Copy-Paste-Syndrome. In: Proceedings of E-Learn 2007, pp. 1071–1079 (2007)
5. Knight, A., Almeroth, K., Bimber, B.: Design, Implementation and Deployment of PAIRwise. Journal of Interactive Learning Research 19(3), 489–508 (2008)
6. iParadigms, TurnItIn (2012), http://turnitin.com/
7. Stetter, M.E.: Plagiarism and the use of Blackboard's TurnItIn. In: Proceedings of ED-MEDIA 2008, pp. 5083–5085 (2008)
8. http://www.ank.co.jp/works/products/copypelna/ (2012)
9. Johnson, D., Patton, R., Bimber, B., Almeroth, K., Michaels, G.: Technology and Plagiarism in the University: Brief Report of a Trial in Detecting Cheating. AACE Journal 12(3), 281–299 (2004)
10. Brown, V., Robin, N., Jordan, R.: A Faculty's Perspective and Use of Plagiarism Detection Software. In: Proceedings of SITE 2008, pp. 1946–1948 (2008)
11. Thomas, M.: Plagiarism Detection Software. In: Proceedings of E-Learn 2008, pp. 2390–2397 (2008)
12. Knight, A., Almeroth, K., Bimber, B.: An Automated System for Plagiarism Detection Using the Internet. In: Proceedings of ED-MEDIA 2004, pp. 3619–3625 (2004)
13. Tang, M., Byrne, R., Tang, M.: University anti-plagiarism efforts versus commercial anti-plagiarism software and services and do online students cheat more? In: Proceedings of E-Learn 2007, pp. 6595–6601 (2007)

14. Pighizzini, G.: How Hard is Computing the Edit Distance. Information and Computation 165(1), 1–13 (2001)
15. Ukkonen, E.: Algorithms for approximate string matching. Information and Control 64, 100–118 (1985)
16. Hyyroe, H.: A Bit-Vector Algorithm for Computing Levenshtein and Damerau Edit Distance. Nordic Journal of Computing 10, 1–11 (2003)
17. Helmer, S.: Measuring the Structural Similarity of Semistructured Documents Using Entropy. In: Proceedings of the 33rd International Conference on Very Large Data Bases, pp. 1022–1032 (2007)
18. Ueta, K., Tominaga, H.: A Development and Application of Similarity Detection Methods for Plagiarism of Online Reports. In: Proceedings of ITHET 2010, pp. 363–371 (2010)
19. Ueta, K., Tominaga, H.: A Prototype and Functions of WebBinder for Student Task Support in Report Processing and Submission. In: Proceedings of ED-MEDIA 2009, pp. 3472–3477 (2009)
20. Kurata, H., Tominaga, H., Hayashi, T., Yamasaki, T.: Contest Style Exercise with Execution Tests for Every Lesson in Introductory C Programming. In: Proceedings of ITHET 2007, pp. 99–102 (2007)
21. Kawasaki, S., Tominaga, H.: Execution Test Series and Partial Scoring in Support Server for Introductory C Programming Exercise. In: Proceedings of ED-MEDIA 2010, pp. 3189–3196 (2011)
22. Nishimura, T., Kawasaki, S., Tominaga, H.: Monitoring System of Student Situation in Introductory C Programming Exercise with a Contest Style. In: Proceedings of ITHET 2011, pp. 1–6 (2011)

Author Index